Lecture Notes in Computer Science 15925

Founding Editors

Gerhard Goos
Juris Hartmanis

Editorial Board Members

Elisa Bertino, *Purdue University, West Lafayette, IN, USA*
Wen Gao, *Peking University, Beijing, China*
Bernhard Steffen, *TU Dortmund University, Dortmund, Germany*
Moti Yung, *Columbia University, New York, NY, USA*

The series Lecture Notes in Computer Science (LNCS), including its subseries Lecture Notes in Artificial Intelligence (LNAI) and Lecture Notes in Bioinformatics (LNBI), has established itself as a medium for the publication of new developments in computer science and information technology research, teaching, and education.

LNCS enjoys close cooperation with the computer science R & D community, the series counts many renowned academics among its volume editors and paper authors, and collaborates with prestigious societies. Its mission is to serve this international community by providing an invaluable service, mainly focused on the publication of conference and workshop proceedings and postproceedings. LNCS commenced publication in 1973.

Jin-song Dong · Jing Sun · Xiaofei Xie ·
Kan Jiang
Editors

Sports Analytics

Second International Conference, ISACE 2025
Shanghai, China, September 26–27, 2025
Proceedings

Editors
Jin-song Dong
National University of Singapore
Singapore, Singapore

Jing Sun
University of Auckland
Auckland, New Zealand

Xiaofei Xie
Singapore Management University
Singapore, Singapore

Kan Jiang
National University of Singapore
Singapore, Singapore

ISSN 0302-9743　　　　　　　ISSN 1611-3349　(electronic)
Lecture Notes in Computer Science
ISBN 978-3-032-06166-9　　　ISBN 978-3-032-06167-6　(eBook)
https://doi.org/10.1007/978-3-032-06167-6

© The Editor(s) (if applicable) and The Author(s), under exclusive license
to Springer Nature Switzerland AG 2026

This work is subject to copyright. All rights are solely and exclusively licensed by the Publisher, whether the whole or part of the material is concerned, specifically the rights of translation, reprinting, reuse of illustrations, recitation, broadcasting, reproduction on microfilms or in any other physical way, and transmission or information storage and retrieval, electronic adaptation, computer software, or by similar or dissimilar methodology now known or hereafter developed.
The use of general descriptive names, registered names, trademarks, service marks, etc. in this publication does not imply, even in the absence of a specific statement, that such names are exempt from the relevant protective laws and regulations and therefore free for general use.
The publisher, the authors and the editors are safe to assume that the advice and information in this book are believed to be true and accurate at the date of publication. Neither the publisher nor the authors or the editors give a warranty, expressed or implied, with respect to the material contained herein or for any errors or omissions that may have been made. The publisher remains neutral with regard to jurisdictional claims in published maps and institutional affiliations.

This Springer imprint is published by the registered company Springer Nature Switzerland AG
The registered company address is: Gewerbestrasse 11, 6330 Cham, Switzerland

If disposing of this product, please recycle the paper.

Preface

This volume contains the papers presented at ISACE 2025: the 2nd International Sports Analytics Conference and Exhibition, held on September 26-27, 2025 in Shanghai.

There were 57 submissions. Each submission was single-blindly reviewed by at least 3 people, and on the average 3.0 program committee members. The committee decided to accept 21 papers.

The International Sports Analytics Conference and Exhibition (ISACE) was established with the vision of advancing the frontiers of sports analytics, which is an interdisciplinary field that integrates artificial intelligence, data science, psychology, and smart devices to enhance athletic performance, strategy, and decision-making.

Sports analytics involves collecting and interpreting data from diverse sources, including video analysis, performance metrics, and scouting reports, to generate actionable insights. These insights support player development, injury prevention, tactical planning, and strategic resource allocation for coaches, analysts, and team managers.

While various seminars, satellite workshops (often held alongside major AI and data science conferences), and regional events such as the MIT Sloan Sports Analytics Conference have contributed to this growing field, ISACE was created to serve as a dedicated international platform. Its goal is to bring together a broad spectrum of stakeholders, such as academics, researchers, technologists, psychologists, coaches, and sports professionals, to share knowledge, foster collaboration, and promote innovation in sports analytics.

The first ISACE successfully laid the foundation for this mission. Building on that momentum, the second ISACE took place in September 2025 in Shanghai, China, just one week before the 2025 Rolex Shanghai Masters tennis tournament. The proceedings are published in the Springer Lecture Notes in Computer Science (LNCS) series.

ISACE aims to become the premier annual event uniting the global sports analytics community.

We would like to express our sincere gratitude to Shanghai Jiao Tong University for serving as the host and local organizer of ISACE 2025. In particular, we extend our heartfelt thanks to Kun Wang for his leadership and coordination in making this event possible.

We also gratefully acknowledge Jin Song Dong, Deputy Head of the School of Computing, National University of Singapore, and Founder of ISACE, for his vision and continued dedication to establishing and advancing the International Sports Analytics Conference and Exhibition series.

The conference was generously supported by the Dependable Intelligence initiative, and we are pleased to recognize the collaboration between Shanghai Jiao Tong University, the National University of Singapore, and Griffith University, whose joint efforts helped bring together a diverse and vibrant community of researchers, practitioners, and thought leaders in the field of sports analytics.

August 2025

Jin Song Dong
Jing Sun
Xiaofei Xie
Kan Jiang

Organization

Honorary Chairs

Jin Song Dong   National University of Singapore, Singapore
Kun Wnag   Shanghai Jiao Tong University, China

General Chairs

Yun Lin   Shanghai Jiao Tong University, China
Tao Huang   Shanghai Jiao Tong University, China
Geguang Pu   East China Normal University, China

Program Committee Chairs

Jing Sun   University of Auckland, New Zealand
Michael Kan Jiang   National University of Singapore, Singapore
Xiaofei Xie   Singapore Management University, Singapore

Organizing Committee

Bimlesh Wadhwa   National University of Singapore, Singapore
Jun Sun   Singapore Management University, Singapore
Yang Liu   Nanyang Technological University, Singapore
Mounir Mokhtari   Institut Mines-Télécom, France
Yamine Aït-Ameur   National Polytechnic Institute of Toulouse, France
Zhiyong Huang   National University of Singapore, Singapore
Teck Khim Ng   National University of Singapore, Singapore
Chunyang Chen   Technical University of Munich, Germany
Ruofan Liu   National University of Singapore, Singapore

Registration Chair

Zhe Hou   Griffith University, Australia

Graduate Symposium Chairs

Wenxi Liu	Shanghai Jiao Tong University, China
Zhaoyu Liu	National University of Singapore, Singapore
Shaobo Cai	Shanghai Jiao Tong University, China

Publicity Chairs

Kailong Wang	Huazhong University of Science and Technology, China
Minghong Zheng	Shanghai Jiao Tong University, China
Hongyi Xie	ShanghaiTech University, China

Web Chair

Mark Huasong Meng	Technical University of Munich, Germany

Program Committee

Bimlesh Wadhwa	National University of Singapore, Singapore
Chunyang Chen	Technical University of Munich, Germany
David Saxby	Griffith University, Australia
Elizabeth Bradshaw	Deakin University, Australia
Hadrien Bride	Dependable Intelligence, France
Jin Song Dong	National University of Singapore, Singapore
John Komar	Nanyang Technological University, Singapore
Kan Jiang	National University of Singapore, Singapore
Lijun Guo	Ningbo University, China
Luke Wildman	RGB Assurance, Australia
Masoumeh Izadi	Television Content Analytics Pte., Ltd, Singapore
Oliver Schulte	Simon Fraser University, Canada
Sebastian Binnewies	Griffith University, Australia
Seungbok Lee	Yonsei University, South Korea
Teck Khim Ng	National University of Singapore, Singapore
Wei-Yao Wang	National Yang Ming Chiao Tung University, Taiwan
Yamine Aït-Ameur	IRIT - National Polytechnic Institute of Toulouse, France
Yun Lin	Shanghai Jiao Tong University, China

Zhaoyu Liu	National University of Singapore, Singapore
Zhe Hou	Griffith University, Australia
Zhiyong Huang	National University of Singapore, Singapore
Xiaofei Xie	Singapore Management University, Singapore
Mehul Raval	Ahmedabad University, India
Srikrishnan Divakaran	Krea University, India
Tolga Kaya	Sacred Heart University, USA
Daniel Bolarinwa	Accenture, UK
Rajdeep Singh Hundal	National University of Singapore, Singapore
Dileepa Fernando	Singapore University of Technology and Design, Singapore
Isuru Supasan	SLTC Research University, Sri Lanka
Karan Gupta	SunPower, USA
Raveendran Paramesran	Monash University Malaysia, Malaysia
Bessam Abdulrazak	University of Sherbrooke, Canada
Steven Lin	Azra Games, USA
Tao Lin	Ethiqly, USA
Robert Moskovitch	Ben-Gurion University of the Negev, Israel
David Clausi	University of Waterloo, Canada
Ruchir Pandya	Moloco, USA
Siqi Hao	Huawei, Finland
Willie Harrison	Brigham Young University, USA
Matthew Caron	VfL Wolfsburg, Germany
Leili Javadpour	University of the Pacific, USA
Jacomine Grobler	Stellenbosch University, South Africa
Charles Danoff	Mr. Danoff's Teaching Laboratory, USA
Klaus Müller	Wictory.ai, Austria
Markus Unterweger	Wictory.ai, Austria
Chean Khim Toa	Xiamen University Malaysia, Malaysia
Divya Mehta	Queensland University of Technology, Australia
Haresh Suppiah	La Trobe University, Australia
Xian Song	Zhejiang University, China
Yu Xin	Ningbo University, China
Ruchika Malhotra	Delhi Technological University, India
Shaobo Cai	Shanghai Jiao Tong University, China
Roland Janos Nemes	Hosei University, Japan

Contents

Fractal Analysis of Ball Movement Maps for Team Performance
Evaluation in Association Football 1
   Ishara Bandara, Sergiy Shelyag, Sutharshan Rajasegarar,
   Daniel B. Dwyer, Eun-jin Kim, and Maia Angelova

YOCO-Sport: An End-to-End Framework for Deep Learning-Based
Camera Calibration from Sports Broadcast Footage 18
   Gerhardt Breytenbach and Jacomine Grobler

One-Shot Team Recognition and 3D Pose Estimation of Cyclists
for Augmented Reality Visualization 36
   Winter Clinckemaillie, Jelle Vanhaeverbeke, Maarten Slembrouck,
   and Steven Verstockt

How Do Football Teams Play? A Deep Embedded Clustering Approach
to Reveal Playing Styles 53
   Ege Demir, Yusuf H. Şahin, and Nazım Kemal Üre

Construction of Sports and Exercise Knowledge Graph 69
   Tao Huang, Zehan Xia, Yangyi Huang, Jiaxin Zheng, Jun Lin,
   and Kun Wang

Personalised Running Coaching with Next-Generation Wearable
Technology 76
   Nathan Hur, Jonathan Soulsby, Zixiao Zhao, and Jing Sun

Does Wellness Predict Performance? Player-Specific Insights from Daily
Monitoring in College Men's Soccer 92
   Leili Javadpour, Ashwinth Reddy Kondapalli, Mehdi Khazaeli,
   and Adam Reeves

Locating Tennis Ball Impact on the Racket in Real Time Using an Event
Camera 99
   Yuto Kase, Kai Ishibe, Ryoma Yasuda, Yudai Washida,
   and Sakiko Hashimoto

An Analysis of Differences in Golf Performance Between Age Groups
for the Development of an XR Metaverse Platform and Content
for Inclusive Digital Leisure 116
   *Yun-hwan Lee, Yeong-hun Kwon, Jin-i Hong, Jongsung Kim,
and Jongbae Kim*

Scalable Tactical Tennis Insights: Hybridizing Automated Reports
and LLM-Powered Analytics 126
   Zizhen Li, Zhaoyu Liu, and Kan Jiang

Analyzing Basketball Lineups with MDP Using NBA Statistics and Player
Tracking Data 142
   Zhaoyu Liu and Shenyi Su

AI and Data Science in Sports Education 155
   Ruchika Malhotra, Bimlesh Wadhwa, Shweta Meena, and Subodh Mor

Examining the Impact of Traffic on Shot Attempts in Ice Hockey 162
   Miles Pitassi, Evan Iaboni, Fauzan Lodhi, and Tim Brecht

Predicting Penalty Kick Direction Using Multi-modal Deep Learning
with Pose-Guided Attention 179
   Pasindu Ranasinghe and Pamudu Ranasinghe

A Bayesian Dual-Skill Framework for Roster-Based Cycling Race
Outcome Prediction 193
   Denis Rize, Paulo Saldanha, and Robert Moskovitch

Horse ReIDing: Addressing Re-Identification in Horse Racing Scenarios 209
   *Luca Francesco Rossi, Andrea Sanna, Federico Manuri,
and Mattia Donna Bianco*

Barbell Trajectory Tracking for Performance Analysis During Snatch
Movement in Weightlifting 218
   *Dhairya Shah, Christopher Taber, Tolga Kaya, Eva Maddox,
and Mehul S. Raval*

A Data-Driven Imputation Scheme for Cohort Studies: A Collegiate
Basketball Casestudy 235
   *Srishti Sharma, Vishal Barot, Srikrishnan Divakaran, Tolga Kaya,
Christopher B. Taber, and Mehul S. Raval*

TANS: A Chess-Inspired Notation System for Strategy Analysis of Tennis Games 253
   Yuexi Song, Chuanfei Li, Hao Cao, Ling Wu, Huanhuan Zheng, and Zhenkai Liang

Agentic Generative AI for Media Content Discovery at the National Football League 268
   Henry Wang, Md Sirajus Salekin, Jake Lee, Ross Claytor, Shinan Zhang, and Michael Chi

Action Sequence Modeling for Tactical Training in Handball 276
   Luke Wildman, Roland Nemes, and Zhe Hou

Author Index 293

Fractal Analysis of Ball Movement Maps for Team Performance Evaluation in Association Football

Ishara Bandara[1,2]($\boxtimes$), Sergiy Shelyag[1,3], Sutharshan Rajasegarar[1], Daniel B. Dwyer[4], Eun-jin Kim[2], and Maia Angelova[5,6]

[1] School of IT, Deakin University, Geelong, VIC 3220, Australia
isharabnd@gmail.com
[2] Research Centre for Fluid and Complex Systems, Coventry University, Coventry CV1 5FB, UK
[3] College of Science and Engineering, Flinders University, Tonsley, Adelaide, SA 5042, Australia
[4] School of Exercise and Nutrition Sciences, Deakin University, Geelong, VIC 3220, Australia
[5] Aston Digital Futures Institute, Aston University, Birmingham B4 7ET, UK
[6] Institute for Biophysics and Bioengineering, Bulgarian Academy of Sciences, Sofia, Bulgaria

Abstract. Spatiotemporal analysis has become a foundation of modern football analytics, particularly in evaluating team performance. However, the complex, dynamic nature of association football makes objective performance evaluation a persistent challenge. While recent studies have explored event distribution randomness and player-to-player interactions, these approaches often overlook the role of ball movement trajectories, which can offer crucial insights into team effectiveness. To address this gap, this study proposes a novel method for quantifying spatial complexity in team ball movement as a measure of offensive performance. A time-series feature extraction approach is introduced, wherein the fractal dimension of 2D ball movement maps are computed to represent spatial complexity across defined time intervals. Correlation analysis reveals a positive association between spatial complexity and match-winning outcomes, particularly during the early phases of play. Furthermore, a Random Forest classification model trained exclusively on spatial complexity features achieved an AUC-ROC of 0.8180 in predicting match winners, underscoring the potential of spatial complexity as a valuable and interpretable time-series metric for evaluating team performance in association football.

Keywords: Football · Soccer · Performance Evaluation · Machine Learning · Fractal Dimension · Complexity · Time-Series

1 Introduction

Performance evaluation in association football remains a complex task, primarily due to the sport's low-scoring nature and the dynamic, fluid structure of gameplay. Unlike high-scoring sports where numerous scoring events allow for clearer outcome attribution, football limits scoring opportunities almost exclusively to goals resulting from successful shots. Moreover, a single event such as an own goal or a deflected shot can decide the match, often misrepresenting the relative quality or dominance of the competing teams over the course of the game. This makes it inherently difficult to assess performance based solely on the final scoreline.

The game of football encompasses a high degree of complexity arising from various sources, such as the tactical decisions of teams [30], the individual and collective abilities of players [22], and the unpredictable nature of factors like chance [23,24]. These diverse and dynamic elements contribute to the inadequacy of traditional score-based metrics in fully capturing the richness of team dynamics. To address these limitations, researchers have explored a wide range of performance indicators. Conventional analyses typically utilize event-based statistics, including possession percentages [2,9,14,21], number of shots, accuracy on target, and successful passes [1,25,31]. Recently, advanced metrics such as expected goals (xG) [3,17] and expected threat (xT) [27] have gained attraction as measures to assess performance in a more comprehensive manner. However, despite their growing popularity, these metrics often overlook the temporal flow and spatial distribution of play. This omission can lead to inconsistent conclusions in research. For instance, while some studies report a positive correlation between possession and match outcomes [9,21], others find no significant relationship [2,14]. Such contradictions highlight the need for time-series performance evaluation frameworks.

Football tactics exhibit significant variability across teams, reflecting a wide range of strategic preferences. Some teams emphasize possession-based styles, aiming to control the game through sustained ball retention, while others adopt more direct approaches, relying on long passes and rapid transitions that de-emphasize possession. Furthermore, tactical decisions are often dynamic and evolve over the course of a match. For example, studies indicate that leading teams shift to a more defensive approach in the latter stages of games [10,11]. Such shifts can result in reduced metrics, such as possession, shot count, and number of passes despite ultimately securing a win. These evolving tactical patterns highlight the importance of capturing temporal variations, which can be effectively analyzed through time series approaches to performance data.

In response to the limitations of conventional metrics, recent advancements in football analytics have increasingly research use of spatiotemporal data to uncover deeper insights into match dynamics. This shift enables analysts to move beyond static statistics and capture the evolving nature of play. Building on this perspective, the authors' previous work introduced the concept of the Ball Carrier Open Space (BCOS) as a time-series metric that reflects the temporal availability of space for the player in possession [5,6]. Predictive models based solely on BCOS derived temporal features demonstrated strong performance, achieving over 82.5% accuracy in forecasting match outcomes underscoring the relevance of BCOS as a meaningful indicator of team effectiveness.

Renowned football manager Pep Guardiola has emphasized that his tactical philosophy centers on creating open spaces to exploit by disrupting and disguising the structure of the opposition's defense [29]. One way attacking teams can achieve this is through unpredictable ball movement, which challenges defensive organization. Reflecting this idea, recent research has proposed unpredictability in ball or player movement as a metric for evaluating team performance. Some studies measure this unpredictability using player-to-player interaction entropy [7,15,19], while others employ event distribution randomness across predefined field regions [4]. These approaches commonly rely on entropy, which is a concept from information theory that quantifies uncertainty to capture the degree of randomness in ball movement. However, player to player interaction entropy can be sensitive to changes in player positioning, and substitutions. To address this, Event Distribution Randomness (EDRan) was proposed, focusing on the spatial and temporal distribution of events rather than player identities or positions. While EDRan mitigates issues tied to dynamic player roles, it does not account for the actual paths of ball movement, potentially omitting important contextual information in its assessment of unpredictability.

To address the limitations of existing performance metrics, this study introduces the temporally extracted fractal dimension of ball movement maps as a novel approach to evaluating team performance. Fractal dimension is a mathematical concept used to quantify the complexity and irregularity of spatial patterns, particularly those that exhibit self-similarity across different scales. Originally formalized by Benoît Mandelbrot in his foundational work on fractal geometry [16], this measure Offers mathematical representation of the complexity and irregularity in spatial patterns, makes it a valuable tool for analyzing tactical behavior and team organization. Existing work using fractal dimension to anyze performance in association football data focus on use of fractal dimension to analyze team formations [8,18], player movements [13] and ball movements [12,26].

Bueno et al. examined different shape descriptors applied to polygonal representations of team formations, aiming to quantify the tactical organization of teams on the pitch [8]. Similarly, Moura et al. investigated how recurrent patterns in team shape relate to technical performance metrics during matches [18]. Both studies utilized multi-scale fractal dimension to capture the complexity of spatial organization [8,18].

In a different context, the motion of the ball in computer-simulated football games was analyzed by extracting two-dimensional coordinates using video processing techniques [12]. The study utilized the box-counting method to compute the fractal dimension, offering insights into the complexity and behavioral patterns of ball movement [12]. In subsequent work, Kim extended this analysis to player movements, evaluating fractal motion behaviors based on positional time series data. The findings indicated that defenders exhibited higher fractal dimensions under sustained pressure, leading to the proposal of a novel metric for assessing relative offensive dominance [13].

Despite these contributions, prior research has not explored the use of team's spatial ball movement complexity as a temporal measure of collective team performance. To address this gap, the present work aims to temporally capture and evaluate team's spatial ball complexity of ball movement and assess their relationship with match outcomes. This is achieved by segmenting each half of a match into equal-duration time intervals, constructing two-dimensional ball movement maps for each team, and quantifying spatial complexity of ball movement by computing the fractal dimension of ball movement using the box-counting method. Applying fractal geometry to spatiotemporal ball movement data enables a novel quantification of how structured or disordered a team's behavior is across different match phases. Such analysis not only deepens the understanding of tactical strategies but also offers new avenues to relate in-game dynamics to final performance outcomes.

2 Methodology

This study investigates the impact of a team's spatial ball movement complexity on match-winning performance by analyzing the fractal dimension of ball movement patterns. Features are extracted across ten predefined time intervals five equal-duration segments from each half of the match. The fractal dimension is computed using the box-counting method applied to two-dimensional ball movement maps generated for each team and time segment. The analysis focuses on understanding how the temporal variation in spatial complexity contributes to match outcomes. Subsequently, a robust prediction model is developed to assess the relationship between these time-dependent metrics and final match results. Figure 1 shows the framework of this study.

2.1 Data

This study utilizes the StatsBomb open-data dataset (as updated on January 01, 2025) [28], the largest publicly available football event-log dataset, which offers detailed event-level logs documenting every recorded in-game action during football matches.

Considering the established differences between men's and women's association football [20], as well as the evolving tactical trends in recent years, the analysis is limited to top-tier men's football matches played after 2010. To ensure consistency across all analyzed games, only matches concluded within regular time were included. The resulting dataset comprises 2,336 matches from premier football competitions, including the top leagues in England, Germany, Spain, Italy, and France, as well as UEFA Champions League, FIFA World Cup, and UEFA Euro Cup fixtures.

Each match is represented as a sequence of named events (e.g., passes, tackles, shots), accompanied by time stamps, team and player identifiers, and spatial

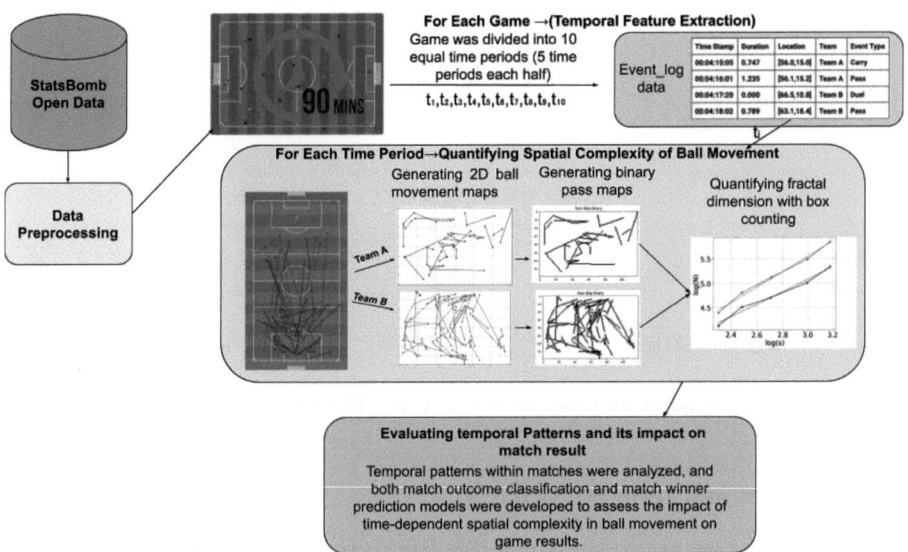

Fig. 1. Framework of the study.

coordinates. Locations are mapped on a 120×80 grid oriented along the direction of attack. For naming consistency, the dataset labels teams as "Home" and "Away," though these do not always reflect actual home-field status. In this study, these teams are referred to as "Team A" and "Team B" respectively. Among the 2,336 matches, 1,042 were won by "Team A", 721 by "Team B", and 573 ended in a draw.

2.2 Temporal Feature Extraction

For temporal feature extraction, each half of a game was partitioned into five equal-duration intervals, resulting in a total of ten time periods for each game (t_i, $1 \leq i \leq 10$) across both halves. When segmenting the time intervals, the total duration of each half, including injury time (i.e., 45 min plus additional stoppage time), was divided equally into five segments. Consequently, the duration of each interval may vary slightly between games and between halves, depending on the specific amount of injury time added.

Within each defined time period, a 2-dimensional ball movement map was generated for each team, based on the start and end locations of every ball movement event (e.g., passes and carries). Off-ball movements were not included in this study, as they fall outside the scope of this analysis. Each resulting ball movement map includes the spatial coordinates of the origins and endpoints, with lines drawn to depict the trajectory of each ball movement, effectively providing a bird's-eye view of all team movements within the specified interval. An example ball movement map of both teams during a t_i time period is depicted in Fig. 2.

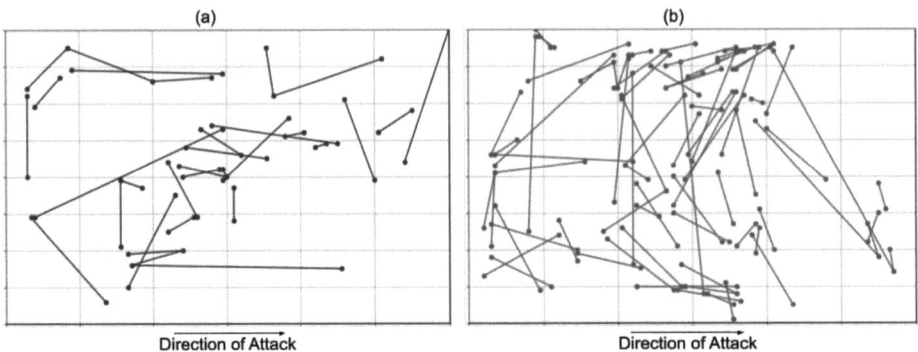

Fig. 2. Ball movement map of (a) Team A and (b) Team B for time period t_i.

2.3 Bresenham's Line Algorithm

As the box-counting approach which this work uses for the estimation of fractal dimension requires a binary representation of spatial data, the two-dimensional ball mvoement maps were converted into binary matrices using Bresenham's line algorithm. This algorithm efficiently traces the discrete grid cells intersected by each pass trajectory, producing a 2D binary map that marks the presence of passes across the field. The resulting binary representation, mapped onto a 120× 80 grid aligned with the direction of play, is essential for accurately applying the box-counting method to estimate the fractal dimension of ball movement.

Bresenham's Line Algorithm is an efficient method for drawing straight lines between two points on a raster grid using only integer arithmetic. It incrementally determines the grid cell closest to the ideal line at each step, eliminating the need for floating-point calculations and thus improving computational performance. This efficiency makes the algorithm well-suited for generating two-dimensional binary ball movement map representations, which are essential for subsequent fractal dimension analysis.

Given two endpoints of a line, (x_0, y_0) and (x_1, y_1), the algorithm identifies the set of discrete grid coordinates that optimally approximate the underlying continuous line. The slope (m) of the line is defined as:

$$m = \frac{\Delta y}{\Delta x} = \frac{y_1 - y_0}{x_1 - x_0} \tag{1}$$

As raster grids are based on integer coordinates, Bresenham's algorithm circumvents the use of floating-point arithmetic by employing an incremental decision parameter. The necessary difference terms are defined as follows:

$$\Delta x = x_1 - x_0, \quad \Delta y = y_1 - y_0 \tag{2}$$

The algorithm uses an initial decision parameter:

$$p_0 = 2\Delta y - \Delta x \tag{3}$$

For each step along the x-axis, the next pixel is chosen based on the decision parameter p_k:

$$p_{k+1} = \begin{cases} p_k + 2\Delta y, & \text{if } p_k < 0 \\ p_k + 2\Delta y - 2\Delta x, & \text{if } p_k \geq 0 \end{cases} \quad (4)$$

Figure 3 presents the binary ball movement maps corresponding to the time interval t_i, generated using Bresenham's line algorithm.

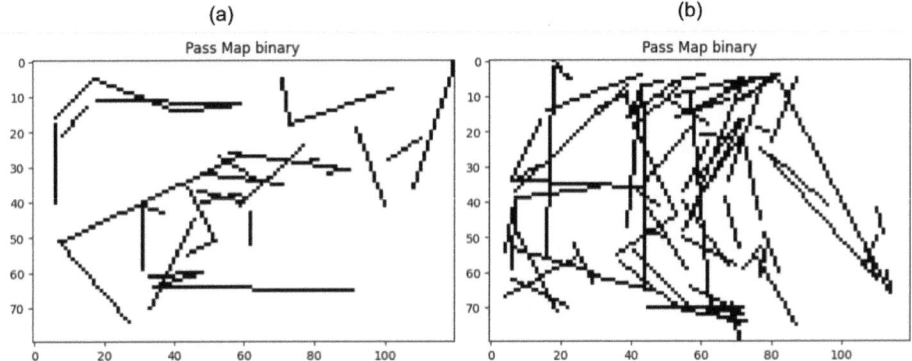

Fig. 3. Binary ball movement maps generated using Bresenham's algorithm for (a) Team A and (b) Team B over the time period t_i.

2.4 Fractal Dimension Calculation with Box Counting

To quantify the spatial complexity of ball movement, this study applies fractal analysis to the binary ball movement maps generated using Bresenham's Line Algorithm. Specifically, the box-counting method is used to estimate the fractal dimension of each ball movement map distribution.

To quantify the spatial complexity of ball and player movement during a match, the fractal dimension of team pass and carry patterns was computed using a two-dimensional box-counting method. The playing field was mapped into a binary grid of size 120 × 80 pixels, representing the height and width of a standard football pitch in meters. Each pass or carry event was mapped by drawing a line between its start and end locations on this grid, with all traversed pixels marked as active.

To estimate the fractal dimension, each binary image was overlaid with grids of varying resolutions, defined by a set of scaling factors. For each scaling factor ϵ, the field was divided into $\epsilon \times \epsilon$ square boxes, and the number of boxes $N(\epsilon)$ containing at least one active pixel was counted. This procedure was repeated across multiple values of ϵ, resulting in a set of data points $(\log(\epsilon), \log(N(\epsilon)))$. The fractal dimension D was estimated as the negative slope of the line fitted

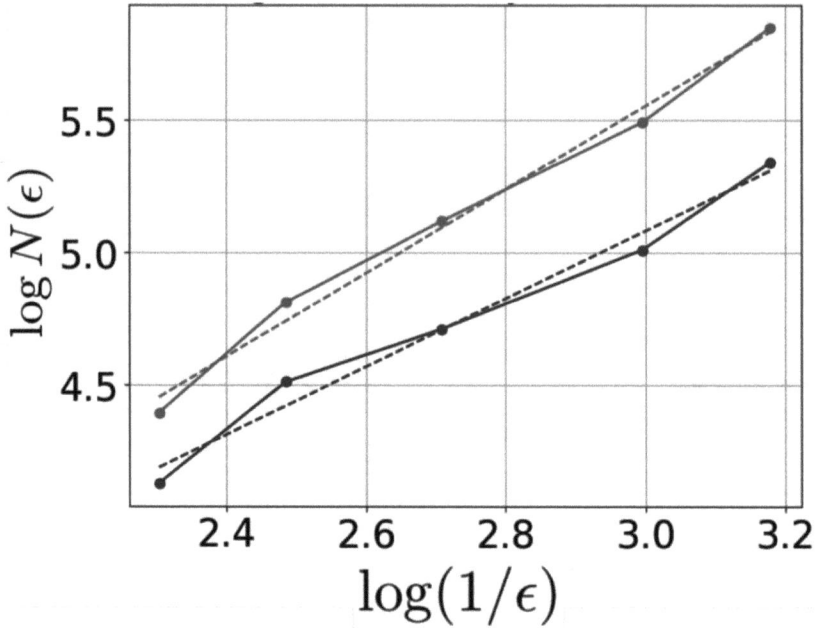

Fig. 4. Linear regression plots to estimate the fractal dimension of two teams.

to these data points using linear regression. Figure 4 presents the linear regression plots(on a log-log plot) used to estimate the fractal dimensions for both teams.

The negative slope of the fitted linear regression line is then interpreted as the fractal dimension of the ball movement for the given team during the time period t_i.

$$D = \lim_{\epsilon \to 0} \frac{\log N(\epsilon)}{\log(1/\epsilon)}. \tag{5}$$

In practice, this limit is approximated using a finite set of scales, and the following expression is computed:

$$D \approx \frac{\Delta \log N(\epsilon)}{\Delta \log(1/\epsilon)}. \tag{6}$$

A higher fractal dimension indicates a more complex and spatially distributed pattern of passes and carries, whereas a lower value suggests a more structured or predictable style of play.

The fractal dimension for each ball movement map, corresponding to each time interval, team, and match, was computed following the described methodology.

As an initial analysis, the 2022 FIFA World Cup final was examined by computing the fractal dimension of ball movement across five equal duration time intervals in each half of the match. Figure 5 illustrates the fractal dimension of ball movement for both teams during the 2022 FIFA World Cup final, shown separately for (a) the first half and (b) the second half.

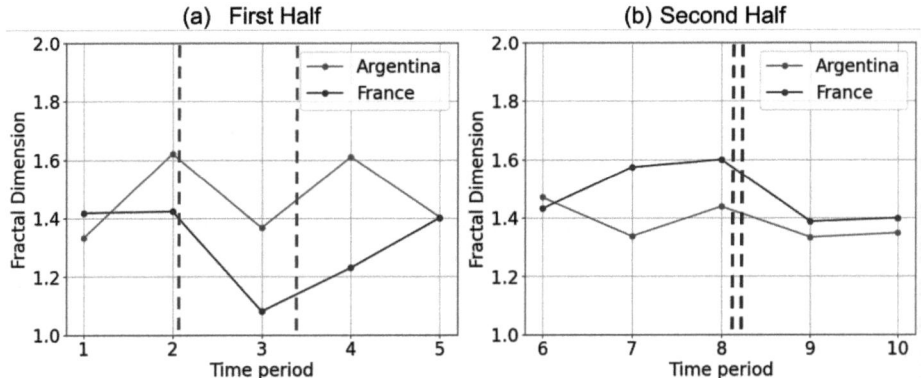

Fig. 5. Fractal dimension of ball movement by France and Argentina during the (a) first and (b) second halves of the FIFA World Cup 2022 final. Green and blue vertical dotted lines represent goals scored by Argentina and France, respectively. (Color figure online)

It was observed that during the first half, Argentina maintained a higher fractal dimension. During this period, they scored two goals in the 23rd and 36th minutes. However, in the second half, France exhibited a higher fractal dimension than Argentina. They have scored two goals during this half in the 80th and 81st minutes of the match.

In this particular game, 7 min were added to the first half as injury time and 9 min were added to the second half as the injury time. Therefore duration of each time period may vary between 1 half to another and one game to another.

To account for variations in the duration of each time interval across games and halves, the raw fractal dimension values were normalized by dividing them by the corresponding interval duration ($Duration_{t_i}$), as shown below:

$$\text{Normalized Fractal Dimension}_{t_i} = \frac{\text{Raw Fractal Dimension}_{t_i}}{\text{Duration}_{t_i}} \qquad (7)$$

The resulting datasets thus contains twenty features per game: ten features for each team, representing the normalized fractal dimension of the ball movement in each distinct time period (t_i; $1 \leq i \leq 10$), along with the corresponding match result. Two separate datasets were constructed to evaluate the impact of spatial complexity on match-winning performance and to develop a match result classification model. For the match-winning performance evaluation, only matches with decisive outcomes were included, excluding draws to specifically analyze the relationship between spatial complexity and victory. In this dataset, the result column is binary, with a value of 1 indicating a win for "Team A" and 0 indicating a win for "Team B". In contrast, the match result classification dataset includes all outcomes and is treated as a three-class classification problem: a value of 1 represents a "Team A" win, 0 denotes a draw, and 2 indicates a "Team B" win.

2.5 Correlation with Match Winning Performances

To assess the significance of spatial complexity and determine whether it contributes positively or negatively to match-winning performance, a correlation analysis was conducted. Matches that ended in a draw were excluded, as the focus of this experiment was specifically on evaluating the influence of spatial complexity in matches with a clear outcome. Given the non-normal distribution of the data, as indicated by the Shapiro-Wilk test, Spearman correlation coefficients were calculated between all 20 features and the match result (binary column).

2.6 Machine Learning Model Development

To investigate the significance of extracted features on match outcomes, a machine learning approach was employed to predict match-winning performances (binary classification of wins and loses) and match results (3 class classification "Team A" win "Team B" win, and draw).

A Random Forest classifier was chosen for this task owing to its capacity to manage a large number of features and maintain robust predictive performance, even with relatively small datasets. In addition, the Random Forest algorithm provides internal feature importance estimates, facilitating the evaluation of each metric's contribution to match-winning performance. As an ensemble learning technique, Random Forest constructs multiple decision trees and aggregates their predictions, thereby enhancing predictive accuracy and mitigating the risk of overfitting. Moreover, Random Forest is inherently resistant to outliers and does not necessitate feature normalization, further supporting its suitability for this analysis.

For a comprehensive evaluation, the model was subjected to fifty rounds of five-fold cross-validation. The match winner classification model was assessed using several performance metrics, including accuracy, F1-score, precision, recall, Matthews Correlation Coefficient (MCC), Area Under the Curve-Receiver Operating Characteristic (AUC-ROC), and per-class accuracy. The mean and standard deviation of each metric across all cross-validation rounds were reported to provide a robust assessment of the predictive performance.

2.7 EPL 2015/2016 Season Analysis

Having identified the contributions in spatial complexity on team ball movement on match winning performances, same approach was used to evaluate team performance in English Premier League (EPL) 2015/2016 season. Leicester City who had narrowly avoided relegation the previous season emerged as league champions with 81 points.

In this study, the spatial complexity of each EPL team in 2015/2016 season was analyzed across all 38 matches of the season, categorized into wins, draws, and losses. To assess each team's performance relative to their opponents, a "relative spatial complexity" measure was considered. This was computed as the difference between a team's spatial complexity in ball movement and that of their opponent at a given time period t_i. Mathematically, the relative spatial complexity in ball movement $RSC(t_i)$ was defined as:

$$RSC(t_i) = SC_{\text{team}}(t_i) - SC_{\text{opponent}}(t_i)$$

where $SC_{\text{team}}(t_i)$ denotes the spatial complexity of the team under consideration at time t_i, and $SC_{\text{opponent}}(t_i)$ represents the spatial complexity of their opponent at the same time instance.

For each team, the average relative spatial complexity $RSC(t_i)$ was computed separately for matches that were won, lost, and drawn, as well as across all matches to obtain a season-wide average. These averages were then plotted to facilitate performance comparison across teams.

3 Results

3.1 Correlation Analysis Results

The correlation analysis reveals a positive association between spatial complexity in Team A's ball movement and match outcomes across the first eight time intervals. The highest correlation is observed during the first time period, gradually decreasing over time, with the final two intervals, t_9 and t_{10}, exhibiting negative correlations. In contrast, for Team B, the first seven time intervals show negative correlations with the match result, with the strongest (most negative) correlation occurring at t_1. The magnitude of these correlations steadily decreases until t_7, after which t_8, t_9, and t_{10} display positive correlations. Table 1 presents the Spearman correlation coefficients of all features with the match result.

Table 1. Spearman Correlation of Features with game Result

Team	Time Periods									
	t_1	t_2	t_3	t_4	t_5	t_6	t_7	t_8	t_9	t_{10}
Team A	0.2934	0.2531	0.2510	0.2082	0.1841	0.1305	0.1101	0.0361	−0.0433	−0.0959
Team B	−0.2597	−0.2324	−0.1663	−0.1767	−0.1741	−0.0672	−0.0372	0.0572	0.1055	0.1839

In the match result column, a value of 1 indicates a victory for "Team A," while 0 denotes a victory for "Team B." As a result, positive correlations for "Team A" and negative correlations for "Team B" imply that greater spatial

Table 2. Comparison of Machine Learning Model Results

Evaluation Metric	Match Winner Prediction 2-class classification	Match Result Prediction 3-class classification
Accuracy	0.7441 ± 0.0215	0.5564 ± 0.0211
F1-score	0.7404 ± 0.0241	0.4955 ± 0.0219
Precision	0.7511 ± 0.0373	0.4985 ± 0.0240
Recall	0.7321 ± 0.0371	0.5055 ± 0.0209
MCC	0.4896 ± 0.0427	0.3020 ± 0.0319
AUC	0.8180 ± 0.0210	-
Per Class Acc.: Team A Win	0.7572 ± 0.0374	0.7396 ± 0.0292
Per Class Acc.: Team B Win	0.7321 ± 0.0371	0.5731 ± 0.0433
Per Class Acc.: Draw	-	0.2039 ± 0.0382

complexity is associated with improved match-winning performance. However, toward the latter stages of the game, this relationship becomes negative, which may be attributed to the defensive strategies often adopted by leading teams in the final phases of play, as noted in prior research [10]. It is also noteworthy that the strongest correlations both in magnitude and direction occur during the early phases of the match, suggesting that spatial complexity during the initial stages plays a more significant role in influencing game outcomes.

3.2 Machine Learning Model Results

The Random Forest model demonstrated strong performance in predicting match winners, achieving a mean AUC-ROC of 0.8180 (±0.0210), f1-score of 0.7404 (±0.0241) and an accuracy of 0.7441 (±0.0215) across fifty rounds of five-fold cross-validation, totaling 250 evaluations. These results highlight the model's effectiveness in capturing factors associated with match-winning performance. For match result prediction, the model achieved a mean accuracy of 0.5564 (±0.0210) and a F1-score of 0.4955 (±0.0219). However, the model exhibited limited ability in correctly identifying drawn outcomes in three class classification. Table 2 provides a detailed summary of the Random Forest model's performance across tasks.

3.3 EPL 2015/2016 Season Analysis Results

The average spatial complexity in ball movement was computed and plotted for each team across matches that resulted in a win, draw, or loss in EPL 2015/2016 season. Interestingly, although Leicester City won the EPL in the 2015/2016 season, they consistently exhibited relatively lower spatial complexity in ball movement compared to the other top five ranked teams. Figure 6 illustrates the mean relative spatial complexity of all EPL teams over the course of the 2015/2016 season.

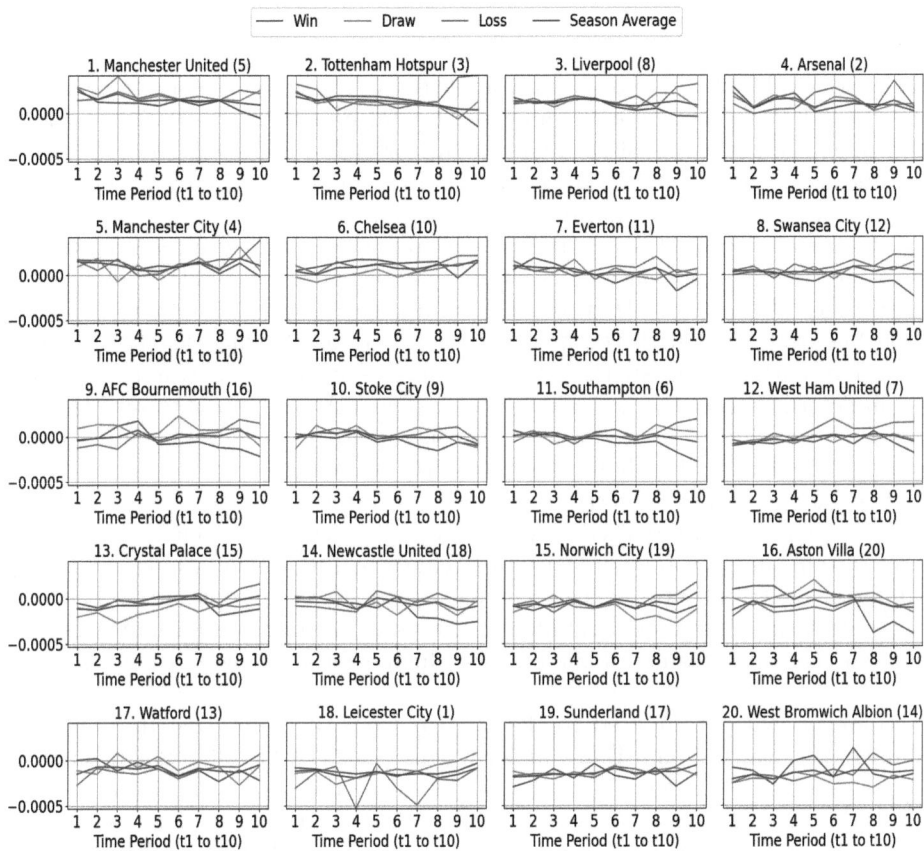

Fig. 6. Team-wise average relative spatial complexity of ball movement exhibited by EPL teams against their opponents during the 2015/2016 season. The average relative spatial complexity was calculated separately for matches resulting in wins, losses, draws, and as a season-wide average across all games. Each plot title includes the respective team's final league position for the 2015/2016 EPL season, indicated in parentheses.

It was observed that traditional "big six" teams in EPL Arsenal, Tottenham Hotspur, Liverpool, Manchester City, Manchester United, and Chelsea maintained the highest average spatial complexity throughout the season, regardless of their final league positions. Overall, there appeared to be a decreasing trend in season-average spatial complexity with lower league rankings. Notably, Leicester City, despite winning the Premier League in the 2015/2016 season, recorded the third-lowest season average in spatial complexity of ball movement, highlighting the effectiveness of their highly defensive tactical approach.

4 Discussion

This work explored how the spatial complexity of a team's ball movement influences match outcomes by examining the fractal dimension of pass patterns of 10 defined time intervals of each game. This proposed approach allowed for the investigation of temporal variations in spatial complexity in ball movement by a team and their contribution to match performance.

The correlation analysis indicates that spatial complexity in ball movement has a positive influence on match-winning performance, particularly during the early stages of the game. However, the significance of this relationship appears to decline over time, with the final two time periods showing negative correlations. This pattern suggests that the initial phases of a match play a more critical role in determining outcomes. The negative correlation observed toward the end of matches may be attributed to the defensive strategies commonly employed by teams in the lead, as highlighted in previous research [10].

The match-winner prediction model demonstrated higher accuracy and AUC-ROC in identifying match-winning performances using only temporal features derived from the spatial complexity of a team's ball movement. This indicates a significant contribution of spatial complexity to successful outcomes. The positive correlations between individual features and model performance further highlight the importance of maintaining high spatial complexity in offensive play.

However, when compared to the contributions of Event Distribution Randomness (EDRan) [4] and Ball Carrier Open-Space (BCOS) [6], which achieved accuracies of 0.7995 and 0.7973 respectively, the correlations with match-winning outcomes and classification performance are relatively lower. This suggests a lesser, though still meaningful, role in determining match success.

Despite utilizing only a single temporal metric for classification, the model achieves notably strong performance. This is particularly impressive when compared to existing models in the literature, such as those by Danisik et al. (70.21% accuracy using 139 features) and Almulla et al. (80.77% accuracy using 396 features), which relied on multiple metrics to develop their match-winner prediction models.

Spatial complexity analysis of 2015/2016 EPL season demonstrated that despite the rank of the team in the season, traditional big six clubs in English football has maintained the highest spatial complexity in ball movement signifying their long term success. Leicester city despite winning the league, have maintained 3rd average lowest spatial complexity during the season. However, Leicester city failed to succeed in following seasons.

This study does not account for spatial complexity across the entire field; however, certain areas may hold more strategic value than others. Investigating spatial complexity by field zones presents a promising direction for future research. Moreover, analyzing how football strategies differ across genders, regions, and teams based on spatial complexity in ball movement offers another potential area for exploration. Additionally, the current work only incorporates the horizontal trajectory of passes, focusing solely on their start and end points.

It excludes vertical movement and other key features, such as the number of defenders bypassed per pass. Incorporating these aspects could enhance future models and deepen our understanding of spatial dynamics in football.

5 Conclusion

In conclusion, this study introduced a novel time-series metric for evaluating team performance based on the spatial complexity of collective ball movement within defined time intervals. By generating 2D ball movement maps and applying a box-counting approach to compute their fractal dimension, the proposed method offers a quantitative means to assess spatial complexity. The use of this metric in a machine learning model to classify match-winning performances yielded promising results, despite relying solely on spatial complexity features. Furthermore, the observed positive correlation between this metric and match outcomes suggests that higher spatial complexity in ball movement is associated with greater chances of success. These findings imply that teams aiming to improve their performance should focus on enhancing the diversity and distribution of their ball movement. Future research could build on this foundation by incorporating additional contextual features, and analyzing how spatial strategies vary across different teams, regions, and genders.

References

1. AlMulla, J., Islam, M.T., Al-Absi, H.R.H., Alam, T.: Soccernet: a gated recurrent unit-based model to predict soccer match winners. PLOS ONE **18**(8), 1–19 (2023). https://doi.org/10.1371/journal.pone.0288933
2. Aquino, R., et al.: Comparisons of ball possession, match running performance, player prominence and team network properties according to match outcome and playing formation during the 2018 FIFA world cup. Int. J. Perform. Anal. Sport **19** (2019). https://doi.org/10.1080/24748668.2019.1689753
3. Bandara, I., Shelyag, S., Rajasegarar, S., Dwyer, D., Kim, E.J., Angelova, M.: Predicting goal probabilities with improved XG models using event sequences in association football. PLOS ONE **19** (2024). https://doi.org/10.1371/journal.pone.0312278
4. Bandara, I., Shelyag, S., Rajasegarar, S., Dwyer, D., Kim, E.J., Angelova, M.: Winning with chaos in association football: spatiotemporal event distribution randomness metric for team performance evaluation. IEEE Access **12**, 83363–83376 (2024). https://doi.org/10.1109/ACCESS.2024.3413648
5. Bandara, I., Shelyag, S., Rajasegarar, S., Dwyer, D.B., jin Kim, E., Angelova, M.: Time-series analysis of ball carrier open-space (BCOS) in association football. SN Comput. Sci. **6**(4), 302 (2025). https://doi.org/10.1007/s42979-025-03815-7
6. Bandara, I., Shelyag, S., Rajasegarar, S., Dwyer, D.B., Kim, E.J., Angelova, M.: Time-series analysis of ball carrier open-space in association football. In: Dong, J.S., Izadi, M., Hou, Z. (eds.) Sports Analytics, pp. 1–17. Springer, Cham (2024)
7. Berman, Y., Mistry, S., Mathew, J., Krishna, A.: Temporal match analysis and recommending substitutions in live soccer games. In: IEEE International Conference on Web Services, pp. 397–404 (2022). https://doi.org/10.1109/ICWS55610.2022.00066

8. Bueno, M., Silva, M., Cunha, S., Torres, R., Moura, F.: Multiscale fractal dimension applied to tactical analysis in football: a novel approach to evaluate the shapes of team organization on the pitch. PLOS ONE **16** (2021). https://doi.org/10.1371/journal.pone.0256771
9. Goral, K.: Passing success percentages and ball possession rates of successful teams in 2014 FIFA world cup. Int. J. Sci. Cult. Sport **3** (2015). https://doi.org/10.14486/IJSCS239
10. Guan, T., Cao, J., Swartz, T.: Should you park the bus? (2021). https://www.sfu.ca/~tswartz/papers/bus.pdf
11. Guan, T., Cao, J., Swartz, T.: Parking the bus. J. Quant. Anal. Sports **19** (2023). https://doi.org/10.1515/jqas-2021-0059
12. Kim, S.: Fractal analysis of the motions of a ball in a computer soccer game. J. Korean Phys. Soc. **44**, 664 (2004). https://doi.org/10.3938/jkps.44.664
13. Kim, S.: Player's positional dependence of fractal behaviors in a soccer game. Fractals **14** (2011). https://doi.org/10.1142/S0218348X06003003
14. Kubayi, A., Toriola, A.: The influence of situational variables on ball possession in the South African premier soccer league. J. Hum. Kinetics **66**, 175–181 (2019). https://doi.org/10.2478/hukin-2018-0056
15. Kusmakar, S., Shelyag, S., Zhu, Y., Dwyer, D., Gastin, P., Angelova, M.: Machine learning enabled team performance analysis in the dynamical environment of soccer. IEEE Access **8**, 90266–90279 (2020). https://doi.org/10.1109/ACCESS.2020.2992025
16. Mandelbrot, B.B.: The Fractal Geometry of Nature. W. H. Freeman and Company (1983)
17. Mead, J., O'Hare, A., McMenemy, P.: Expected goals in football: improving model performance and demonstrating value. PloS one **18**, e0282295 (2023). https://doi.org/10.1371/journal.pone.0282295
18. Moura, F., Bueno, M., Caetano, F., Silva, M., Cunha, S., Torres, R.: Exploring the recurrent states of football teams' tactical organization on the pitch during Brazilian official matches. PLOS ONE **19** (2024). https://doi.org/10.1371/journal.pone.0308320
19. Neuman, Y., Israeli, N., Vilenchik, D., Cohen, Y.: The adaptive behavior of a soccer team: an entropy-based analysis. Entropy **20**(10) (2018). https://doi.org/10.3390/e20100758,
20. Pappalardo, L., Rossi, A., Natilli, M., Cintia, P.: Explaining the difference between men's and women's football. PLOS ONE **16**(8), 1–17 (2021). https://doi.org/10.1371/journal.pone.0255407
21. Peñas, C., Lago Ballesteros, J., A, D., Gómez López, M.: Game-related statistics that discriminated winning, drawing and losing teams from the Spanish soccer league. J. Sports Sci. Med. **9**, 288–93 (2010)
22. Plakias, S., Armatas, V., Mitrotasios, M.: Influence of tactics and situational variables on goal scoring in European football. Proc. Inst. Mech. Eng. Part P J. Sports Eng. Technol. (2025). https://doi.org/10.1177/17543371241313252
23. Reep, C., Pollard, R., Benjamin, B.: Skill and chance in ball games. J. Roy. Stat. Soc. Ser. A (General) **134**, 623 (1971). https://doi.org/10.2307/2343657
24. Reep, C., Benjamin, B.: Skill and chance in association football. J. R. Stat. Soc. Ser. A **131**(4), 581–585 (1968)
25. Rocha-Lima, E., Tertuliano, I., Fischer, C.: Determinant football elements for euro16 match results. Revista Inteligência Competitiva **13**, e0428 (2023). https://doi.org/10.24883/eagleSustainable.v13i.428

26. Simon, L., Soós, A.: Fractal analysis on the football pitch. In: 2019 22nd International Conference on Control Systems and Computer Science (CSCS), pp. 583–585 (2019). https://doi.org/10.1109/CSCS.2019.00106
27. Singh, K.: Introducing expected threat (XT) (2018). https://karun.in/blog/expected-threat.html
28. StatsBomb: Statsbomb open data (2022). https://github.com/statsbomb/open-data.git
29. Swaby, S.: Pep Guardiola Calls Tiki-Taka 'Rubbish,' Admits That He Loathes All the Passing — bleacherreport.com. https://bleacherreport.com/articles/2234051-pep-guardiola-calls-tiki-taka-rubbish-admits-that-he-loathes-all-the-passing. Accessed 21 Nov 2023
30. Utama, C., Maksum, A., Kristiyandaru, A.: The influence of physical condition, skill, and mental factors on the ability to play football. COMPETITOR: Jurnal Pendidikan Kepelatihan Olahraga **16**, 466 (2024). https://doi.org/10.26858/cjpko.v16i2.63597
31. Versic, S., Modric, T., Ćorluka, M., Zaletel, P.: The comparison of position-specific match performance between the group and knockout stage of the UEFA champions league. Sport Mont J. **22** (2024). https://doi.org/10.26773/smj.240702

YOCO-Sport: An End-to-End Framework for Deep Learning-Based Camera Calibration from Sports Broadcast Footage

Gerhardt Breytenbach[(✉)] and Jacomine Grobler

Department of Industrial Engineering, Stellenbosch University,
Stellenbosch, Western Cape, South Africa
breytenbachgerdo@gmail.com, jacominegrobler@sun.ac.za

Abstract. Camera calibration is often a necessary requirement for advanced computer vision-based sports analytics, such as player tracking, augmented reality, or movement-based performance analysis. In this paper, a novel end-to-end framework for automatic camera calibration is proposed. The specific novel aspect concerns the intentional design for sports with minimal to no pre-labelled camera calibration data. The proposed framework is tailored for sports with strictly defined rectangular sports fields, *i.e.* soccer, badminton, rugby, or volleyball, and delineates the design of deep learning-based algorithms to solve camera calibration from single-feed broadcast footage. The deep learning paradigms described in this framework are search-based and prediction-based camera calibration. Prediction-based camera calibration, however, is preferred due to the ease of data labelling and scalability of the solution. The application of the framework is verified by an implementation thereof for badminton, a sport without a public calibration dataset. For ease of use, the popular pose estimation algorithm by Ultralytics is utilised, namely YOLO11x-pose. The selected algorithm is validated independently with respect to two of the established benchmark datasets for soccer, namely the World Cup 2014 dataset and the time-sequence World Cup dataset. Comparative results were achieved with median intersection over union accuracies.

Keywords: Camera calibration · Keypoint detection · Framework · Sports broadcast footage

1 Introduction

Many sports analytics tasks—such as player tracking, augmented reality, ball (or object) tracking, and movement-based performance analysis—may necessitate interpreting pixel-based image data into physical units relative to the playing field [11,13]. For example, suppose that a ball is identified in an image at pixel coordinate (u, v), it is often necessary to express its position in world coordinates with respect to the known geometry of the field, represented in physical units such as meters or yards. This transformation from image space to world space is defined by a set of parameters, namely, the camera parameters [16]. Camera calibration can be utilised to determine these camera parameters, using only broadcast footage in which the playing field is visible [6].

Camera calibration is a principal computer vision task concerned with estimating the geometric relationship between image and world coordinates [40]. The world coordinate system is traditionally modelled as a three-dimensional (3D) Cartesian space, denoted by (x, y, z) and expressed in physical units [16]. The world coordinate system is typically defined with respect to a known object in the scene—in the context of sports broadcast footage, this object is commonly the sports field itself [25].

The implementation of camera calibration from sports broadcast footage dates back to 1998, where traditional computer vision techniques, like the Hough transform, were implemented [2, 25]. In 2017, however, Homayounfar *et al.* [19] proposed the first deep learning-based solution in which they emphasised the robustness of deep learning solutions in changing line visibility and weather conditions. Their paper is one of the seminal research works in the field and has led to many automatic deep learning-based solutions. For the same reason, the proposed framework in this paper is specifically focused on deep learning-based solutions.

Beyond developing individual calibration algorithms, several frameworks have been proposed to structure camera calibration solutions. These frameworks, however, primarily emphasise algorithm design rather than the data acquisition and preparation needed to adapt them to new sports [25]. In 2022, Zhang and Izquierdo formalised a framework for algorithmic design that has since frequently appeared in literature [32, 37, 39]. Their framework defined three modules: semantic segmentation, homography estimation, and homography refinement [39]. In a previous paper, Zhang and Izquierdo [38] also explicitly categorised the two main paradigms of deep learning-based camera calibration solutions, namely, *search-based* and *prediction-based* solutions. The principal difference between the two paradigms concerns the proposed method to interpret the input image. Search-based methods are characterised by identifying the most similar image in a database of images with corresponding calibration solutions [6, 34]. Prediction-based solutions, on the other hand, refer to methods that use the latent representation of a deep learning pipeline to estimate the solution directly [21, 39]. A popular use of the prediction-based paradigm is to predict visual cues in an image, after which an optimisation algorithm is deployed to estimate the calibration solution [29, 37]. Visual cues in an image may refer to keypoints, lines, or any visual information needed for the specific camera calibration approach.

Regardless of the selected paradigm, the most cited and accurate solutions to date are notably characterised by supervised learning methods. Accordingly, data availability, acquisition, and preparation are critical factors in enabling robust, transferable camera calibration pipelines. Unfortunately, the only sport with well-documented data for camera calibration is soccer [7, 14, 19]. Although other sports have also been investigated, publicly available datasets remain limited. Badminton, for example, is popular in the domain of computer vision-based sports analytics, yet to the best of the author's knowledge, a public dataset for camera calibration is not available—despite relevant literature using traditional camera calibration techniques [24]. This lack of data motivated the selection of badminton as a case study to validate the proposed framework.

Recently, in August of 2024, Roboflow published an article on their website detailing the use of a *you only look once* (YOLO) model variant for prediction-based camera calibration [35]. They utilised YOLOv8 to predict the intersection of soccer field lines but lacked validation against benchmark datasets, and their analysis was limited to scenes with abundant visible field lines [30]. Moreover, they did not address zoomed-in broadcast shots, a common occurrence in real-world footage. These limitations are addressed in this paper whilst using the latest Ultralytics model, YOLO11 [23], which is validated on standardised benchmark datasets for soccer, before extending it to badminton.

The framework proposed in this paper addresses the limitations of prior work by introducing an end-to-end design that spans from data acquisition to automated homography (or projection) matrix estimation. The main contributions of the paper are threefold: (1) A novel end-to-end framework for automatic camera calibration from sports broadcast footage, (2) a new camera calibration solution validated by popular soccer benchmark datasets, and (3) the application of this framework and solution to badminton. A more detailed discussion of the related work is written in Sect. 2 followed by the proposed framework in Sect. 3, after which its adaptation to badminton is described in Sect. 4, and the numerical results are discussed in Sect. 5.

2 Related Work

Historically, traditional computer vision techniques were used to solve camera calibration from broadcast footage [25]. In 2017, Homayounfar *et al.* [19] initiated a fundamental change in the solution approach by incorporating a deep segmentation network to interpret scene context and robustly extract the field markings. In 2018, Sharma *et al.* published the first search-based camera calibration solution [34]. Their data acquisition approach included labelling four points in an image, estimating the homography matrix via the *direct linear transform* (DLT) [1], and using the matrix to generate a synthetic edge image. Subsequently, Chen and Little [6] proposed an alternative method that defined statistical distributions over camera parameters, enabling the automatic generation of any number of synthetic edge images through sampled sets of camera parameters. The remainder of this section is focused on search-based approaches in Sect. 2.1, prediction-based approaches in Sect. 2.2, the scalability of the paradigms in Sect. 2.3, and formulating camera calibration as a keypoint prediction task in Sect. 2.4

2.1 Search-Based Approaches

Within the paradigm of search-based camera calibration, synthetic edge images are assembled into a database, each paired with its corresponding projection (or homography) matrix and a latent encoding. The fundamental idea of this paradigm is to process an input image to obtain its latent representation, and then *search* the database for the most similar encoding. Various matching and encoding methods have been proposed, however, all require a training dataset composed of pairs of broadcast and synthetic edge images. Consequently, search-based approaches inherently depend on acquiring these matching pairs. Since the generation of the synthetic edge image relies on a known homography matrix [6, 33], the exact homography must first be identified for each broadcast frame.

2.2 Prediction-Based Approaches

Prediction-based approaches, on the other hand, can be grouped into two subsets, namely, *direct* and *indirect* prediction-based approaches. Direct prediction-based approaches directly predict the calibration solution as the latent representation of a deep learning pipeline [21, 38]. Indirect prediction-based approaches, instead, predict visual cues in an input image upon which optimisation techniques can be deployed to define the calibration solution [29, 37]. In 2024, Gutiérrez-Pérez and Agudo proposed a third paradigm, termed *optimisation-based* approaches, describing methods utilising segmentation networks (such as U-Net [31]) to extract visual cues as unstructured image data, after which optimisation techniques estimate the calibration solution [15]. Based on the above-mentioned definition of prediction-based approaches, however, it can be argued that optimisation-based approaches can form a subclass of indirect prediction-based approaches, because they first predict visual cues before adopting an optimisation approach. Therefore, upon considering the paradigm differences, distinctions can be made between the different deep learning paradigms based on the fundamental processes from the initial unstructured (image) data to structured data.

2.3 Scalability of the Different Paradigms

Given these distinctions, it becomes important to understand the influences of each paradigm regarding the complexity and flexibility of data acquisition. The critical advantage of indirect prediction-based approaches, and the main motivation for distinguishing them clearly, lies in their fundamentally simpler requirements for data acquisition and adaptability. Specifically, indirect prediction-based approaches require only annotated point correspondences between broadcast images and known field templates [28, 29]. Unlike search-based and direct prediction-based methods, indirect prediction-based methods can utilise simple keypoint identifiers without explicitly needing homography matrices or predefined calibration parameters. While all paradigms benefit from homography-supervised labels when available, indirect prediction-based methods uniquely possess the ability to be developed entirely without such explicit supervision, significantly simplifying data acquisition. Furthermore, because indirect prediction-based models are trained to detect general visual cues rather than learn specific calibration parameters, their training process does not inherently encode precise field dimensions. This independence greatly facilitates adaptation to varying field sizes without extensive retraining. Conversely, search-based methods are inherently tied to exact field dimensions, as field templates directly define the synthetic training database. Similarly, direct prediction-based approaches implicitly encode specific dimensional information through their explicitly defined calibration labels. Such dimensional dependence constitutes a notable limitation for sports like soccer, where variations in field dimensions lead to non-linear scaling of geometric features [4,20]. Consequently, indirect prediction-based methods offer a potentially more scalable solution across sports with diverse and variable field dimensions.

2.4 Camera Calibration as Keypoint Prediction

These advantages, particularly simplified data acquisition and adaptability, motivate the adoption of indirect prediction-based approaches for developing scalable, transferable calibration frameworks, such as the one presented in this paper. A recent practical demonstration by Roboflow provides a compelling approach that treats camera calibration as a keypoint detection task, deploying YOLOv8x-pose to identify line intersections as keypoints [35]. Specifically, they employ OpenCV's homography estimation function, which first constructs a system of equations using the DLT and, subsequently, applies the least squares method to robustly estimate the homography parameters from point correspondences between their field template and the predicted keypoints. While their results are visually convincing, no quantitative validation against established benchmarks was presented. A further notable advantage of YOLO is its accessibility—despite its underlying architectural complexity, it is easily deployable as a plug-and-play model provided by Ultralytics [23].

3 Framework

The general methodology from the Roboflow article [35] is adopted in this study, utilising the most recent version of YOLO by Ultralytics—YOLO11x-pose—which is explicitly validated using two well-known benchmark datasets: the *World Cup 2014* (WC14) dataset [19] and the *Time-Sequence World Cup* (TSWC) dataset [7]. Building on the advantages of indirect prediction-based approaches outlined in the preceding section, a sequential, four-module framework is presented for the development of automated camera-calibration solutions directly from raw broadcast footage. The paradigm of choice is indirect prediction-based camera calibration, with the use of YOLO11x-pose to predict keypoints in a frame, after which the DLT is deployed to estimate the homography matrix. The framework, *you only calibrate once* (YOCO)-Sport, is comprised of four consecutive modules: (i) the *field template*, (ii) *data preparation*, (iii) *keypoint prediction*, and (iv) *matrix formation*. Each module is described in detail, according to the pipeline displayed in Fig. 1.

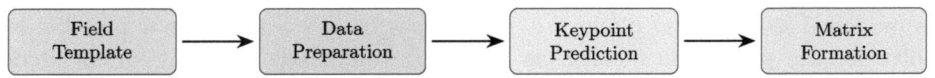

Fig. 1. Overview of the YOCO-Sport framework's four modules.

3.1 Field Template

The field template is a parametric virtual representation of all field markings, in which the intersections of boundary and line markings (vertices) and the segments between them (edges) are defined with explicit coordinates. Each vertex defined within the field template must be treated as a distinct and uniquely identifiable keypoint. To ensure consistent training and inference, a fixed numerical identifier (ID) is assigned to every keypoint, starting from 1 up to the total number of vertices. These IDs must remain consistent across all annotations and data-processing steps, as the prediction model is trained to detect specific keypoints by their assigned identities, rather than arbitrary field intersections. In practice, this data may be stored in a configuration file [19], but it is often preferable to encapsulate it within a dedicated class that provides specific methods to extract the vertices and edges [14]. For sports with flexible dimensions, e.g. soccer or rugby, a class-based approach avoids the need for multiple files covering every possible field size.

Accurate field templates are essential, as even minor discrepancies can lead to noticeable homography matrix estimation errors. First, ensure all units are consistent, noting that some publicly available templates use yards and others use meters. Second, verify that vertex locations conform to the sport's official geometry. For instance, Roboflow's template erroneously positions the penalty-arc semicircle so that it intersects the penalty area at the same horizontal level as the goal-area line. In reality, the arc should meet the penalty area closer to the horizontal axis of the field [20,36].

For the purpose of training a keypoint prediction algorithm, however, the above-mentioned error will not be particularly noticeable, because the network will be trained to predict the pixel coordinates of line intersections without necessarily learning the underlying structure of the field template. Only when the homography matrix is estimated and applied, based on the incorrect field template, will the projection errors be visible. Fortunately, within this paradigm of solutions, the field template can be adjusted at any time without retraining the network. Furthermore, based on a thorough study of the use of a generic virtual field template [4], it is nearly impossible to attain perfect calibration solutions if the field template does not perfectly represent the true field. Consequently, this paradigm is potentially the most accurate when accounting for varying field sizes.

In addition to defining the coordinates and connectivity of vertices, it is important to consider the dimensionality of the field template in relation to the target camera model. If the aim is to estimate a projection matrix, the field template must provide 3D world coordinates for each keypoint, with variation across at least two distinct planes in 3D space. For example, if the soccer field is modelled as a flat 2D plane, *i.e.* all vertices have $z = 0$, it is not possible to solve for the projection matrix, as the DLT would lack sufficient independent constraints to estimate the full 3D-to-2D mapping. In such cases, only a homography matrix can be estimated, capturing the projective relationship between two planes. Therefore, when constructing the field template, the intended calibration objective—homography matrix or projection matrix—must inform whether 2D or 3D coordinates are assigned to the vertices.

3.2 Data Preparation

The first stage of data preparation involves sourcing broadcast footage of a selected sport. While most sports have publicly available broadcast footage, few have datasets that are pre-labelled for camera calibration. If calibration labels are available, they can be directly leveraged to generate keypoint annotations for the associated frames. Otherwise, raw broadcast footage must be manually sampled and annotated to create the necessary supervision.

Broadcast footage may be obtained from public sources, *e.g.* YouTube, or through proprietary datasets. It is unnecessary to use every frame, as most consecutive frames are highly similar. Instead, frames may be sampled periodically, for example, every 10th frame, to maximise diversity while maintaining temporal relevance. After sampling, any frame with fewer than four visible vertices is discarded, as the DLT requires at least four point correspondences to estimate the homography matrix [16]. These frames are then annotated by marking the pixel locations and visibility flags of each keypoint according to the Ultralytics YOLO format, the preferred annotation scheme for YOLO-based keypoint detection tasks [26].

Within this annotation scheme, each visible object is associated with a class label, a bounding box, and a set of ordered keypoints. Each object is described in a single line within a text file corresponding to an image in the dataset. Consequently, the field template itself is seen as an object, where the bounding box encompasses the visible portion of the field, and the keypoints correspond to the vertices defined in the field template. Each label follows the format:

$$\{c, x, y, w, h, p_{x_1}, p_{y_1}, v_1, p_{x_2}, p_{y_2}, v_2, \ldots, p_{x_n}, p_{y_n}, v_n\},$$

where c denotes the class index, (x, y) is the normalised bounding box center, (w, h) are the normalised width and height, and each (p_{x_i}, p_{y_i}, v_i) triplet defines the normalised pixel coordinates and visibility of keypoint i. All coordinates, as well as the width and height of the bounding box, are normalised between 0 and 1. Visibility is modelled as a discrete variable $v \in \{0, 1, 2\}$, where $v = 0$ indicates an absent keypoint, $v = 1$ an occluded keypoint, and $v = 2$ a fully visible keypoint. If a keypoint is absent, it must still be included in the annotation, however, it can be assigned a default location of $(0, 0)$ and visibility flag of zero, to facilitate robust handling of missing data during training.

The dataset configuration is specified through a `data.yaml` file, which defines the structure and location of the training resources. The file includes the relative paths to the training, validation, and, optionally, the testing datasets. The class index and class name must also be specified. Furthermore, the expected keypoint array shape should be defined, e.g. $(n, 3)$, where n refers to the number of keypoints and 3 corresponds to the dimensions per keypoint, i.e. (p_{x_i}, p_{y_i}, v_i). An optional inclusion is the "flip_idx" parameter, which defines pairs of symmetric keypoints under horizontal flipping, however, since each vertex is explicitly represented in the virtual field template, symmetry mapping is not required and generally not recommended in this framework.

3.3 Keypoint Prediction

The keypoint prediction module is implemented using the Ultralytics YOLO Python interface, which provides easily accessible software for training keypoint detection models. In particular, an instance of the YOLO class from the ultralytics Python package is initialised with the desired model, after which the training procedure can be launched by specifying the dataset and hyperparameters. All training artefacts, including detailed results as well as the best and last model weights, are automatically saved upon completion. Further documentation and examples for configuring the training process are available in the Ultralytics guide [22].

It is critical to employ a YOLO-pose model, as standard YOLO object-detection models predict only bounding boxes without keypoint annotations, whereas the pose models also output keypoint locations. Different pose models can be explored depending on the complexity of the task, however, YOLO11x-pose is recommended in this framework to maximise robustness across varying line visibility and weather conditions. This specified model requires approximately 194.9 *giga floating operations per second* (GFLOPS), for an input resolution of 640 × 640. This computational complexity is roughly 24% more efficient than the YOLOv8x-pose network reported in [35], with a computational complexity of 257.8 GFLOPS [27].

In real-time processing scenarios, a throughput of at least 25 *frames per second* (FPS) is required. Given that YOLO11x-pose demands approximately 194.9 GFLOPS per image, achieving 25 FPS necessitates a sustained computational performance of roughly 4.9 *tera* FLOPS (TFLOPS). Modern laptop GPUs, *e.g.*, the NVIDIA RTX 4050 and RTX 4060 Laptop GPUs, offer approximately 9 and 11.6 TFLOPS of theoretical performance, respectively [17,18]. Consequently, real-time deployment is feasible on mid-range to high-end consumer hardware. Practical inference speeds, however, may vary depending on factors such as memory bandwidth, thermal throttling, and software stack efficiency.

3.4 Matrix Formation

The matrix formation module estimates the homography matrix that maps the virtual field template to the broadcast frame. Following keypoint prediction, the set of predicted keypoints must be filtered to extract only those deemed visible and reliable. Specifically, any keypoint predicted on the image boundaries, *i.e.*

with normalised coordinates equal to 0 or 1, is excluded from the correspondence set. The resulting filtered set of predicted keypoints is then matched to the corresponding vertices of the field template.

To solve the homography matrix, the DLT algorithm is employed, using a least squares solution to minimise the reprojection error. In practice, OpenCV's `findHomography` function is utilised, which implements the DLT with least squares optimisation by default. This framework supports both least squares and robust estimators such as *random sample consensus* (RANSAC) [12] for solving the DLT, depending on the characteristics of the data. Least squares is preferred when accurate keypoints are reliably available, while RANSAC is used to mitigate the impact of spurious keypoints in scenarios with limited keypoint visibility. If the field template provides 3D coordinates for the vertices, and sufficient variation exists across multiple planes, the full camera projection matrix can be estimated instead of a homography matrix. In this case, the DLT is adapted to solve the linear system relating 3D world coordinates to 2D image coordinates. OpenCV's `solvePnP` function provides a suitable implementation for estimating the projection matrix under such conditions, utilising methods such as iterative least squares.

Although matrix refinement steps are often recommended in the literature, they are not necessary within this framework. When the network predicts keypoint locations accurately, the DLT reliably estimates high-quality homography matrices. Since this framework treats the problem as a keypoint detection task, rather than a pose estimation task, occasional anomalous keypoint predictions may occur. Unlike pose estimation, keypoint detection does not inherently enforce the predefined structural relationships of the object on the image plane. To address such inconsistencies, an optional refinement step can be introduced, imposing the field template's structure onto the predicted keypoints through a heuristic adjustment. As noted, such refinement remains an optional addition to the framework.

4 Badminton Implementation

YOCO-Sport is applied to develop a camera calibration solution for badminton from broadcast footage. A field template is constructed in Python as a class according to official court dimensions [8], while a 13-minute compilation of professional match highlights, covering multiple camera angles, is curated and downloaded from YouTube[1]. In Fig. 2, an illustration of the field template (left) is provided to scale in meters, along with an example frame from broadcast footage (right).

Roboflow's Universe platform [10] is used to sample the broadcast footage and label the keypoints for each frame. A keypoint detection project is created, incorporating a class *skeleton* that defines the connections between keypoints as edges. This skeleton serves as a visual guide during labelling but does not preserve metric properties such as edge lengths. The labelling tool on the Roboflow

---

[1] https://youtu.be/6D7RtmlyIq0?si=aRo5zzLAHqgI5kGT.

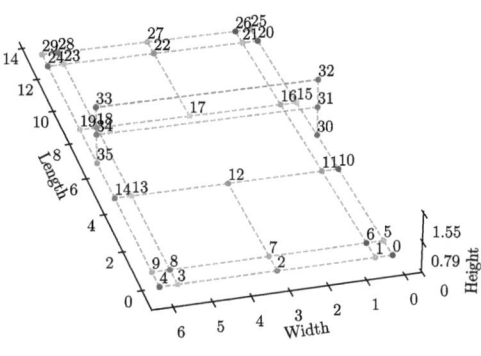 

(a) Field template in meters.   (b) Example broadcast frame.

Fig. 2. The inputs used for badminton camera calibration: (a) the field template constructed in Python, and (b) a representative frame from broadcast footage.

platform is employed to annotate the dataset. The annotation tool allows labels from previous frames to be loaded and refined, which is advantageous for video footage where consecutive frames are highly similar. Vertices may be slightly repositioned, and bounding boxes adjusted through the drag-and-drop interface.

After labelling, the dataset is formatted and divided into 204 training images, 40 validation images, and 20 testing images. All images are resized to 640 × 640 pixels to standardise input dimensions for training, while no augmentations are applied. The dataset is publicly available on the Roboflow Universe platform, along with a pretrained model [3]. This online model, however, is not the solution reported in this paper—it was trained in the Roboflow Universe using the "Roboflow 3.0 Keypoint Detection (Fast)" architecture.

After downloading the dataset, the data.yaml file is modified by removing the "flip_idx" field and correcting the directory structure to match the local project. A YOLO11x-pose model is trained for 300 epochs on the curated badminton dataset, with images resized to 512 × 512 and processed in batches of three. The loss function weights are adjusted to reflect the relative importance and difficulty of keypoint prediction versus object detection. Three specific loss function weights are changed—the box loss weight is reduced from 7.5 to 3, while the pose loss and keypoint objectness loss weights are each set to 10.

Following keypoint prediction, homography matrix estimation is performed to relate the predicted keypoints with the field template. The methodology outlined in Sect. 3.4 is adopted, where predicted keypoints located at the image boundaries are excluded, and the remaining correspondences are used to solve the homography matrix via the DLT algorithm. OpenCV's findHomography function is utilised with least squares optimisation to compute the final transformation. Through this procedure, the field template is mapped accurately to each broadcast frame, completing the implementation of the framework.

5 Results

The following results encompass the training of the keypoint detection model in Sect. 5.1, the accuracy of the homography matrices estimated via the DLT for badminton broadcast footage in Sect. 5.2, and lastly, a validation of the algorithmic pipeline used on a comparative study with soccer benchmark data in Sect. 5.3.

5.1 Keypoint Prediction Training Results

The keypoint detection model is evaluated by considering the set of predicted keypoints per frame, referred to as the predicted *pose*. To assess the quality of each pose, the *object keypoint similarity* (OKS) score is computed between 0 and 1, based on the Euclidean distances between corresponding predicted and ground-truth keypoints. These distances are normalised by the object scale and adjusted according to keypoint visibility. To summarise the model's precision and recall performance, the *average precision* (AP) is computed as the area under the precision-recall curve, indicating both accuracy and completeness of the predictions.

The mAP@50 metric reports the mean AP across all pose predictions, where a prediction is considered correct only if its OKS exceeds 50%. In contrast, mAP@50-95 provides a more comprehensive evaluation by averaging the AP across ten OKS thresholds, ranging from 50% to 95%, in 5% increments. This stricter metric evaluates the model's robustness under increasingly stringent matching criteria. Both metrics, computed on the validation set across 300 training epochs, are illustrated in Fig. 3, where the solid line denotes mAP@50 and the dotted points indicate mAP@50-95.

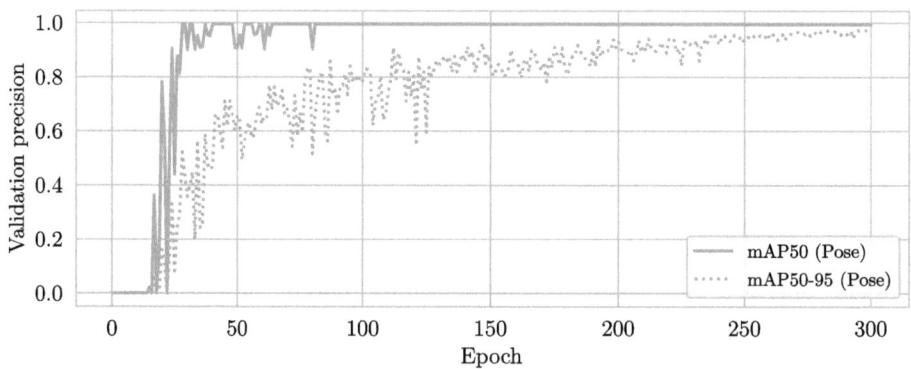

Fig. 3. Validation average precision curves for keypoint detection over training epochs. The solid line represents mAP@50, and the dots represent mAP@50-95.

The learning trends illustrated in Fig. 3 indicate that the model progressively improves its ability to predict keypoint poses with higher precision. The gap between mAP@50 and mAP@50-95 highlights the increased challenge of maintaining high prediction accuracy across stricter OKS thresholds. This difference supports the extended training duration, as the more stringent mAP@50-95 metric continues to improve steadily, particularly between epochs 250 and 300.

While mAP metrics offer insight into keypoint prediction accuracy, they do not directly reflect the quality of the resulting camera calibration, because predicted keypoints may lie close to the ground truth yet still fail to preserve the correct spatial configuration of the field template. To address this potential misalignment, a homography matrix is estimated from the predicted keypoints and used to project the camera's viewpoint onto the field template. The alignment between the predicted and ground-truth projections onto the field template is then evaluated using the *intersection over union* (IoU) metric, specifically referred to as IoU_{part}.

5.2 Badminton Homography Estimation Accuracy

The metric IoU_{part} is particularly useful in sports such as hockey, where only a portion of the field is typically visible in the broadcast view. In badminton, however, most of the court is usually in view. Therefore, two additional metrics are introduced: IoU_{whole} and IoU_{mask}. The IoU_{whole} metric evaluates how accurately the homography projects the entire field template, including regions not visible in the image. In contrast, IoU_{mask} measures the overlap when projecting the field template onto the image plane, constrained to the visible area. All three metrics are computed on the test set of 20 images and summarised in Table 1.

Table 1. IoU metrics to assess the predicted homography matrices on the 20 test images for badminton camera calibration.

Metric	Mean	Std	Median	Min	Max
IoU_{part}	97.72%	2.21%	98.67%	92.55%	99.73%
IoU_{whole}	94.33%	6.22%	96.86%	78.21%	98.52%
IoU_{mask}	97.50%	3.41%	98.99%	88.60%	99.59%

Among the three metrics, IoU_{whole} shows the greatest variability, indicating increased difficulty in estimating the complete field projection. Two illustrative examples are presented in Fig. 4: the best and worst-performing test images according to the IoU_{part} metric. In each image, the true field projection derived from ground-truth keypoints is shown in green (bottom), the predicted projection is shown in red (top), and the detected keypoints are marked in black. For the best example, the model achieved accuracies of 99.73%, 98.40%, and 99.37% for IoU_{part}, IoU_{whole}, and IoU_{mask}, respectively. The right-hand image represents the most challenging case in the dataset, with this example producing the lowest recorded scores across all three metrics as listed in Table 1.

Fig. 4. Example overlays of the field template warped onto broadcast frames using (left) the best homography matrix and (right) the worst homography matrix, according to the IoU$_{part}$ metric. The green (bottom) lines represent the true template, whereas the red (top) lines represent the predicted template, and the black markers represent the keypoint predictions. (Color figure online)

The visual examples in Fig. 4 indicate that the central region of the court is consistently well predicted, whereas distortions are more common near the far and near ends. In the right-hand image, the predicted keypoints in the top left corner appear to misinterpret the boundary between the green court outer area and the adjacent red stadium flooring as part of the white court lines. This confusion is likely due to a distributional bias in the training data, where scenes with similar red backgrounds were under-represented in the relatively small training set of 204 images. Further analysis revealed that the sharp contrast between colours in these areas is often misinterpreted as structural edges by traditional computer vision algorithms such as the Canny edge detector [5] and Hough Transform [9]. Increasing both the size and visual diversity of the training set is therefore expected to improve the model's robustness in such challenging regions. Additionally, expanding the dataset may support distortion estimation by enabling the model to detect deviations from geometric expectations. For instance, if a predicted set of points is noticeably curved despite corresponding to a known straight field line, the deviation between the predicted and expected geometries could be used to estimate distortion parameters.

Since the physical dimensions of a badminton court are fixed, it is unnecessary to compute segment-wise overlap metrics. The overall IoU measures already account for geometric consistency across the field, due to the uniform dimensions of badminton courts. Segment-level IoU would be more appropriate in cases where the true and predicted field templates differ structurally, such as in settings with uncertain or variable field geometries, for example, soccer.

Field template prediction in badminton is relatively simple due to the mostly visible and well-defined court. Soccer footage, by contrast, poses greater challenges: substantially limited keypoint visibility, varying camera angles, and player or weather-related occlusions. To evaluate robustness under such conditions, the YOLO11x-pose model is evaluated on public soccer benchmark datasets.

5.3 Soccer Benchmark Results

The soccer field template employed in this study is based on the Roboflow configuration and conforms to the official *International Football Association Board* (IFAB) laws [20, 36]. Three public datasets are incorporated: WC14 [19], TSWC [7], and the Roboflow calibration dataset [30]. To train the model evaluated on the WC14 benchmark, the TSWC training set is downsampled to 300 images and combined with 209 WC14 training images and 255 Roboflow images. In contrast, the TSWC benchmark model is trained exclusively on the temporally consistent TSWC dataset to preserve its sequential structure. In both cases, true keypoint projections are generated using the provided ground-truth homography matrices. The predicted homography matrices are estimated via DLT, using RANSAC to improve robustness against keypoint prediction noise. The resulting calibration performance is summarised in Table 2 and compared against state-of-the-art methods to assess the validity of the proposed pipeline.

Table 2. Comparison of homography matrix estimation performance on WC14-test and TSWC-test datasets, evaluated using IoU_{part}.

Dataset	Approach	Mean (%)	Median (%)
WC14-test	Jiang et al. [21]	95.1	96.7
	Nie et al. [28]	95.9	97.5
	Chu et al. [7]	96.0	97.0
	Zhang et al. [39]	95.9	97.3
	Oo et al. [29]	96.9	97.9
	Gutiérrez-Pérez et al. [15]	**97.0**	**98.2**
	WC14- YOLO11x-pose (Proposed)	90.6	95.1
TSWC-test	Nie et al. [28]	97.4	97.8
	Chu et al. [7]	98.1	98.3
	Oo et al. [29]	98.5	98.7
	Gutiérrez-Pérez et al. [15]	**98.6**	**98.9**
	YOLO11x-pose (Proposed)	96.1	97.1

While the proposed approach does not outperform state-of-the-art methods on WC14 or TSWC, it demonstrates competitive accuracy with a notably simpler design. Rather than leveraging tailored architectures or curated pipelines, the method capitalises on standard models and broadly available data. This contributes to a scalable and reproducible calibration pipeline with strong baseline performance. Importantly, the implementation omits homography refinement—considered essential in prior work [15]—and avoids augmentation techniques such as greyscaling or synthetic noise. Lower accuracy samples typically coincide with fewer predicted keypoint predictions, especially where the template underrepresents curved line regions.

6 Conclusion

A novel end-to-end framework for automatic camera calibration from sports broadcast footage was presented in this paper. Designed specifically for rectangular sports fields, the framework integrates deep learning-based keypoint detection with homography matrix estimation using the DLT algorithm. The framework consists of four essential modules: field template design, data preparation, keypoint prediction, and matrix formation. Particular emphasis was placed on the first two modules, which are typically underexplored in the literature. By explicitly detailing the design of a field template and proposing a practical, reproducible data acquisition process, this work addresses a critical gap in the literature and enables the scalable generation of annotated datasets for underrepresented sports.

The latter two modules, namely Keypoint Prediction and Matrix Formation, comprise the algorithmic pipeline designed to predict the homography matrix from a given input image. This pipeline, built on the YOLO11x-pose architecture, was evaluated using two established soccer benchmark datasets (TSWC and WC14). To further demonstrate the framework's generalisability and robustness, it was also applied to badminton, a sport for which no pre-labelled calibration dataset currently exists. The resulting fully automatic solution highlights the adaptability of the proposed framework to new sports domains without requiring ground-truth homography matrices, nor the development of advanced, task-specific deep learning models.

The results confirm the viability of indirect prediction-based camera calibration for sports where annotated data is limited or unavailable. By framing keypoint detection as a pose estimation problem and leveraging modern deep learning models, the proposed approach simplifies data acquisition and enhances scalability. The successful application across both benchmarked and unlabelled sports demonstrates the framework's potential as a scalable, general-purpose tool for automated camera calibration in sports with standardised field geometries.

Despite its demonstrated effectiveness, the framework has several limitations that can inform avenues for future research. The accuracy of the estimated homography matrices remains sensitive to motion blur, occlusion, poor lighting, sparse keypoint predictions, and lens distortion. These limitations highlight the need for high-quality data acquisition and the potential value of adding noise to training datasets. In addition to enhanced datasets, further work may include adding multiple keypoints along predicted field lines to estimate potential distortion—by deriving deviations from expected straight lines. Incorporating temporal information across consecutive frames and applying homography refinement techniques may further improve the reliability and stability of the predicted matrices.

Overall, the proposed framework offers a scalable, reproducible, and adaptable solution for automated camera calibration across varied sports field geometries. It eliminates the reliance on homography-labelled datasets and provides a foundation for integrating distortion correction to enhance calibration accuracy.

References

1. Abdel-Aziz, Y.I., Karara, H.M.: Direct linear transformation from comparator coordinates into object space coordinates in close range photogrammetry. In: Proceedings of the Symposium on Close-Range Photogrammetry, Urbana (IL), USA, pp. 1–19 (1971)
2. Bebie, T., Bieri, H.: SoccerMan-reconstructing soccer games from video sequences. In: Proceedings of the 1998 International Conference on Image Processing (ICIP), vol. 1, pp. 898–902. IEEE, Chicago (1998). https://doi.org/10.1109/ICIP.1998.723665
3. Breytenbach, G.: Badminton-Keys dataset. Roboflow Universe. https://universe.roboflow.com/gerdos-computer-vision-games/badminton-keys. Accessed 29 Apr 2025
4. Breytenbach, G., Grobler, J.: Evaluating the accuracy of a generic field template for camera calibration in soccer broadcast footage. SN Comput. Sci. 6(2), 107 (2025). https://doi.org/10.1007/s42979-024-03636-0
5. Canny, J.: A computational approach to edge detection. IEEE Trans. Pattern Anal. Mach. Intell. PAMI 8(6), 679–698 (1986). https://doi.org/10.1109/TPAMI.1986.4767851
6. Chen, J., Little, J.J.: Sports camera calibration via synthetic data. In: Proceedings of the IEEE/CVF Conference on Computer Vision and Pattern Recognition Workshops (CVPRW), pp. 2497–2504. IEEE/CVF, Long Beach (2019). https://doi.org/10.1109/CVPRW.2019.00305
7. Chu, Y.J., et al.: Sports field registration via keypoints-aware label condition. In: Proceedings of the IEEE/CVF Conference on Computer Vision and Pattern Recognition Workshops (CVPRW), pp. 3523–3530. IEEE/CVF, New Orleans (2022). https://doi.org/10.1109/CVPRW56347.2022.00396
8. Dimensions.com: Badminton court dimensions. https://www.dimensions.com/element/badminton-court. Accessed 29 Apr 2025
9. Duda, R.O., Hart, P.E.: Use of the Hough transformation to detect lines and curves in pictures. Commun. ACM 15(1), 11–15 (1972). https://doi.org/10.1145/361237.361242
10. Dwyer, B., Nelson, J., Hansen, T., et al.: Roboflow (Version 1.0) [Software]. https://roboflow.com. Accessed 29 Apr 2025
11. Działowski, K., Forczmański, P.: Football players movement analysis in panning videos. In: Paszynski M., Kranzlmüller D., Krzhizhanovskaya V.V., Dongarra J.J., Sloot P.M.A. (eds.) Computational science – ICCS 2021, vol 12746, pp. 193–206. Springer, Cham (2021). https://doi.org/10.1007/978-3-030-77977-1_15

12. Fischler, M.A., Bolles, R.C.: Random sample consensus: a paradigm for model fitting with applications to image analysis and automated cartography. Commun. ACM **24**(6), 381–395 (1981). https://doi.org/10.1145/358669.358692
13. Ghassab, V., Maanicshah, K., Bouguila, N., Green, P.: REP-MODEL: a deep learning framework for replacing ad billboards in soccer videos. In: 2020 IEEE International Symposium on Multimedia, pp. 149–153. IEEE (2020). https://doi.org/10.1109/ISM.2020.00032
14. Giancola, S., Cioppa, A., Deliège, A., Magera, F., Somers, V., Kang, L.: SoccerNet 2022 challenges results. In: Proceedings of the 5th International ACM Workshop on Multimedia Content Analysis in Sports, pp. 75–86. ACM, New York (2022). https://doi.org/10.1145/3552437.3558545
15. Gutiérrez-Pérez, M., Agudo, A.: PnLCalib: sports field registration via points and lines optimization. arXiv preprint arXiv:2404.08401 (2024). https://doi.org/10.48550/arXiv.2404.08401
16. Hartley, R., Zisserman, A.: Multiple view geometry in computer vision (Chapter 3: Projective Geometry and Transformations of 3D), 2nd edn. Cambridge University Press, Cambridge (2003)
17. Hinum, K.: NVIDIA GeForce RTX 4050 laptop GPU benchmarks and specs. Notebookcheck. https://www.notebookcheck.net/NVIDIA-GeForce-RTX-4050-Laptop-GPU-Benchmarks-and-Specs.675695.0.html. Accessed 28 Apr 2025
18. Hinum, K.: NVIDIA GeForce RTX 4060 laptop GPU benchmarks and specs. Notebookcheck. https://www.notebookcheck.net/NVIDIA-GeForce-RTX-4060-Laptop-GPU-Benchmarks-and-Specs.675692.0.html. Accessed 28 Apr 2025
19. Homayounfar, N., Fidler, S., Urtasun, R.: Sports field localization via deep structured models. In: Proceedings of the IEEE Conference on Computer Vision and Pattern Recognition (CVPR), pp. 4012–4020. IEEE, Honolulu (2017). https://doi.org/10.1109/CVPR.2017.427
20. IFAB Law 1 - The Field of Play. https://www.theifab.com/laws/latest/the-field-of-play/#field-surface. Accessed 24 Mar 2025
21. Jiang, W., Gamboa Higuera, J.C., Angles, B., Sun, W., Javan, M., Yi, K.M.: Optimizing through learned errors for accurate sports field registration. In: Proceedings of the IEEE/CVF Winter Conference on Applications of Computer Vision (WACV), pp. 201–210. IEEE, Snowmass (2020). https://doi.org/10.1109/WACV45572.2020.9093581
22. Jocher, G.: Ultralytics YOLO documentation. https://docs.ultralytics.com/modes/train/. Accessed 28 Apr 2025
23. Jocher, G., Qiu, J.: Ultralytics YOLO11 (Version 11.0.0) [Software]. https://github.com/ultralytics/ultralytics. Accessed 25 Apr 2025
24. Ma, H., Ding, X.: Robust automatic camera calibration in badminton court recognition. In: Proceedings of the 2022 IEEE Asia-Pacific Conference on Image Processing, Electronics and Computers (IPEC), pp. 893–898. IEEE, Dalian (2022). https://doi.org/10.1109/IPEC54454.2022.9777532
25. Manafifard, M.: A review on camera calibration in soccer videos. Multimed Tools Appl. **83**, 18427–18458 (2024). https://doi.org/10.1007/s11042-023-16145-8
26. Munawar, R., Jocher, G., Noyce, M.: YOLO keypoint annotation format. Ultralytics Documentation. https://docs.ultralytics.com/datasets/pose/. Accessed 28 Apr 2025
27. Munawar, R., Jocher, G.: YOLOv8 performance metrics. Ultralytics Documentation. https://docs.ultralytics.com/models/yolov8/#performance-metrics. Accessed 28 Apr 2025

28. Nie, X., Chen, S., Hamid, R.: A robust and efficient framework for sports-field registration. In: Proceedings of the IEEE/CVF Winter Conference on Applications of Computer Vision (WACV), pp. 1936–1944. IEEE, Virtual Event (2021). https://doi.org/10.1109/WACV48630.2021.00198
29. Oo, Y.M., Jamsrandorj, A., Chao, V., Mun, K.R., Kim, J.: A residual attention-based EfficientNet homography estimation model for sports field registration. In: Proceedings of the 49th Annual Conference of the IEEE Industrial Electronics Society (IECON), pp. 1–7. IEEE, Singapore (2023). https://doi.org/10.1109/IECON51785.2023.10312494
30. Roboflow: football-field-detection dataset. Roboflow Universe. https://universe.roboflow.com/roboflow-jvuqo/football-field-detection-f07vi. Accessed 25 Apr 2025
31. Ronneberger, O., Fischer, P., Brox, T.: U-Net: convolutional networks for biomedical image segmentation. In: Proceedings of the International Conference on Medical Image Computing and Computer-Assisted Intervention (MICCAI), pp. 234–241. Springer, Munich (2015). https://doi.org/10.1007/978-3-319-24574-4_28
32. Shang, J.C., Chen, Y., Shafiee, M.J., Clausi, D.A.: Rink-agnostic hockey rink registration. In: Proceedings of the 6th International Workshop on Multimedia Content Analysis in Sports, pp. 73–81. ACM, Ottawa (2023). https://doi.org/10.1145/3606038.3616161
33. Sha, L., Hobbs, J., Felsen, P., Wei, X., Lucey, P., Ganguly, S.: End-to-end camera calibration for broadcast videos. In: Proceedings of the IEEE/CVF Conference on Computer Vision and Pattern Recognition (CVPR), Seattle, WA, USA, pp. 13624–13633 (2020). https://doi.org/10.1109/CVPR42600.2020.01364
34. Sharma, R.A., Bhat, B., Ghandi, V., Jawahar, C.V.: Automated top view registration of broadcast football videos. In: Proceedings of the IEEE/CVF Winter Conference on Applications of Computer Vision (WACV), pp. 305–313. IEEE, Lake Tahoe (2018). https://doi.org/10.1109/WACV.2018.00040
35. Skalski, P.: Camera calibration in sports with keypoints. Roboflow Blog. https://blog.roboflow.com/camera-calibration-sports-computer-vision/. Accessed 25 Apr 2025
36. Skalski, P.: Soccer field configuration script. GitHub. https://github.com/roboflow/sports/blob/main/sports/configs/soccer.py. Accessed 27 Apr 2025
37. Theiner, J., Ewerth, R.: TVCalib: camera calibration for sports field registration in soccer. In: Proceedings of the IEEE/CVF Winter Conference on Applications of Computer Vision (WACV), pp. 1166–1175. IEEE/CVF, Waikoloa (2023). https://doi.org/10.1109/WACV56688.2023.00122
38. Zhang, N., Izquierdo, E.: A high accuracy camera calibration method for sport videos. In: Proceedings of the 2021 International Conference on Visual Communications and Image Processing (VCIP), pp. 1–15. IEEE, Munich (2021). https://doi.org/10.1109/VCIP53242.2021.9675379
39. Zhang, N., Izquierdo, E.: A fast and effective framework for camera calibration in sport videos. In: Proceedings of the IEEE International Conference on Visual Communications and Image Processing (VCIP), pp. 1–5. IEEE, Suzhou (2022). https://doi.org/10.1109/VCIP56404.2022.10008882
40. Zhang, Z.: A flexible new technique for camera calibration. IEEE Trans. Pattern Anal. Mach. Intell. **22**(11), 1330–1334 (2000). https://doi.org/10.1109/34.888718

One-Shot Team Recognition and 3D Pose Estimation of Cyclists for Augmented Reality Visualization

Winter Clinckemaillie$^{(\boxtimes)}$, Jelle Vanhaeverbeke , Maarten Slembrouck , and Steven Verstockt

IDLab, Ghent University-imec, Ghent, Belgium
winter.clinckemaillie@ugent.be

Abstract. Advanced computer vision and machine learning technologies transform how we experience sports events. This research focuses on enhancing the viewing experience of cycling races by automatically identifying teams from helicopter footage. It employs a multi-stage pipeline that tackles challenges such as rapid motion and similar team uniforms. Initially, cyclists are detected and tracked. Team recognition is then performed using a one-shot learning approach based on Siamese neural networks, achieving a classification accuracy of 85% on a test set composed of previously unseen teams. This method reduces the need for extensive labeling. Additionally, temporal post-processing techniques, such as applying a moving average to confidence scores, further enhance classification performance. These methods ensure reliable identification of teams and track their presence throughout the race footage. Furthermore, we integrate 3D pose estimation to generate augmented reality (AR) overlays that display rider-specific information, such as names and speeds, enhancing the broadcast's informational value. The combination of advanced computer vision and AR showcases new possibilities for improving live sports broadcasts, particularly in challenging environments like road cycling.

Keywords: Computer vision · Object detection · One-shot classification · 3D Pose Estimation · Augmented Reality · Sports Broadcasting

1 Introduction

Cycling, in particular, has lagged behind other sports in adopting interactive and dynamic broadcast technologies and faces growing pressure to enhance its traditional broadcast model to meet modern viewer demands. Despite advancements in broadcasting technology, coverage of cycling events remains largely manual. Key information, such as rider names and team affiliations, is manually added to the broadcast, limiting automation and the depth of insights provided to viewers. In contrast, sports like football and basketball utilize augmented reality (AR) for dynamic overlays and real-time statistics [7]. For example, in football, virtual advertising dynamically adjusts digital banners based on the viewer's location or the target market of the broadcast. Technologies like these enable broadcasters to tailor advertising content during live broadcasts, providing localized and personalized experiences for viewers [23]. However, applying similar AR concepts to cycling is much more challenging due to the unique nature of its broadcasting environment. Unlike stadium sports with controlled conditions, cycling races take place outdoors over vast terrains with constantly changing viewpoints and varying environmental factors, such as weather and lighting. These challenges complicate the implementation of automated, dynamic overlays that remain consistent throughout the race.

This paper aims to address these challenges by exploring how advanced computer vision and artificial intelligence (AI) techniques can enhance cycling broadcasts, making them more interactive and informative. The first part focuses on team recognition in aerial cycling footage, a task complicated by occlusions from varying camera angles, such as side or frontal views, and visual obstructions like trees. Additionally, changing lighting conditions and the visual similarity of team jerseys make this task even more complicated. This work proposes an automated computer vision approach that leverages deep learning to improve accuracy and efficiency in team recognition. By applying one-shot learning techniques, the system is designed to generalize across different races and team jerseys without the need for extensive retraining.

The second part investigates the use of AR to dynamically visualize rider-specific data, such as names and speeds, directly on broadcast footage. We develop methods to stabilize and position these overlays based on riders' movements and camera perspectives, enhancing the informativeness of cycling broadcasts. Figure 1 illustrates examples of these AR elements, including dynamic overlays that follow riders and display relevant data during the race.

By combining team recognition with AR, this work aims to enhance cycling broadcasts, reducing manual input and improving race coverage with contextual visualizations. In Sect. 2, we provide an overview of related work on cyclist recognition and augmented reality applications in sports broadcasts. Next, in

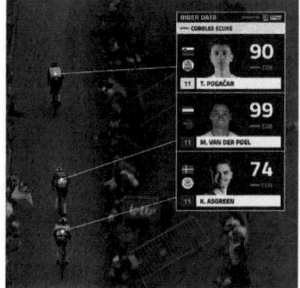  

Fig. 1. Examples of augmented reality overlays applied to helicopter footage during cycling races. The static card overlay (left) displays rider data. The dynamic name card (middle) follows the rider's movement and orientation. The ground marker (right) projects a visual indicator on the ground beneath the cyclist.

Sect. 3 we describe our proposed pipeline, which includes detection, tracking, team recognition, 3D bounding box calculation, and dynamic AR visualizations. Finally, in Sect. 4 we summarize the key findings and discuss potential directions for future research.

2 Related Work

This section provides an overview of the scientific context for this research, focusing on two main areas: identifying cycling teams and utilizing dynamic AR elements in sports broadcasts. This sets the foundation for the research questions addressed in this work, highlighting existing knowledge and ongoing challenges.

2.1 Cyclist Detection and Team Recognition

Cyclist Detection. Accurate cyclist detection is crucial for analyzing race footage, as it forms the foundation for tracking and team recognition. This requires high accuracy and speed, especially for real-time use. Challenges include distinguishing cyclists from motorcycles and spectators, managing occlusions within the peloton, and dealing with varying camera perspectives [17]. Recent deep learning approaches, particularly convolutional neural networks (CNNs), have significantly improved detection accuracy. Among these, the YOLO (You Only Look Once) framework stands out for its ability to process entire images in a single forward pass, making it well-suited for real-time applications. YOLOv8 introduced an anchor-free architecture and improved feature extraction, enhancing small-object detection [9]. More recently, YOLOv11 further advanced performance by integrating efficient attention mechanisms and optimized architectural components, achieving state-of-the-art accuracy for real-time detection [8].

Team Recognition. After detecting cyclists, identifying them into their respective teams is essential for enriching live broadcasts and improving race analyses. Various methods have been explored, each with specific applications and limitations in cycling. Liu and Bhanu introduced a pose-guided R-CNN for recognizing jersey numbers. While effective under controlled conditions, it is less suitable for aerial footage, where jersey numbers are rarely visible [14].

Verstockt et al. proposed a methodology using skeleton-based pose detection for identifying cyclists, combining pose orientation with jersey number recognition to improve identification accuracy [25]. By leveraging both the structural data from cyclists' poses and visual cues from their jerseys, this dual focus allows for more reliable team recognition, especially in dynamic scenes where traditional methods might struggle. Lavent's research demonstrated robust rider identification by detecting bib numbers using YOLOv8, leveraging asynchronous processing to enable real-time operation [13]. However, this method relies on close-range views and is less effective when bib numbers are occluded, which often occurs in aerial footage.

Recent work on team recognition in soccer has shown promising results using Siamese neural networks [21]. These networks effectively extract feature vectors from jerseys and compare them to a database of known jerseys. They are advantageous when there is a small amount of data for each class, employing a one-shot learning approach [5,12]. Siamese networks use loss functions like contrastive loss or triplet loss to minimize the distance between similar images and maximize the distance between dissimilar ones. This capability allows for the identification of new cycling teams with just one reference image, making the method highly adaptable and efficient for team recognition [22].

Tracking. Accurate tracking is essential for real-time cyclist monitoring. Traditional methods like SORT use a Kalman filter to estimate the position of cyclists across frames, allowing it to bridge short detection gaps caused by obstacles like poles or trees [2]. Deep SORT enhances this by adding deep appearance descriptors to reduce identity switches [26]. Lighter alternatives such as ByteTrack [27] and BoT-SORT [1] avoid deep embeddings, focusing instead on motion cues and detection confidence. ByteTrack achieves impressive speeds (1265 fps) by linking low-confidence detections. BoT-SORT further improves this by introducing enhanced motion modeling and outlier filtering, running at 46 fps [4]. This combination of high accuracy and speed makes BoT-SORT ideal for real-time cyclist tracking.

Research Gap. While techniques such as YOLO, Siamese networks, and BoT-SORT show promise, their performance in live cycling broadcasts is limited. Most were designed for ground-level or controlled settings and struggle with aerial views, small objects, rapid scene changes, and occlusion. This paper adapts detection and one-shot team recognition methods to these broadcast conditions and integrates tracking to improve temporal consistency. Addressing these challenges is crucial for enhancing automated team recognition and providing a more engaging viewing experience. This work aims to bridge the gap between current state-of-the-art methods and their practical use in real-world broadcasts.

2.2 AR Visualization

AR has transformed sports broadcasting by enhancing viewer experiences with dynamic overlays. For instance, in Formula 1, AR shows real-time data such as driver names and positions directly above cars [18]. Similarly, American football uses a digital first-down line for clear indicators of game progress [16]. In tennis, systems like Hawk-Eye project ball trajectories using precisely calibrated, high-speed cameras in controlled environments [19].

However, these systems are not directly applicable to cycling broadcasts. Unlike stadium sports or events in confined, well-defined areas, cycling races take place over long, open courses with varying terrain and dynamic camera views. Traditional AR alignment that relies on fixed reference points is less effective. Instead, AR overlays must adapt to continuously changing perspectives and athlete movements.

3D pose estimation offers a promising alternative by reconstructing an athlete's spatial position and orientation from video footage, enabling better alignment of overlays with moving cyclists. 3D Human Pose Estimation (HPE) models predict 3D joint coordinates, allowing for near real-time use with decent accuracy [15]. More recent transformer-based approaches, such as MotionBERT [28], further improve accuracy by modeling temporal and spatial dependencies across frames. However, these methods lack a holistic body model, so they can output anatomically implausible poses when joints are occluded or misdetected, causing jitter or limb errors in outdoor scenes like road cycling.

Human mesh recovery (HMR) addresses these issues by estimating full-body 3D meshes using parametric models. Early models like HMR minimized 2D reprojection loss [10], while PARE added part-based attention for better occlusion handling [11]. More recently, TokenHMR (Fig. 2) introduced a discrete token-based pose representation that enhances robustness in complex, real-world scenes [6]. Although computationally intensive, mesh models offer anatomically consistent outputs, making them suitable for cycling broadcasts. Further optimization is needed to make them viable for AR overlay placement during live races.

Research Gap. While existing AR systems perform well in controlled environments, cycling broadcasts require more adaptable methods. This work applies mesh-based 3D pose estimation to enable more robust, anatomically coherent overlay alignment under real-world conditions. This improves the accuracy and temporal stability of visualizations, even under motion and occlusion. Addressing these challenges is crucial for deploying AR in cycling and offering viewers a clearer, more informative, and engaging broadcast experience.

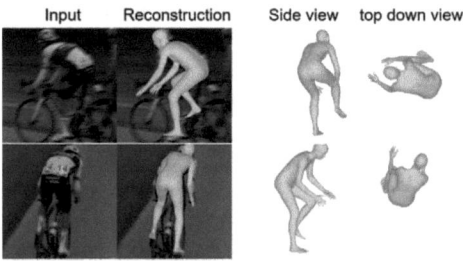

Fig. 2. Examples of reconstructed 3D meshes using TokenHMR from single RGB images.

3 Methodology and Results

This section presents the pipeline for team recognition and AR visualization in cycling race footage. As shown in Fig. 3, the pipeline consists of several key components, including cyclist detection, team recognition, 3D bounding box calculation, temporal postprocessing, and AR visualization.

The pipeline starts with per-frame cyclist detection using the YOLO model. This is followed by BoT-SORT tracking, which assigns unique IDs to each cyclist and maintains their identity across frames. This ensures consistency even during temporary occlusions or rapid camera movements. The persistent IDs allow detections to be linked over time, aiding in temporal postprocessing and enabling semi-automatic annotation, where metadata is assigned once per tracked cyclist and automatically propagated throughout the track.

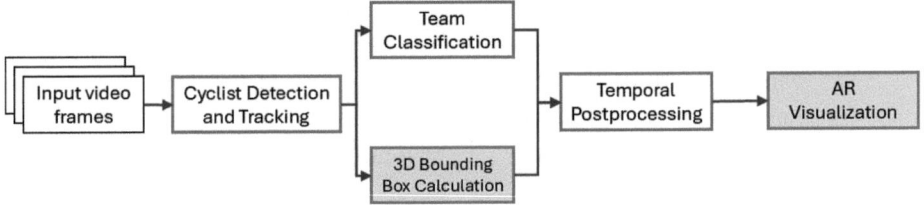

Fig. 3. Overview of the proposed pipeline for cyclist detection, tracking, team recognition, and AR visualization.

Team recognition employs a one-shot learning approach using Siamese neural networks. To align AR overlays with each cyclist's body orientation, TokenHMR is used to extract 3D keypoints. These keypoints are then used to compute a 3D bounding box, ensuring that overlays remain dynamically and accurately positioned across varying camera perspectives. Temporal postprocessing further enhances the pipeline by smoothing predictions, handling occlusions, and interpolating missing data, leading to stable outputs. Finally, AR visualization dynamically places text overlays, such as rider names and speeds, based on the 3D bounding boxes. These elements adjust according to cyclists' movements, providing viewers with dynamic information. The following sections provide a detailed breakdown of the pipeline's core components.

3.1 Cyclist Detection

We created a custom dataset of annotated images for detecting cyclists and motorcycles from helicopter footage, available on Roboflow [24]. This dataset contains 660 images with 20,111 annotations, focusing on cyclists' upper bodies to recognize teams by jerseys and logos. It also includes motorcycles to help differentiate cyclists from other objects commonly present in race footage. The dataset is split into training (70%), validation (15%), and test (15%) sets for thorough evaluation. We trained YOLOv8 and YOLOv11 models at two input resolutions: 640×640 (referred to as S for small) and 1920 × 1080 (referred to as L for large), across five model sizes ranging from nano to extra-large. To improve robustness, we applied data augmentation techniques such as horizontal flipping, rotation, brightness adjustment, and noise addition, along with early stopping to prevent overfitting. Models were assessed using mAP@0.5 on cyclists and motorcycles. In addition, inference time was measured on an NVIDIA GeForce RTX 4090 GPU, including pre- and post-processing steps, to reflect realistic performance in real-time settings. Results are summarized in Table 1.

Table 1. mAP@0.5 and inference time (ms) for YOLOv8 and YOLOv11 across different model sizes at two input resolutions: large (1920 × 1080) and small (640 × 640).

(a) YOLOv8

Size	mAP@0.5		Time (ms)	
	L	S	L	S
Nano	0.86	0.72	5.8	2.4
Small	0.86	0.75	7.5	2.6
Medium	0.89	0.80	14.9	4.5
Large	0.89	0.82	21.0	6.8
X-Large	0.90	0.83	33.1	9.5

(b) YOLOv11

Size	mAP@0.5		Time (ms)	
	L	S	L	S
Nano	0.88	0.73	5.5	2.8
Small	0.90	0.81	7.7	2.8
Medium	0.91	0.83	14.9	4.2
Large	0.92	0.85	18.9	5.4
X-Large	0.92	0.85	32.1	8.3

The results indicate that YOLOv11 consistently outperforms YOLOv8 in both accuracy and inference speed across nearly all configurations. Consequently, we have chosen the YOLOv11-large model at a resolution of 1920 × 1080 as the default detector in our pipeline. This model best balances high accuracy (0.92 mAP) and an acceptable processing time (18.9 ms) for real-time applications, which is vital for reliable tracking and team recognition.

In situations where additional downstream modules, such as 3D pose estimation or full AR rendering, increase the computational load, a smaller variant of YOLOv11 may be preferable. For instance, the nano model is about three times faster (5.5 ms compared to 18.9 ms at 1920 × 1080) while maintaining competitive accuracy (0.88 mAP), making it a viable fallback option for stricter real-time constraints. Overall, YOLOv11 enhances detection quality and allows for flexible trade-offs between speed and accuracy based on system requirements. Although there were fewer annotations for motorcycles, their inclusion highlighted opportunities for improving the dataset to further reduce false positives, particularly in distinguishing cyclists from similar non-cyclist objects.

3.2 One-Shot Team Recognition

The team recognition module is designed to identify cyclists' teams based on the visual appearance of their jerseys, even when encountering new or previously unseen teams. This capability is essential because jersey designs often change throughout the season, such as when riders wear national champion kits or special editions. It is not practical to retrain a classification model each time a design changes; therefore, a scalable solution that generalizes across seasons and variations is necessary. To address this, we employ a Siamese neural network approach, which is effective for one-shot learning. This setup processes images through a shared encoder to generate feature embeddings. The network is trained to minimize the distance between embeddings for the same team while maximizing the distance for different teams.

Dataset and Preprocessing. We created a custom dataset of cropped images of cyclist jerseys. Each image measures at least 35 × 35 pixels and is resized to 105 × 105 pixels. Each image is labeled by team, and we applied data augmentation techniques, such as rotation and brightness adjustments, to enhance robustness, as illustrated in Fig. 4. The primary dataset includes an average of 80 images per team across 24 teams (with 70% for training and 30% for validation). To evaluate the one-shot capability, we established a separate test set comprising 15 new teams, each with an average of 10 images. This set assesses the recognition of unfamiliar jerseys, reflecting the frequent outfit changes in cycling.

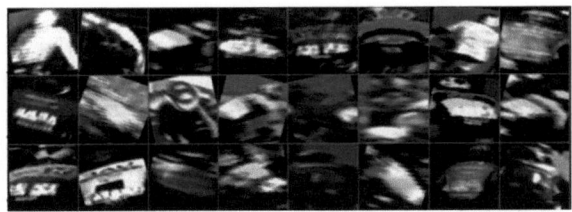

Fig. 4. Examples of cropped rider jersey images used for team recognition.

Network Architecture. The Siamese network, shown in Fig. 5, consists of two identical encoders producing 256-dimensional embeddings, based on the design by Koch et al. [12]. This structure includes five convolutional layers with ReLU activations and 2 × 2 max-pooling, followed by two fully connected layers. We reduced the final embedding size from 4096 to 256 dimensions to lower computational costs and improve classification efficiency during inference. Using the contrastive loss formulation by Pei-Xia et al. [20], the network is trained to minimize the distance between embeddings of positive pairs (same team) and ensure negative pairs (different teams) are separated by a margin m. The loss is defined as:

$$\mathcal{L}(y, d) = (1 - y)\frac{1}{2}d^2 + y\frac{1}{2}\max(0, m - d)^2 \qquad (1)$$

where $y \in \{0, 1\}$ indicates whether a pair is dissimilar (1) or similar (0), d is the Euclidean distance between embeddings, and m is the margin.

To support interpretability and provide a binary similarity score during training, a dense layer is added on top of the computed distance to predict whether a given pair represents the same team or not. This allows for a direct performance metric alongside the embedding space learning.

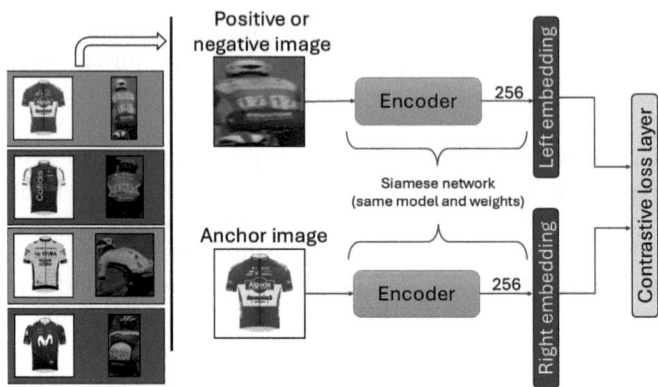

Fig. 5. Siamese network architecture and examples of positive (green border) and negative (red border) pairs used in training. (Color figure online)

Training and Validation Performance. To evaluate pairwise classification, we utilized the output from the dense layer. Initially, each team was represented by only one online-sourced anchor image (as shown in Fig. 5), and negative sampling was performed randomly, which resulted in a validation accuracy of 71%.

We then enhanced our approach by replacing the single anchor image with three images from actual race footage, each from a different viewpoint (front, side, above). This allowed for a more comprehensive similarity model that better reflects how jerseys might appear in real conditions, rather than relying on a single reference that may not generalize well. Furthermore, using three anchors for each team allowed us to generate multiple positive (one for each anchor) and negative pairs (from other teams) for each training sample. Additionally, we employed hard negative mining to target visually similar jerseys by explicitly sampling negative examples from teams with nearly identical color schemes. This approach reduced the number of false positives associated with near-identical apparel. We empirically selected the margin hyperparameter used in the contrastive loss by conducting a grid search over values from 0.5 to 1.5 (with step size 0.2). The best validation performance was achieved at $m = 0.7$. Overall, these refinements increased the dense layer's validation accuracy to 83%, significantly improving our ability to distinguish between teams.

Model Usage and One-shot Capability. After training the Siamese network, we utilize the shared encoder to map newly detected jerseys and anchor images into a common embedding space. We employ a k-Nearest Neighbors (kNN) algorithm with cosine distance for team label assignment: each new crop is embedded, and the closest anchor determines its team. This approach allows for easy integration of new teams with minimal anchor images.

To enhance prediction reliability, we compute confidence scores from the inverse distances to all anchors and normalize them with a softmax function:

$$s_i = \frac{e^{d_i^{-1}}}{\sum_{j=1}^{N} e^{d_j^{-1}}} \qquad (2)$$

Here, s_i is the normalized confidence score for anchor i, d_i is the cosine distance between the input crop and anchor i, and N is the total number of anchors. The inverse distance ensures that closer anchors contribute more to the final score.

To classify the team, scores of all anchors belonging to the same team class C are summed:

$$S_C = \sum_{i \in C} s_i \qquad (3)$$

where S_C is the total confidence score for class C, and the sum is taken over all anchors i that represent team C. The team with the highest S_C is selected. This method improves interpretability and accuracy, especially when combined with race context. For example, broadcasters typically know which riders or teams are being filmed; by restricting recognition to those specific teams, we reduce the likelihood of misidentifications.

The model achieved a mean accuracy of 94% (minimum 67%) on known teams in the validation set, confirming its effectiveness for previously encountered jerseys. Crucially, it maintained strong performance on the unseen test set, with an 85% mean accuracy (minimum 60%), demonstrating robust one-shot generalization.

We observed that classification accuracy decreases when the jersey images are smaller, as these contain less visual detail, reducing model confidence and recognition accuracy. However, temporal tracking across frames can partially compensate by accumulating predictions, improving stability. Overall, Siamese networks offer effective, scalable one-shot recognition, enabling quick adaptation to new jersey designs without extensive retraining.

3.3   3D Bounding Box Calculation

A key challenge in cycling broadcasts is aligning AR elements, such as name cards and speed readouts, with each rider's orientation and movement. Traditional 2D bounding boxes are insufficient for achieving dynamic alignment in uncontrolled environments. We propose a 3D bounding box method utilizing Human Mesh Recovery (TokenHMR) for accurate 3D keypoint estimation. This approach ensures AR overlays remain visually aligned with each cyclist's posture, regardless of changes in camera perspective or orientation.

TokenHMR requires a full-body bounding box as input. However, the YOLO model discussed in Sect. 3.1 detects only upper bodies for team recognition. Currently, we employ a two-step detection process: an upper-body YOLO model first identifies cyclists, followed by a second general YOLO model detecting full-body bounding boxes. We apply an Intersection over Union (IoU) threshold of

0.3 to obtain cyclist-specific full-body bounding boxes, filtering out non-cyclist detections.

Figure 6 illustrates the procedure for calculating a cyclist's 3D bounding box. Key points on the cyclist are first extracted (see Fig. 6a) to define orientation axes. The x-axis is drawn from the left to the right hip, the y-axis is the average of vectors from the hips to the hands and along the spine, and the z-axis is the cross product of the x- and y-axes. The y-axis is then recalculated using the new z- and x-axes to maintain orthogonality, as illustrated in Fig. 6b. Key points are projected onto these axes to find extreme values, from which the bounding box corners are derived (see Fig. 6c). The resulting 3D bounding boxes accurately capture cyclist translation and rotation, enabling precise 2D projection.

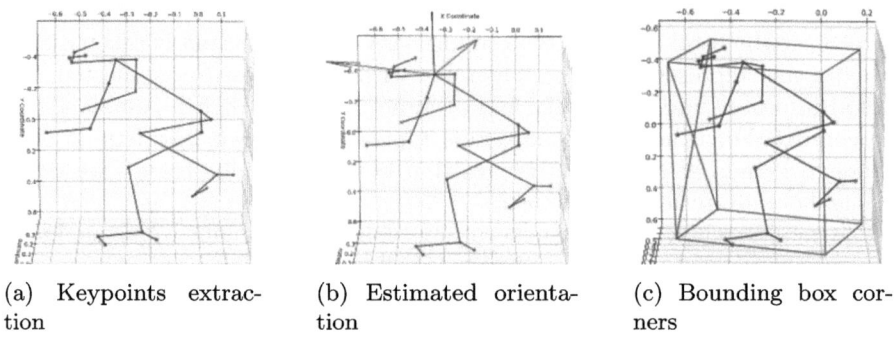

(a) Keypoints extraction (b) Estimated orientation (c) Bounding box corners

Fig. 6. Steps for calculating a 3D-oriented bounding box.

3.4 Temporal Postprocessing

Temporal postprocessing enhances stability and accuracy by leveraging tracking IDs from detection to maintain consistency across frames. Key techniques include:

- Enhancing cyclist detection reliability: The BoT-SORT tracker, which internally uses a Kalman filter, predicts the future positions of cyclists when they are temporarily occluded, such as when passing behind obstacles like poles. This helps maintain consistent tracking IDs across frames, even when cyclists are briefly out of view. Additionally, the system filters out false positives, such as when spectators or motorcycles are mistakenly detected as cyclists, by confirming detected cyclists across multiple consecutive frames (e.g. five frames) before assigning an ID.
- Improve team recognition accuracy: A moving average of confidence scores smooths short-term fluctuations in team recognition, reducing the impact of errors from individual frames.
- Interpolate missing 3D pose data: Linear interpolation fills gaps in the 3D bounding boxes for frames where human mesh recovery is not possible. This ensures that virtual elements remain aligned with cyclists, even when pose estimation data is missing due to detection errors or occlusions.

– Reduce jitter in AR overlays: Simple exponential smoothing functions are applied to 3D bounding box data to minimize jitter in AR overlays, ensuring stable and accurate alignment with cyclists' movements, even during rapid motion or changes in camera angles.

3.5 AR Visualization

This section explores the integration of AR elements into our cycling broadcast pipeline. We use 3D bounding boxes to position virtual overlays, such as rider names and race statistics, around individual cyclists. By focusing on smaller groups, such as breakaways, we can prevent visual overload, as applying AR to the entire peloton could become confusing and quickly exceed current technical limits. Also, breakaways typically involve riders from different teams, simplifying the identification process, as it relies on recognizing teams and linking them to the correct riders.

Our system leverages broadcasters' data to identify which groups and riders are being filmed. Incorporating this information allows us to accurately assign recognized teams to the right cyclists, ensuring precise overlays. Although this semi-automated process already provides flexibility, future iterations could further enhance it with more sophisticated data integration from broadcast systems. For identifying individual cyclists, soft biometrics such as helmet designs, body posture, or unique riding styles could improve accuracy. Meanwhile, recognizing bib numbers is more suitable for close-up footage; from aerial or long-distance angles, the numbers are generally too small to detect reliably.

AR elements are positioned using the 3D bounding boxes described in Sect. 3.3. The area allocated for these virtual elements is determined by the dimensions of the bounding boxes. The height of the text area corresponds to the height of the bounding box, while the width is based on the aspect ratio of the text or image. There are multiple ways to position the text area, as illustrated in Fig. 7. Three examples of possible placements include: above the rider, aligned with their body orientation (Fig. 7a); on the ground ahead of the rider (Fig. 7b); and perpendicular to the rider's direction (Fig. 7c).

Currently, placement is selected manually, but future iterations could automate this process by incorporating artificial intelligence (AI) techniques to detect available space within the frame. This would involve identifying road surfaces, calculating rider directions, and ensuring that text does not overlap with other riders in the broadcast image.

Text overlays are projected onto the original image by mapping 3D corner points to 2D image coordinates using OpenCV [3], as illustrated in Fig. 8. Alpha compositing is utilized to make the text semi-transparent; however, blending operations may introduce processing overhead. Optimized GPU support helps mitigate some of these extra costs.

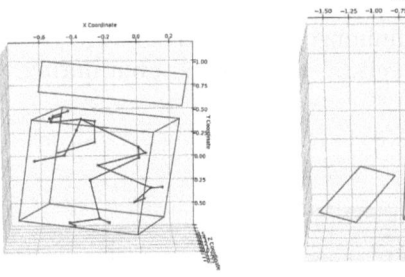

(a) Positioned above the rider

(b) On the ground ahead of the rider

(c) Perpendicular to the rider's path

Fig. 7. Several options for positioning the text area.

Fig. 8. Examples of AR visualizations.

Improvements were made to stabilize overlays and reduce the chaotic visualizations caused by noisy bounding box positions and orientations across frames. First, a mean orientation axis for all cyclists was calculated for each frame, rather than determining each cyclist's orientation independently. This adjustment ensured that the text angles within a frame aligned uniformly, resulting in a more stable view. Figure 9a illustrates the original approach, which used individual orientations, while Fig. 9b shows the uniform visualization achieved through averaging the orientation axes.

A second improvement involved smoothing the corner points of the text projection using tracked data from previous frames. Variations in bounding box orientation and text area positions across consecutive frames often caused jitter. By applying simple exponential smoothing to key points and a moving average to orientation axes over a buffer of recent frames, the text display became more stable and fluid. A promising next step would be integrating a road segmentation model to ensure that text planes are projected only onto road pixels, avoiding overlap with riders or distracting background elements.

A demo showcasing both 3D-aligned text and static name overlays is available as supplementary material at: https://youtu.be/0BbghrJUJmM.

 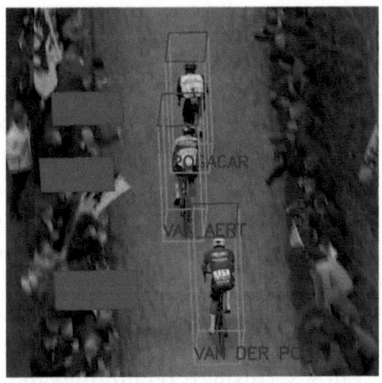

(a) Text placement using individual cyclist's orientation.

(b) Text placement using averaged orientation.

Fig. 9. Comparison of text placement strategies.

Despite these improvements, the computational cost of TokenHMR remains a significant challenge, averaging 90 ms per frame on an NVIDIA GeForce RTX 4090 limiting real-time performance. Hardware-specific optimizations like TensorRT acceleration, could substantially reduce inference time.

Beyond hardware-level enhancements, several pipeline-level strategies could further reduce inference time. Adopting asynchronous team recognition, triggered only once sufficient evidence per tracked cyclist is available, would increase reliability and reduce unnecessary computation. Furthermore, simplifying the current two-step detection process (upper-body and full-body) into a unified cyclist detection model could eliminate redundant computations. Extraction of upper-body regions for team recognition could then rely on 2D keypoints derived from the 3D pose estimation, streamlining the pipeline and enhancing efficiency.

Additionally, frame skipping and interpolation techniques could significantly decrease computational demands. By processing fewer frames directly and interpolating results from periodically computed keyframes, minimal latency is introduced, which remains acceptable in broadcast scenarios. This strategy also ensures smooth visual continuity, particularly when camera movement is relatively stable.

Overall, these strategies provide a clear path toward achieving real-time operation at 25 fps, highlighting the practical feasibility of the proposed pipeline for live cycling broadcasts.

4 Conclusion

In this research, we presented a comprehensive pipeline for enhancing live cycling broadcasts through advanced computer vision and augmented reality techniques. The proposed system integrates YOLO-based cyclist detection, BoT-SORT tracking, Siamese network-based team recognition, and 3D pose estimation using TokenHMR, enabling dynamic, context-aware AR visualizations from helicopter footage.

To detect cyclists, a YOLOv11 object detection model was trained on a dataset focusing on the cyclists' upper bodies. The YOLOv11-large model, trained on 1920 × 1080 pixel images, achieved a mean Average Precision (mAP) of 92% at an Intersection over Union (IoU) of 0.5, and a mean inference time of 19 ms per frame. This demonstrated its robustness and efficiency in diverse conditions, making it suitable for real-time applications. BoT-SORT was then utilized to track cyclists across multiple frames, ensuring accurate tracking and maintaining the identities of detected cyclists.

For team recognition, a Siamese neural network was employed, achieving an 85% accuracy through techniques like hard negative mining and multiple anchor images. The model demonstrated strong generalization by accurately recognizing unseen jerseys in the test set. Temporal data from tracking further improved recognition reliability, forming a comprehensive system for detecting, tracking, and identifying cycling teams in real-time.

AR components were developed to project visualizations based on 3D poses of cyclists, determined using the TokenHMR method. This approach allowed for the creation of 3D bounding boxes and dynamic text overlays, which were then applied to the footage using perspective transformations and alpha compositing. While the system demonstrated the feasibility of AR integration, optimizing processing speed remains essential for live broadcast applications.

Future work could explore integrating sensor data, such as bike positions, to enable individual rider identification and more personalized insights during broadcasts. Automating text placement using road segmentation models could further reduce visual clutter and enhance readability. Additionally, advancements in hardware or lightweight neural networks could improve processing times, making real-time AR visualizations more viable. Employing interpolation techniques for frame prediction could also reduce computational demands while ensuring smooth and stable overlays. An additional improvement could involve accounting for lighting conditions and shadows during AR overlay rendering to enhance visual realism and coherence in broadcast footage.

In conclusion, this work demonstrates the potential of integrating computer vision and AR technologies to transform the viewing experience of cycling races. By advancing team recognition accuracy and developing AR visualizations, this research lays the groundwork for more engaging and informative sports broadcasts. Continued exploration and refinement of these technologies will be crucial in addressing current limitations and expanding their applicability, ultimately contributing to a more immersive viewing experience for cycling enthusiasts.

References

1. Aharon, N., Orfaig, R., Bobrovsky, B.Z.: Bot-sort: robust associations multi-pedestrian tracking (2022). https://arxiv.org/abs/2206.14651
2. Bewley, A., Ge, Z., Ott, L., Ramos, F., Upcroft, B.: Simple online and realtime tracking. CoRR arxiv:1602.00763 (2016)

3. Bradski, G.: The OpenCV Library. Dr. Dobb's J. Softw. Tools (2000)
4. Broström, M.: BoxMOT: pluggable SOTA tracking modules for object detection, segmentation and pose estimation models. https://zenodo.org/record/7629840. https://github.com/mikel-brostrom/boxmot
5. Duque Domingo, J., Medina Aparicio, R., González Rodrigo, L.M.: Improvement of one-shot-learning by integrating a convolutional neural network and an image descriptor into a siamese neural network. Appl. Sci. **11**(17) (2021). https://doi.org/10.3390/app11177839. https://www.mdpi.com/2076-3417/11/17/7839
6. Dwivedi, S.K., Sun, Y., Patel, P., Feng, Y., Black, M.J.: Tokenhmr: advancing human mesh recovery with a tokenized pose representation (2024). https://arxiv.org/abs/2404.16752
7. Goebert, C.: Augmented reality in sport marketing: uses and directions. Sports Innov. J. **1**, 134–151 (2020). https://doi.org/10.18060/24227
8. Jocher, G., Qiu, J.: Ultralytics yolo11 (2024). https://github.com/ultralytics/ultralytics
9. Jocher, G., Qiu, J., Chaurasia, A.: Ultralytics YOLO (2023). https://github.com/ultralytics/ultralytics
10. Kanazawa, A., Black, M.J., Jacobs, D.W., Malik, J.: End-to-end recovery of human shape and pose (2018). https://arxiv.org/abs/1712.06584
11. Kocabas, M., Huang, C.H.P., Hilliges, O., Black, M.J.: Pare: part attention regressor for 3d human body estimation (2021). https://arxiv.org/abs/2104.08527
12. Koch, G.R.: Siamese neural networks for one-shot image recognition. In: ICML Deep Learning Workshop (2015). https://api.semanticscholar.org/CorpusID:13874643
13. Lavent, Q.: Bib Number-Based Cyclist Identification in Motorcycle Video Footage for Race Reporting. Master's thesis, Ghent University (2024)
14. Liu, H., Bhanu, B.: Pose-guided r-cnn for jersey number recognition in sports. In: 2019 IEEE/CVF Conference on Computer Vision and Pattern Recognition Workshops (CVPRW), pp. 2457–2466 (2019). https://doi.org/10.1109/CVPRW.2019.00301
15. Martinez, J., Hossain, R., Romero, J., Little, J.J.: A simple yet effective baseline for 3d human pose estimation (2017). https://arxiv.org/abs/1705.03098
16. Mashable: The yellow first-down line: an oral history of a game changer (2022). https://mashable.com/archive/yellow-first-down-line
17. Naik, B.T., Hashmi, M.F., Bokde, N.D.: A comprehensive review of computer vision in sports: open issues, future trends and research directions. Appl. Sci. **12**(9) (2022). https://doi.org/10.3390/app12094429. https://www.mdpi.com/2076-3417/12/9/4429
18. Noble, J.: Ai replays and augmented reality: What's new for f1's tv coverage in 2023 (2023). https://www.autosport.com/f1/news/ai-replays-and-more-augmented-reality-whats-new-for-f1s-tv-coverage-in-2023/10439342/
19. Owens, N., Harris, C., Stennett, C.: Hawk-eye tennis system. In: International Conference on Visual Information Engineering, pp. 182–185 (2003). https://doi.org/10.1049/cp:20030517
20. Pei-Xia, S., Hui-Ting, L., Luo, T.: Learning discriminative cnn features and similarity metrics for image retrieval, pp. 1–5 (2016). https://doi.org/10.1109/ICSPCC.2016.7753634
21. Santos, P.D., Jerri, V.: Team Recognition in Sports Events. Master's thesis, Polytechnic University of Catalonia (2023)
22. Schroff, F., Kalenichenko, D., Philbin, J.: Facenet: a unified embedding for face recognition and clustering. CoRR arxiv:1503.03832 (2015)

23. Supponor: Cutting through the regulation of the virtual advertising landscape. Technical report. Supponor Ltd. (2020). https://supponor.com/whitepaper-cutting-through-the-regulation-of-the-virtual-advertising-landscape/. Accessed 14 Jan 2025
24. UGent: Ventoux dataset (2024). https://universe.roboflow.com/ugent-hkozj/ventoux-chxhg
25. Verstockt, S., Van den broeck, A., Van Vooren, B., De Smul, S., De Bock, J.: Data-driven summarization of broadcasted cycling races by automatic team and rider recognition. In: Proceedings of the 8th International Conference on Sport Sciences Research and Technology Support - icSPORTS, pp. 13–21. INSTICC, SciTePress (2020). https://doi.org/10.5220/0010016900130021
26. Wojke, N., Bewley, A., Paulus, D.: Simple online and realtime tracking with a deep association metric. CoRR arxiv:1703.07402 (2017)
27. Zhang, Y., et al.: Bytetrack: multi-object tracking by associating every detection box (2022). https://arxiv.org/abs/2110.06864
28. Zhu, W., et al.: Motionbert: a unified perspective on learning human motion representations (2023). https://arxiv.org/abs/2210.06551

How Do Football Teams Play? A Deep Embedded Clustering Approach to Reveal Playing Styles

Ege Demir[(✉)], Yusuf H. Şahin, and Nazım Kemal Üre

Istanbul Technical University, Istanbul, Turkey
{demireg20,sahinyu,ure}@itu.edu.tr

Abstract. This study proposes a novel approach to clustering football teams' playing styles using a Deep Embedded Clustering (DEC) algorithm applied to large-scale event data. The dataset comprises over 3 million events from 1,826 matches played in the top-tier leagues of England, Spain, Italy, Germany, and France during the 2016/2017 season. Each match event, such as passes, shots, and duels, contributes to a comprehensive representation of team behavior on the field. To capture tactical nuances, the dataset is segmented into four distinct phases of play, and each phase is clustered independently. These intermediate clustering results are aggregated to create a feature representation for each team, which is subsequently clustered to reveal dominant playing styles. A detailed feature engineering process, inspired by recent literature, incorporates spatial and temporal elements of play, including pass motifs, positional tendencies, and graph-based metrics. The resulting clusters are evaluated in terms of their clustering quality, measured by Silhouette, $A(C)_1$, and $A(C)_2$ scores, their predictive utility for match outcomes, and their tactical interpretability. The analysis demonstrates clear performance disparities among styles, offering insights into the effectiveness of specific tactical schemas. This methodology enables data-driven tactical analysis and benchmarking in football, highlighting the potential of unsupervised learning, particularly deep clustering, to inform strategic decision-making in sports analytics. The source code of this study is available at: https://github.com/egecjdemir/how_football_teams_play.

Keywords: Football analytics · Deep clustering · Playing styles · DEC · Unsupervised learning · Sports data mining

1 Introduction

The strategic analysis of football has experienced a paradigm shift with the rise of data-driven methodologies enabled by the increasing availability of detailed event data. These datasets comprised of time-stamped, spatially tagged actions such as passes, shots, and duels offer a high-resolution lens into the tactical patterns and behaviors of teams. As a result, clustering-based playing style identification has emerged as a prominent research frontier in sports analytics.

Among the most notable contributions to this field is the work by Moffatt et al. [14], which establishes a benchmark framework for identifying team styles across four game phases using traditional clustering algorithms. Their study introduces robust evaluation metrics Adjusted Cluster Accuracy 1 ($A(C)_1$) and Adjusted Cluster Accuracy 2 ($A(C)_2$) designed to assess both statistical coherence and expert-aligned interpretability of discovered clusters. However, their framework relies on static clustering algorithms (K-Means, K-Medoids, hierarchical agglomeration clustering using Ward's linkage) and does not incorporate representation learning, limiting its capacity to model complex, high-dimensional tactical behaviors.

To address these limitations, our study introduces a novel deep clustering pipeline based on Deep Embedded Clustering (DEC) [22], which integrates representation learning and clustering in a unified architecture. Unlike traditional methods that operate directly on raw features, DEC learns a compact latent space tailored for clustering, enabling it to capture nuanced team behavior patterns that are otherwise difficult to distinguish.

Our pipeline segments data into four distinct game phases—In-Possession, Out-of-Possession, Positive Transition, and Negative Transition—reflecting the dynamic structure of football gameplay. A rich feature set is engineered for each phase, incorporating spatial zoning, pass motifs, graph connectivity, and contextual metadata. DEC is applied independently to each game phase, and team-level cluster assignments are determined through majority voting across match-level representations.

Beyond internal validity metrics such as silhouette score [19], we evaluate our clusters using $A(C)_1$ and $A(C)_2$, ensuring comparability with prior work. In our experiments, DEC consistently outperforms benchmark algorithms across all metrics and game phases. Furthermore, we provide actionable tactical insights by examining the average playing characteristics of each cluster, their win rates against other styles, and their distribution across Europe's top five leagues.

The main contributions of our work can be summarized as:

- We introduce DEC to football analytics as a robust alternative to traditional clustering algorithms for style identification.
- We provide a comprehensive multi-phase framework combining unsupervised learning with feature-rich representations and evaluation through both statistical and outcome-based metrics.
- We offer interpretable, practical outputs—such as tactical matchup outcomes and league-specific cluster distributions—that can support data-informed coaching and strategic planning. Identifying the dominant playing style of football teams can help coaches tailor their game plans to exploit the strategic tendencies and weaknesses of their opponents.

By combining deep learning and tactical analysis, our work demonstrates the power of modern unsupervised techniques to uncover interpretable and effective playing styles in football.

2 Related Work

Recent work in football analytics has seen growing adoption of unsupervised learning to categorize team playing styles from event or tracking data [2,5]. Among the most influential contributions is Moffatt et al. [14], who proposed a flexible clustering pipeline applied separately to four game phases. Their framework introduced composite clustering quality indices $A(C)_1$ and $A(C)_2$, integrating statistical and expert-weighted scores to assess clustering outcomes. To compute these metrics, the authors combined indicators such as within-cluster sum of squares (I_{wcss}) [9], separation index (I_{sep}) [2], distance to the cluster centroid (I_{distcc}) [1], and density index (I_{dens}) [9], using either equal weighting or expert-informed importance weights as follows:

$$A(C)_1 = \frac{I_{wcss} + I_{sep} + I_{distcc} + I_{dens}}{4} \tag{1}$$

$$A(C)_2 = \frac{I_{wcss} + 0.5 \cdot I_{sep} + I_{distcc} + 0.25 \cdot I_{dens}}{2.75}. \tag{2}$$

This work benchmarked various algorithms, including K-Means, K-Medoids, and Ward's method, and provided a replicable basis for evaluating tactical clustering frameworks. We use this as the main baseline for comparison and apply their composite scores to assess the effectiveness of our approach. While their method offers a solid foundation, it remains reliant on shallow representations and classical clustering techniques. To address these limitations and enhance clustering quality, we introduce a deep learning-based alternative.

To improve upon this benchmark, our study employs Deep Embedded Clustering (DEC) [22], which integrates deep representation learning with unsupervised clustering via an autoencoder architecture. DEC jointly optimizes feature compression and cluster assignment by minimizing a Kullback-Leibler divergence loss between soft and target distributions. Unlike traditional methods that operate on raw or dimensionally reduced data (e.g., PCA [6]), DEC learns task-specific embeddings, which lead to better separation and more semantically meaningful clusters. In our experiments across all four game phases, DEC consistently outperforms classical algorithms on all three metrics: Silhouette score, $A(C)_1$, and $A(C)_2$, where the Silhouette score represents the cohesion and separation of cluster assignments and is given as:

$$s(i) = \frac{b(i) - a(i)}{\max\{a(i), b(i)\}} \tag{3}$$

Here, $a(i)$ is the average distance between sample i and all other points in the same cluster, while $b(i)$ is the minimum average distance between sample i and all points in the nearest neighboring cluster. The final score is computed as the mean of $s(i)$ across all data points.

To situate our contribution within the broader landscape of football analytics, we refer to the taxonomy proposed by Plakias et al. [18], which organizes the literature into three categories: playing style recognition, contextual analysis, and style effectiveness. Most studies in the first category rely on static features or aggregate match-level statistics, such as possession percentages or shots taken. By contrast, our method captures more nuanced dynamics through a rich feature set constructed from event-level spatiotemporal data.

Several prior works inspired our feature engineering strategy. Diquigiovanni and Scarpa [7] and Peña and Touchette [17] advocate for network-based representations of team behavior using passes and duels. We incorporate their ideas by modeling pass interactions as directed graphs and computing connectivity scores. In addition, we extend the concept of pass motifs introduced by Gyarmati et al. [8] as indicators of short-passing structure and build-up complexity. These approaches are complementary to our event-level feature engineering, which synthesizes both structural and temporal aspects of play.

Earlier methods such as PCA-based clustering [12,20] have been used to extract latent tactical dimensions, but often sacrifice interpretability and coherence. Moreover, many studies rely on tracking data to assess formations and player movements [4], which offer granular insights but are typically inaccessible for public research. Our method, in contrast, relies solely on open-source event data, making it more scalable and reproducible.

Emerging work also investigates football through the lens of broader AI frameworks. For instance, Yildiz et al. [10] propose using probabilistic model checkers to simulate match outcomes in a transparent, interpretable fashion, diverging from conventional black-box approaches. Meanwhile, Bandara et al. [3] analyze open-space dynamics using time-series models derived from television broadcasts, introducing a novel spatial-temporal metric to assess performance. Additionally, recent efforts such as [21] demonstrate how large language models (LLMs) can be integrated with football event data via retrieval-augmented generation pipelines, highlighting new directions for accessibility and narrative analysis in the sport.

Finally, we go beyond existing studies by evaluating not only the coherence of discovered playing styles but also their relative effectiveness in real matchups. This mirrors the under-explored third branch of tactical analysis highlighted by Plakias et al. [18]. Our win-percentage matrix shows how styles perform against each other, offering potential tactical applications for match planning.

By integrating phase-specific modeling, rich feature construction, deep clustering, and outcome-based validation, our study provides a comprehensive and novel solution for discovering tactical typologies in football. It addresses key gaps in the literature by combining methodological rigor with practical applicability and advances the state-of-the-art in unsupervised football analytics.

3 Methodology

Our full pipeline for feature engineering and cluster assignment is given in Fig. 1, which outlines the complete flow from raw event data preprocessing to feature extraction, latent representation learning, and final clustering using DEC.

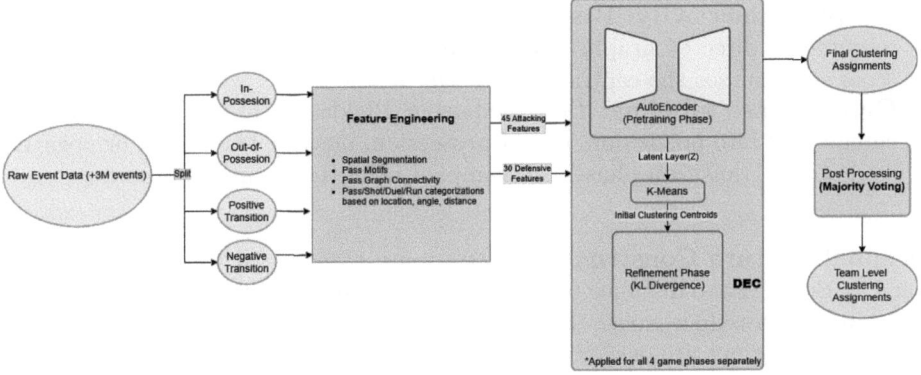

Fig. 1. Diagram of the proposed method.

3.1 Data Collection and Preprocessing

The dataset used in this study was provided by Massucco and Pappalardo [16], containing over 3 million event records from 1,826 matches played in the top-tier football leagues of France, England, Germany, Italy, and Spain during the 2016–17 season. Events include actions such as passes, duels, shots, and runs, recorded with spatiotemporal precision.

To mirror real tactical diversity and temporal dynamics, the data was segmented into four distinct **game phases** based on possession dynamics:

- **In-Possession (IP):** Attacking behavior while holding the ball.
- **Out-of-Possession (OP):** Defensive activity while the opponent is in control.
- **Positive Transition (PT):** Immediate phase after gaining possession.
- **Negative Transition (NT):** Immediate phase after losing possession.

To enrich the dataset beyond the 98 teams, each match played by a team was treated as an individual sample, resulting in 3,652 team-match entries per game phase. This approach allows for granular tactical analysis and facilitates generalization across varying match contexts.

3.2 Feature Engineering

Adopted Feature Concepts. We incorporate several established ideas from the literature:

- **Spatial segmentation:** Inspired by Diquigiovanni and Scarpa [7], the pitch is divided into three zones to preserve information about action location while avoiding feature explosion.
- **Pass motifs:** Following Gyarmati et al. [8], pass sequences are labeled (e.g., ABAB, ABCD) to infer playing style from pass patterns.

- **Graph connectivity:** Based on Peña and Touchette [17], passes are represented as directed graphs with players as nodes. The average connectivity score summarizes the centrality and ball distribution.
- **Game phase labeling:** Borrowed from Plakias et al. [18], each event is categorized into possession, out-of-possession, positive transition, or negative transition. Transition phases are sampled from their parent datasets.

Original Feature Construction. Passes are categorized by angle (side, forward, backward) and distance (short, mid, long), and subtypes (e.g., high, smart) are grouped to reduce noise. These are normalized to obtain frequency ratios. Similar calculations are performed per zone. Shots are classified according to the distance from the goal.

Defensive data includes duels, runs, and fouls. Runs are categorized by direction and distance; duels and fouls are broken down by context (e.g., aerial, ground, defensive). In total, 30 features are extracted for attacking phases and 45 for defensive ones.

3.3 Deep Embedded Clustering (DEC)

To overcome the limitations of traditional clustering algorithms such as K-Means in high-dimensional and non-linearly separable spaces, we adopt the **Deep Embedded Clustering (DEC)** model introduced by Xie et al. [22]. DEC unifies representation learning and clustering into a joint optimization problem by leveraging a deep autoencoder network with a clustering objective based on the Kullback-Leibler (KL) divergence.

Autoencoder Pretraining. The DEC pipeline begins by training a symmetric autoencoder on the normalized feature matrix. The encoder network reduces the input to a low-dimensional latent space $z \in \mathbb{R}^{d_z}$, while the decoder reconstructs the input from this embedding. The autoencoder is trained to minimize the mean squared reconstruction error:

$$\mathcal{L}_{AE} = \frac{1}{N} \sum_{i=1}^{N} \|x_i - \hat{x}_i\|_2^2$$

where x_i is the input feature vector, and $\hat{x}_i$ is the reconstructed output.

We use fully connected layers with ReLU [15] activations in both encoder and decoder, and train the model for 3,000 epochs using the Adam [11] optimizer with a learning rate of 10^{-3}. The latent vector size is set to 10 to balance representational compactness and information preservation.

Clustering Initialization. Once pretraining converges, the encoder is used to transform the entire dataset into the latent space. Initial cluster centroids are obtained by applying the K-Means algorithm on these latent vectors:

$$\{\mu_j^{(0)}\}_{j=1}^{k} = \text{KMeans}(\{z_i\}_{i=1}^{N})$$

These centroids serve as the starting point for the DEC optimization.

KL-Based Optimization. DEC refines the encoder to jointly minimize a clustering loss function based on the KL divergence between a soft assignment distribution Q and a sharpened target distribution P. Specifically, the soft assignment of a point z_i to cluster j is defined as:

$$q_{ij} = \frac{(1 + \|z_i - \mu_j\|^2/\alpha)^{-\frac{\alpha+1}{2}}}{\sum_{j'}(1 + \|z_i - \mu_{j'}\|^2/\alpha)^{-\frac{\alpha+1}{2}}}$$

where α is the Student's t-distribution [13] degree of freedom, typically set to 1.

The target distribution P is computed to emphasize confident assignments and normalize cluster frequency:

$$p_{ij} = \frac{q_{ij}^2/\sum_i q_{ij}}{\sum_{j'}(q_{ij'}^2/\sum_i q_{ij'})}$$

The final objective function is the KL divergence between P and Q:

$$\mathcal{L}_{KL} = \mathrm{KL}(P \parallel Q) = \sum_i \sum_j p_{ij} \log \frac{p_{ij}}{q_{ij}}$$

DEC is trained for another 3,000 epochs using the same optimizer, during which the encoder is continuously updated to minimize $\mathcal{L}_{KL}$, while cluster centroids are kept fixed.

Model Architecture. The DEC framework is built around a symmetric autoencoder architecture composed of fully connected layers with ReLU activations. The encoder maps the input $x \in \mathbb{R}^D$ to a latent space $z \in \mathbb{R}^{d_z}$, and the decoder attempts to reconstruct the input. In our implementation:

- The **encoder** has three dense layers: 128 → 64 → 10 dimensions.
- The **decoder** mirrors the encoder: 10 → 64 → 128 → input dimension.
- ReLU activations follow each hidden layer.

Formally, the encoder $f_\theta(x)$ is defined as:

$$f(x) = \mathrm{ReLU}(W_3(\mathrm{ReLU}(W_2(\mathrm{ReLU}(W_1 x + b_1)) + b_2)) + b_3)$$

where W_1, W_2, W_3 and b_1, b_2, b_3 are learnable parameters. The decoder follows the same structure in reverse.

Phase-Specific Training. This procedure is repeated independently for each of the four game phases: In-Possession (IP), Out-of-Possession (OP), Positive Transition (PT), and Negative Transition (NT), ensuring phase-specific clustering behavior.

For each phase, we set $k = 4$, since it consistently yielded the best results across all evaluation metrics (silhouette score, $A(C)_1$, and $A(C)_2$) in the in-possession data, as shown in Fig. 2.

Clustering Output. Upon convergence, final soft assignments q_{ij} are used to assign discrete labels by:

$$\hat{y}_i = \arg\max_j q_{ij}$$

These labels are then used for downstream analyses such as team-level aggregation via majority voting, visualization via PCA, and evaluation using clustering metrics.

3.4 Post-processing and Cluster Assignment

As each match is treated as a distinct instance, a given team may appear in multiple clusters throughout the season. To derive a single representative cluster for each team per game phase, we apply **majority voting** over all of its match-level assignments. This ensures that the final label reflects the team's dominant tactical behavior throughout the season.

The entire methodology, including preprocessing, feature extraction, DEC clustering, and majority voting, is conducted **independently for each game phase**, ensuring that phase-specific insights are preserved.

4 Evaluation

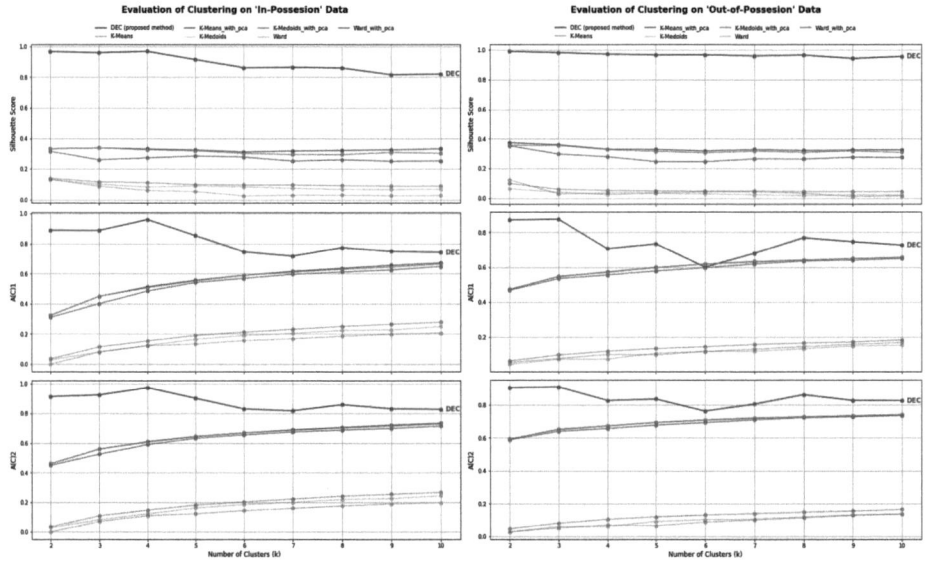

Fig. 2. Evaluation of clustering algorithms across various 'k' values for In-Possession (left) and Out-of-Possession (right) game phases.

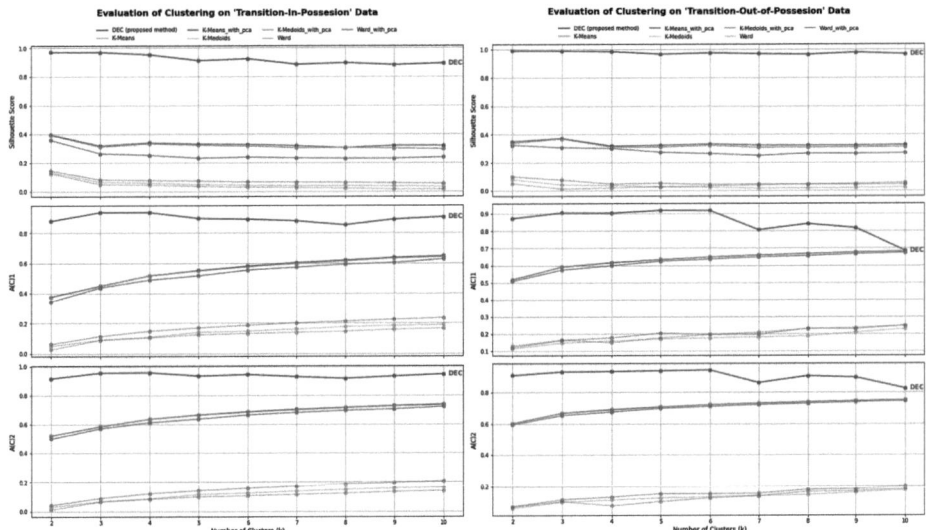

Fig. 3. Evaluation of clustering algorithms across various 'k' values for Positive Transition (left) and Negative Transition (right) game phases.

To assess the quality of the discovered clusters, we evaluate the performance of our proposed method: Deep Embedded Clustering (DEC) [22] against three classical clustering algorithms: K-Means, K-Medoids, and Ward's hierarchical clustering. These methods were chosen based on their usage in the benchmark study by Moffatt et al. [14], which also introduced two of the three evaluation metrics: Adjusted Cluster Accuracy 1 ($A(C)_1$), and Adjusted Cluster Accuracy 2 ($A(C)_2$).

As illustrated in Figs. 2 and 3, the DEC method consistently outperforms the baseline methods at nearly all values of k and in all phases of the game. It achieves a higher clustering quality as measured by all three metrics indicating that the clusters it produces are both internally coherent and externally meaningful.

Another notable finding is the performance improvement observed when K-Means, K-Medoids, and Ward are applied after reducing the dataset to two dimensions via PCA. This phenomenon highlights the curse of dimensionality faced by traditional clustering algorithms. The high number of input features can dilute distance-based similarity calculations, degrading cluster quality. In contrast, the DEC model is inherently robust to high-dimensional spaces, as it jointly optimizes feature representation and cluster assignment.

The loss curves and latent space visualizations in Figs. 4 and 5 support the stability and effectiveness of DEC training. Both the Mean Squared Error (MSE) loss of the autoencoder and the Kullback-Leibler (KL) divergence loss decrease rapidly and plateau near zero, indicating fast and stable convergence. This is expected behavior in well-regularized unsupervised learning tasks and confirms

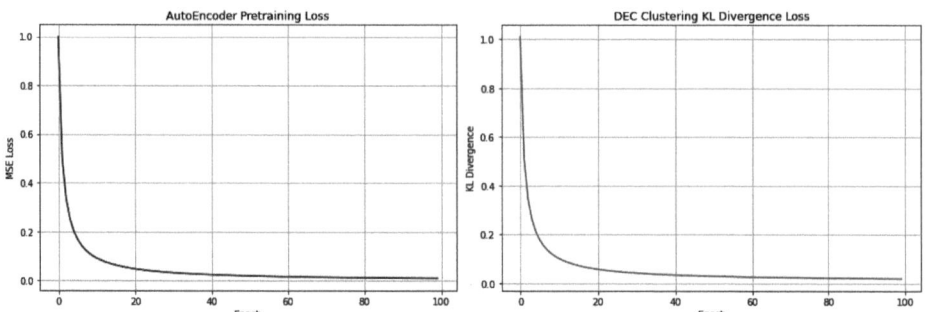

Fig. 4. Autoencoder pretrain and KL Divergence losses of DEC for In-Possession game phase (first 100 epochs).

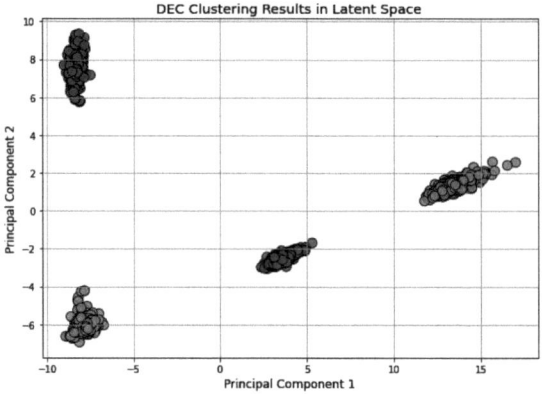

Fig. 5. Latent space representation of DEC's clustering results for In-Possession game phase.

that the encoder has successfully captured compact, meaningful representations for clustering.

Furthermore, the PCA projection of the latent space (Fig. 5) shows that the final representations are well separated and clusterable, further validating the discriminative power of the DEC-encoded features.

In summary, the DEC method offers a significant methodological advantage over traditional clustering approaches, producing more interpretable and effective tactical groupings in football data. These findings validate the utility of deep clustering techniques for complex multi-phase sports analytics tasks.

5 Results and Discussions

*Note: All findings reported in this section are based on the **In-Possession** game phase. While the proposed approach is applied to all four game phases independently, we focus on In-Possession here for simplicity and clarity.*

Additionally, the number of clusters $k = 4$ was selected based on the evaluation metrics—Silhouette Score, Adjusted Cluster Accuracy 1 $(A(C)_1)$, and Adjusted Cluster Accuracy 2 $(A(C)_2)$—all of which reach their highest values at $k = 4$ for the In-Possession data, as shown in Fig. 2.

5.1 Cluster Playing Characteristics

To better understand the stylistic tendencies of each cluster, we analyzed the average values of five representative features: *Passes per Game*, *High Pass Ratio*, *Shots per Game*, *Pass Network Connectivity*, and the frequency of the triangular pass motif $ABCB$. Table 1 summarizes these statistics.

Table 1. Playing Characteristics of Clusters (Average Values)

Cluster Name	Passes/Game	High Pass Ratio	Shots/Game	Pass Network Connectivity	ABCB Motif Freq.
Reactive Direct (C2)	302.7	0.21	9.1	6.7	0.087
Highly Proactive (C1)	576.9	0.10	14.3	8.1	0.098
Moderately Reactive (C0)	375.8	0.18	10.1	7.3	0.093
Balanced Proactive (C3)	453.9	0.14	10.8	7.7	0.096

From these results, four distinct tactical profiles emerge:

- **Highly Proactive (Cluster 1)**: This group demonstrates dominant ball control and offensive intensity. It records the highest average *Passes per Game* and *Shots per Game*, the lowest *High Pass Ratio* (favoring short, controlled passing), and the strongest *Pass Network Connectivity*. The elevated *ABCB Motif Frequency* indicates structured and multi-directional ball movement, typical of possession-oriented teams.
- **Reactive Direct (Cluster 2)**: In contrast, this group reflects a reactive and vertical playstyle. It has the fewest *Passes per Game* and *Shots per Game*, the highest *High Pass Ratio* (indicating long, direct play), and the lowest values for *Pass Network Connectivity* and *ABCB Motif Frequency*. This pattern suggests teams that prioritize quick transitions and counter-attacks over possession.
- **Moderately Reactive (Cluster 0)**: Positioned between the extremes, this cluster shows moderate pass counts and shot attempts with a relatively high *High Pass Ratio*. It may represent teams that adapt their style based on match context, balancing structured build-up with occasional directness.

- **Balanced Proactive (Cluster 3):** This group is similar to Cluster 1 in proactive tendencies but with slightly less intensity. It exhibits strong *Pass Network Connectivity* and *ABCB Motif Frequency*, yet with a moderately higher *High Pass Ratio* than Cluster 1. These teams may blend positional dominance with strategic long passes to stretch the defense.

Overall, the clusters capture a clear spectrum of tactical preferences, ranging from possession-heavy, build-up focused teams to direct, transition-based strategies. These insights provide interpretable labels for downstream match preparation and comparative tactical analysis.

5.2 Win Percentages

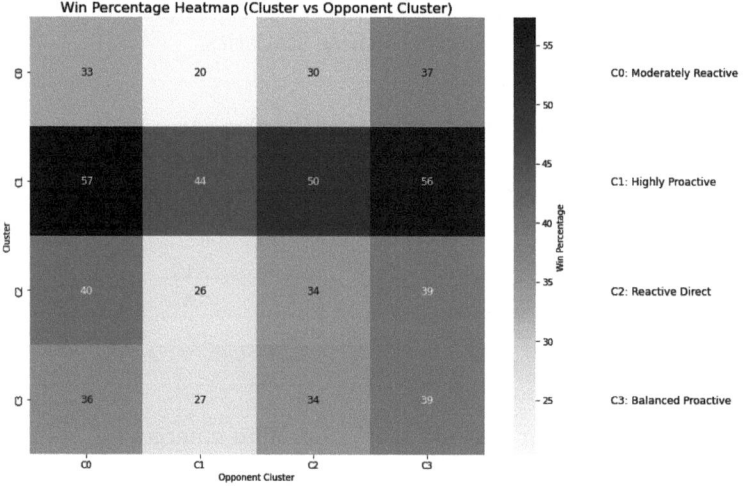

Fig. 6. Win percentages of each cluster against one another.

To assess the effectiveness of each tactical cluster, we analyzed match outcomes based on the cluster affiliations of competing teams. Figure 6 shows the win percentages of each cluster against all others in a 4 × 4 heatmap format. The results provide further insight into the strengths and weaknesses associated with each style of play.

Importantly, this subsection highlights the practical utility of our framework: coaches and analysts can use this data to identify tactical matchups and plan accordingly. By choosing a playing style with higher win probability against an opponent's dominant cluster, decision-makers can improve match preparation and strategic planning.

- **Highly Proactive (C1)** teams stand out as the most dominant. They boast the highest win rates against all other clusters, ranging from 44% to 57%, and

even maintain a relatively high intra-cluster win rate (44%). Their tactical intensity—evidenced by superior connectivity, frequent attacking motifs, and volume of passes—translates into consistent success on the pitch regardless of the opponent's strategy.
- **Reactive Direct (C2)** teams surprisingly perform better than their moderate or balanced counterparts, despite having the lowest overall shot and pass metrics. Their win rates are above 34% in most matchups, suggesting that their conservative and counter-attacking approach is difficult to exploit, especially by teams that overcommit offensively. Notably, they outperform both Moderately Reactive (C0) and Balanced Proactive (C3) clusters.
- **Balanced Proactive (C3)** teams show middling performance. While their style is well-rounded, it lacks the overwhelming pressure of Highly Proactive teams or the deep defensive structure of Reactive Direct teams. As a result, they struggle to establish superiority, with win rates generally below 39%.
- **Moderately Reactive (C0)** teams appear the least effective overall. Their win percentages are consistently the lowest across nearly all matchups. Their mid-range characteristics may lack the decisive traits needed to outperform more specialized styles, making them vulnerable against both proactive and reactive opponents.

Overall, these findings highlight the importance of tactical extremity. Both Highly Proactive and Reactive Direct teams—despite being opposites in style—achieve greater success than those in more balanced or moderate clusters.

5.3 League-Wise Distribution of Clusters

To understand the strategic diversity across top European leagues, we aggregated team-level cluster assignments using majority voting across all their match representations. That is, each team in each match is treated as a separate instance in clustering, and final team assignments are determined by the most frequently assigned cluster over the season.

Table 2 shows the number of teams falling into each cluster by league, along with example teams per cluster.

Table 2. Cluster Distribution Across Leagues (Majority Vote Based)

League	C0	C1	C2	C3	C0 Teams	C1 Teams	C2 Teams	C3 Teams
Bundesliga	4	4	2	8	Augsburg	Bayern Munich	Union Berlin	Leipzig
La Liga	4	4	6	6	Osasuna	Barcelona	Celta Vigo	Atletico Madrid
Ligue 1	0	6	4	9	-	PSG	Metz	Marseille
Premier League	5	6	8	1	Everton	Manchester City	Sheffield Utd	Arsenal
Serie A	1	9	7	3	Salernitana	Napoli	Verona	Lazio

Several insights emerge from this distribution:

- **La Liga** exhibits a relatively even distribution across all clusters, suggesting a high degree of tactical diversity in the Spanish league.
- **Serie A** has the highest number of Highly Proactive teams (9 out of 20), which is counterintuitive given the league's reputation for defensive rigor. One possible explanation is the high number of Reactive Direct teams (7), potentially forcing other teams to adopt proactive styles to break them down.
- **Premier League** shows a near absence of Balanced Proactive teams, with most teams skewed toward Reactive Direct or Highly Proactive. This polarized distribution may explain the dynamic nature of English football.
- **Ligue 1** and **Serie A** each have only one or zero Moderately Reactive teams, indicating a potential lack of hybrid or adaptable tactical identities.
- **Bundesliga** is more balanced but also shows a skew toward Balanced Proactive playstyles, likely influenced by the structured and pressing-heavy tendencies of several German teams.

This league-wise view supports the notion that broader tactical trends exist within domestic competitions and that cluster membership can provide useful summaries of league-level playing identities.

6 Conclusion

This study introduces a novel application of Deep Embedded Clustering (DEC) to the domain of football analytics for the purpose of identifying interpretable and effective team playing styles. Building upon recent advances in tactical clustering frameworks, such as Moffatt et al. [14], we segment matches into four distinct game phases and treat each team-match instance independently to enhance both granularity and generalization.

We engineered a diverse set of spatial, network-based, and motif-driven features to capture the complexity of tactical behaviors. The DEC model [22] was then applied separately to each game phase, outperforming traditional clustering techniques—including K-Means, K-Medoids, and Ward's hierarchical clustering—across all evaluation metrics: Silhouette Score, Adjusted Cluster Accuracy 1 ($A(C)_1$), and Adjusted Cluster Accuracy 2 ($A(C)_2$).

Our results show that the DEC method not only achieves superior clustering quality but also leads to highly interpretable clusters that correspond to intuitive tactical profiles, such as Highly Proactive and Reactive Direct. A focused analysis on the In-Possession game phase revealed that both stylistic extremities—possession-dominant and counter-attacking—are more effective in terms of match outcomes than moderate or hybrid styles.

Moreover, the league-wise distribution analysis provides additional insight into how playing styles differ across Europe's top five leagues. Tactical imbalances and style concentrations—such as Serie A's surprising abundance of proactive teams—demonstrate that cluster identities are both team-specific and league-contextual.

The framework presented here not only serves as a methodological contribution to the clustering of sports event data but also has practical implications for match preparation, opponent scouting, and long-term tactical planning. Coaches and analysts can use these clusters to identify opponent tendencies and optimize their game plans accordingly.

Limitations and Future Work. This work is limited by its reliance on event data from a single season and the absence of tracking information. Future research could integrate positional tracking data for richer spatial context or develop game phase-aware supervised models using DEC-generated labels. Additionally, evaluating clusters with expert tactical annotations would further validate their semantic coherence.

In conclusion, deep clustering methods such as DEC offer a powerful, scalable, and interpretable alternative to traditional unsupervised techniques for football tactical analysis capturing not just who a team is, but how and why they play the way they do.

References

1. Agarwal, S.: Data mining: data mining concepts and techniques. In: 2013 International Conference on Machine Intelligence and Research Advancement, pp. 203–207. IEEE (2013)
2. Akhanli, S.E., Hennig, C.: Comparing clusterings and numbers of clusters by aggregation of calibrated clustering validity indexes. Stat. Comput. **30**(5), 1523–1544 (2020)
3. Bandara, I., Shelyag, S., Rajasegarar, S., Dwyer, D.B., Kim, E.J., Angelova, M.: Time-series analysis of ball carrier open-space in association football. In: International Sports Analytics Conference and Exhibition, pp. 1–17. Springer, Heidelberg (2024). https://doi.org/10.1007/978-3-031-69073-0_1
4. Bialkowski, A., Lucey, P., Carr, P., Yue, Y., Sridharan, S., Matthews, I.: Identifying team style in soccer using formations learned from spatiotemporal tracking data. In: 2014 IEEE International Conference on Data Mining Workshop, pp. 9–14. IEEE (2014)
5. Born, Z.: Tactical Performance Insights for Australian Rules Football Using Deep Learning. Master's thesis, The University of Western Australia (2022)
6. Chapman, R.M., McCrary, J.W.: Ep component identification and measurement by principal components-analysis. Brain Cogn. **27**(3), 288–310 (1995)
7. Diquigiovanni, J., Scarpa, B.: Analysis of association football playing styles: an innovative method to cluster networks. Stat. Model. **19**(1), 28–54 (2019)
8. Gyarmati, L., Kwak, H., Rodriguez, P.: Searching for a unique style in soccer (2014)
9. Hennig, C.: An empirical comparison and characterisation of nine popular clustering methods. Adv. Data Anal. Classif. **16**(1), 201–229 (2022)
10. Hundal, R.S., Liu, Z., Wadhwa, B., Hou, Z., Jiang, K., Dong, J.S.: Soccer strategy analytics using probabilistic model checkers. In: International Sports Analytics Conference and Exhibition, pp. 249–264. Springer, Heidelberg (2024). https://doi.org/10.1007/978-3-031-69073-0_22
11. Kingma, D.P., Ba, J.: Adam: a method for stochastic optimization. arXiv preprint arXiv:1412.6980 (2014)

12. Lopez-Valenciano, A., et al.: Association between offensive and defensive playing style variables and ranking position in a national football league. J. Sports Sci. **40**(1), 50–58 (2022)
13. Van der Maaten, L., Hinton, G.: Visualizing data using t-sne. J. Mach. Learn. Res. **9**(11) (2008)
14. Moffatt, S.J., Gupta, R., Rakshit, S., Keller, B.S.: Identifying team playing styles across phases of play: a user-specific cluster framework. In: International Sports Analytics Conference and Exhibition, pp. 129–136 (2024)
15. Nair, V., Hinton, G.E.: Rectified linear units improve restricted Boltzmann machines. In: Proceedings of the 27th International Conference on Machine Learning (ICML-10), pp. 807–814 (2010)
16. Pappalardo, L., et al.: A public data set of spatio-temporal match events in soccer competitions (2019)
17. Pena, J.L., Touchette, H.: A network theory analysis of football strategies (2012)
18. Plakias, S., et al.: Identifying soccer teams' styles of play: a scoping and critical review. J. Funct. Morphol. Kinesiol. **8**(2), 39 (2023)
19. Rousseeuw, P.J.: Silhouettes: a graphical aid to the interpretation and validation of cluster analysis. J. Comput. Appl. Math. **20**, 53–65 (1987)
20. Ruan, L., Ge, H., Shen, Y., Pu, Z., Zong, S., Cui, Y.: Quantifying the effectiveness of defensive playing styles in the chinese football super league. Front. Psychol. **13**, 899199 (2022)
21. Schilling, A., et al.: Querying football matches for event data: towards using large language models. In: International Sports Analytics Conference and Exhibition, pp. 216–227. Springer, Heidelberg (2024). https://doi.org/10.1007/978-3-031-69073-0_19
22. Xie, J., Girshick, R., Farhadi, A.: Unsupervised deep embedding for clustering analysis. In: International Conference on Machine Learning, pp. 478–487. PMLR (2016)

Construction of Sports and Exercise Knowledge Graph

Tao Huang[1]([✉]), Zehan Xia[1], Yangyi Huang[1], Jiaxin Zheng[2], Jun Lin[3], and Kun Wang[1]([✉])

[1] Department of Physical Education, Shanghai Jiao Tong University, Shanghai, China
{taohuang,wangkunz}@sjtu.edu.cn
[2] School of Physical Education and Sport Science, Fujian Normal University, Fuzhou, China
[3] Department of Physical Education, Wuyi University, Nanping, China

Abstract. The growing demand for precise and scientific exercise knowledge and personalized training guidance has highlighted the urgent need for standardized organization and intelligent expression of sports and exercise knowledge. This study addresses these needs by utilizing ontology theory to collect and mine multi-source, complex sports and exercise knowledge at a large scale. We establish a hierarchical classification system for sports and exercise knowledge and develop an ontology-based knowledge representation model to structure the data semantically. By integrating and deeply mining fragmented and cross-disciplinary sports and exercise knowledge, we construct a sports and exercise knowledge corpus that facilitates centralized integration, shared analysis, and full utilization of sports and exercise knowledge data. Furthermore, using this knowledge corpus, we explore key technologies for an intelligent query system, enabling intelligent management, querying, and Q&A functionalities for sports and exercise knowledge. This knowledge graph can meet the public's diverse and personalized needs for sports and exercise knowledge, thereby supporting targeted and effective exercise and workouts.

Keywords: Sports and exercise knowledge · Knowledge corpus · Knowledge graph · Intelligent query

1 Introduction

Currently, physical activity and exercise are becoming a healthy lifestyle choice for many people. More and more individuals are pursuing scientific exercise guidance and balanced nutritional strategies to improve their overall well-being. Physical exercise is widely recognized for its potential to enhance physical health by improving cardiovascular function, strengthening bones, managing weight, and increasing both muscle strength and endurance, as well as enhancing posture and coordination [1, 2]. Additionally, it plays a crucial role in promoting mental health by alleviating stress and improving mood, and it also helps improving social adaptability [3].

The rapid advancement of the Internet has led to an exponential increase in information related to sports and exercise. This surge has made it increasingly difficult to accurately identify valuable insights from the vast sea of available literature [4]. The absence of standardized concept descriptions and semantic links between them reduces the accuracy of traditional search engines. Common problems include inconsistent use of terms, vague concept definitions, and unclear term relationships. This increases the challenge of obtaining scientific sports and exercise knowledge, causing issues like "information overload" [5] and sports injuries due to lack of knowledge and improper methods.

Consequently, there is a urgent need for a structured approach to data representation that can filter out irrelevant information and effectively organize valuable data, thereby simplifying the processes of information retrieval, analysis, and application [6, 7]. In this context, knowledge graph can be used as an important tool for managing and applying knowledge in the field of sports and exercise.

In this paper, we described the construction of a knowledge graph database in the field of sports and exercise. Combined with large language models (LLM), it aims to achieve intelligent management, querying, and Q&A functionalities for sports and exercise knowledge.

2 Classification Methods for Sports and Exercise Knowledge

To build a sports and exercise knowledge base, the data must be complete, contain minimal redundancy, and exhibit clear correlations among attributes. Redundant information in the knowledge base can severely impact the operational efficiency of the sports and exercise knowledge query system platform. Additionally, semantic ambiguity and polysemy can negatively affect the learning, execution, and output of large language models. Therefore, it is essential to establish a unified and consensus-based hierarchical classification system for sports and exercise knowledge at the early stages of building the knowledge base and query system platform. This will effectively promote the dissemination and development of scientific, systematic, and standardized sports and exercise knowledge.

The classification hierarchy of sports and exercise knowledge (Fig. 1) is constructed with exercise modalities at the core. Focusing on the concepts and attributes of exercise modalities, and considering the elements involved in the exercise process, three additional major categories can be extended: exercise populations, exercise purposes, and exercise norms. In summary, this study divides the sports and exercise knowledge hierarchy mainly into four major categories: exercise populations, exercise modalities, exercise purposes, and exercise norms. Each of these categories is further divided into several levels of subcategories.

3 Ontology-Based Knowledge Representation Techniques for Sports and Exercise

Based on the process of ontology-based knowledge representation methods and the characteristics of sports and exercise knowledge [8, 9], we constructed the ontology representation framework for sports and exercise knowledge. This framework essentially

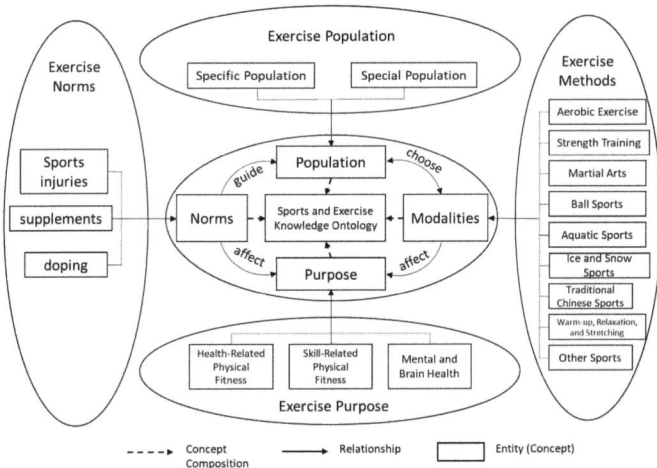

Fig. 1. Ontology Model of Sports and Exercise Knowledge

provides a logical abstraction and ontology modeling process summary of the four elements of sports and exercise, divided into three layers: the descriptive layer, the abstraction layer, and the ontology layer [10]. The descriptive layer defines the scope of the sports and exercise knowledge covered, determined by the range of exercise populations, exercise modalities, exercise purposes, and exercise norms, but is only described in natural language. The abstraction layer abstracts the four elements of sports and exercise into a set of concepts, relationships between concepts, and attributes. The ontology layer further extracts information from the abstraction layer and uses Protégé (http://protege.stanford.edu) software for ontology modeling, with formal representation in OWL language.

4 Construction of Sports and Exercise Knowledge Corpus

In previous studies, the sports and exercise knowledge ontology defined the "skeleton" of the graph database (concepts and concept attributes), while the entities and entity attributes in the sports and exercise knowledge base constitute the actual components of the graph database. The graph database is a structured semantic knowledge base that symbolicly describes concepts and their interrelationships in the physical world. Its basic building blocks are "entity-relationship-entity" triples and entity-attribute-value pairs. Entities are interconnected through relationships, forming a web-like knowledge structure. Given this, the data sources for the sports and exercise knowledge graph database are the sports and exercise knowledge corpora based on the ontology.

This study constructed a knowledge graph which covers diverse exercise populations based on age and health status, exercise modalities (e.g., aerobic exercise, strength training), and workouts purpose, including improving cardiovascular endurance, muscular strength, and mental well-being etc. It also provides knowledge related to preventing sports injuries, the rational use of sports supplements, and anti-doping. The descriptions of the relationships between the above-mentioned ontologies reveal their interactions, which enable the sports and exercise knowledge graph database to provide a comprehensive and detailed reference for the construction of a sports and exercise knowledge query system platform.

There are usually two approaches to extract corpora, namely manual acquisition and web crawling. Because the sports and exercise ontology is still in its early development and the corpus is plagued by noise (polysemy, terminological inconsistency, and undefined relations), we decided to manually organize the data to ensure maximal relevance and quality. Raw corpora were converted into plain text using OCR (Optical Character Recognition) technology, after which classes, relations, attributes, and instances were manually extracted and cleaned to ensure data accuracy.

In this study, the corpus is constructed by dividing it into ontology layer, an abstraction layer and an instance layer. After referring to core literature, authoritative books, official guidelines, and expert consensus in the field of sports and exercise as initial corpora, the core knowledge sources for sports and exercise are determined. Based on the hierarchical classification system of sports and exercise knowledge, Protégé ontology modeling software is used to establish corresponding major categories and their refined subcategories in a top-down manner (Fig. 2). Subsequently, relationships and attributes are added to the categories of exercise populations, exercise modalities, exercise purposes and exercise norms. This process involves the participation of experts who thoroughly read through the textual materials and annotate the conceptual terms within the sports and exercise field. Then, by analyzing the lexical, grammatical, and syntactic aspects, the relationships between these conceptual terms are extracted. After refinement, standardization, and validation, the relationships between the conceptual terms are ultimately established. Finally, based on the abstraction layer of the sports and exercise knowledge corpus, the Protégé tool is utilized to add instances of the sports and exercise knowledge ontology, thereby completing the construction of the instance layer.

5 Construction of Sports and Exercise Knowledge Graph Database

The storage of sports and exercise knowledge data is the fundamental purpose of constructing a sports and exercise knowledge graph database. It is used to organize and manage sports and exercise knowledge data that are multi-sourced, fragmented, and cross-disciplinary. The Neo4j graph database provides a powerful solution for the storage of sports and exercise knowledge data. Compared with traditional relational databases, Neo4j (https://neo4j.com/) is an efficient tool for handling highly interconnected data and can easily manage complex networks of relationships. Applied to this study, it enables flexible storage of knowledge related to exercise selection, movement steps, technical requirements, relevant norms, injury prevention, nutritional supplements, anti-doping, and other aspects used by exercise enthusiasts. It also visually represents the relationships between different categories of knowledge. This storage method not only provides a foundation for the construction of a sports and exercise knowledge query system platform but also ensures the continuous updating and scalability of sports and exercise knowledge data.

In the sports and exercise knowledge graph database, visualization is a key part of knowledge data communication (Fig. 2). It transforms abstract knowledge data into easily understandable graphics or charts. Through the visualization of sports and exercise knowledge data, complex relational data within the core knowledge sources of the exercise domain, such as exercise modalities, exercise outcomes, target populations, and their connections to individual health, can be converted into intuitive graphical interfaces. This not only enhances the richness of user interaction but also makes the experience of searching for sports and exercise knowledge more intuitive and efficient for the general public. Meanwhile, the visualization function supports the dynamic display of sports and exercise knowledge data. Users can observe changes and comparison results of sports and exercise knowledge data based on personalized query conditions. This real-time dynamic visualization method improves user experience and provides the public with a new interactive way to learn and explore sports and exercise knowledge.

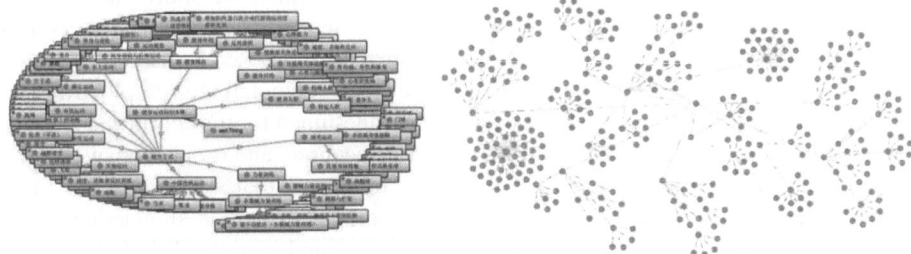

Fig. 2. Data Storage Instances in Protégé(left) and Graph Database Instances in Neo4j(right)

6 Intelligent Query Platform Based on Large Language Models

The intelligent Q&A platform for sports and exercise knowledge is designed to address open-ended questions within the sports and exercise domain. This functionality is built upon the sports and exercise knowledge ontology that was developed in the early stages of this study. By using the semantic expression of knowledge data in the sports and exercise domain, the system effectively retrieves relevant concept classes, attributes, and attribute values. Furthermore, by integrating large language models and the constructed sports and exercise knowledge base, the system applies natural language processing (NLP) techniques to deeply identify, process, and analyze the textual content of the core knowledge sources in the sports and exercise domain.

To enhance the accuracy of the intelligent Q&A, we incorporate two techniques: retrieval-augmented generation (RAG) and fine-tuning of large language models. The RAG framework integrates a knowledge base into the retrieval and generation process, allowing the system to retrieve relevant information from the knowledge base and generate more accurate and contextually appropriate answers. This integration ensures that the responses are based on verified knowledge, thereby improving the reliability and accuracy of the Q&A system.

Additionally, we employ fine-tuning techniques to adapt the large language models specifically to the sports and exercise domain. By fine-tuning the models on domain-specific data, the understanding of the terminology, concepts, and context relevant to sports and exercise is enhanced. This fine-tuning process improves the model's ability to generate responses that are not only accurate but also tailored to the specific needs and nuances of the sports and exercise domain.

Owing to the continual updating of knowledge in sports and exercise, the platform must be persistently updated and refined to keep pace with emerging insights and user expectations. On one hand, based on user feedback and data analysis, the platform's functions and performance are continuously refined to better meet user needs. On the other hand, by utilizing machine learning techniques, the large language models are constantly trained and adjusted. This ongoing process enhances the platform's understanding and response capabilities regarding sports and exercise knowledge, improving the accuracy and efficiency of intelligent queries.

The functional modules of the sports and exercise knowledge intelligent query platform include sports and exercise knowledge Q&A, sports and exercise knowledge search, and sports and exercise knowledge visualization. Specifically, the sports and exercise knowledge Q&A module aims to answer various open-ended questions that users have about the sports and exercise domain. Based on the questions posed by users, the system platform uses a knowledge base to initially filter relevant content. It then employs a large-scale model, enhanced by the RAG framework and fine-tuned for the sports and exercise domain, to conduct in-depth reasoning and generate accurate and reliable answers. This approach ensures that users receive high-quality responses that are both contextually relevant and scientifically accurate. Tests on the sports and exercise knowledge corpus showed that all query results were displayed correctly and completely. Moreover, the knowledge graph-based recommender and Q&A service achieved an overall accuracy of over 90% in this study.

7 Conclusion

In this study we constructed a sports and exercise knowledge graph to address key challenges in organizing and accessing sports and exercise knowledge effectively. Using ontology-based methods, we have been able to standardize and structure a large amount of sports and exercise knowledge. The constructed knowledge corpus serves as a robust foundation for the knowledge graph database, which not only centralizes the data but also enables efficient sharing and analysis. Moreover, the implementation of an intelligent query platform based on knowledge graph and large language models supports the creation of a more personalized and intelligent exercise guidance system. The knowledge graph has the potential to support the future innovations in sports and health science.

Acknowledgments. This study was funded by National Key Research and Development Program of China (grant number 2022YFC3600401).

Disclosure of Interests. The authors have no competing interests to declare that are relevant to the content of this article.

References

1. Warburton, D.E.R., Bredin, S.S.D.: Health benefits of physical activity: a systematic review of current systematic reviews. Curr. Opin. Cardiol. **32**(5), 541–556 (2017)

2. Dhuli, K., et al.: Physical activity for health. J. Prev. Med. Hyg. **63**(2 Suppl 3), E150–E159 (2022)
3. Harvey, S.B., Øverland, S., Hatch, S.L., Wessely, S., Mykletun, A., Hotopf, M.: Exercise and the prevention of depression: results of the HUNT cohort study. Am. J. Psychiatry **175**(1), 28–36 (2018)
4. Li, L., et al.: Real-world data medical knowledge graph: construction and applications. Artif. Intell. Med. **103**, 101817 (2020)
5. Kamilaris, A., Yumusak, S., Ali, M.I., (eds.) WOTS2E: a search engine for a semantic web of things. In: 2016 IEEE 3rd World Forum on Internet of Things (WF-IoT) (2016)
6. Xu, J., et al.: Building a PubMed knowledge graph. Sci. Data **7**(1), 205 (2020)
7. Lin, J., Zhao, Y., Huang, W., Liu, C., Pu, H.: Domain knowledge graph-based research progress of knowledge representation. Neural Comput. Appl. **33**(2), 681–690 (2021)
8. Studer, R., Benjamins, V.R., Fensel, D.: Knowledge engineering: Principles and methods. Data Knowl. Eng. **25**(1), 161–197 (1998)
9. Maedche, A., Staab, S.: Ontology learning for the semantic web. IEEE Intell. Syst. **16**(2), 72–79 (2001)
10. Cardoso, S.D., Da Silveira, M., Pruski, C.: Construction and exploitation of an historical knowledge graph to deal with the evolution of ontologies. Knowl.-Based Syst. **194**, 105508 (2020)

Personalised Running Coaching with Next-Generation Wearable Technology

Nathan Hur, Jonathan Soulsby, Zixiao Zhao(✉)[ID], and Jing Sun(✉)[ID]

School of Computer Science, University of Auckland, Auckland, New Zealand
{zixiao.zhao,jing.sun}@auckland.ac.nz

Abstract. The growing adoption of fitness wearables has resulted in the collection of extensive exercise and biometric data, with numerous studies highlighting the health benefits of such technologies. However, a persistent challenge remains: many commercial wearables lack long-term efficacy and motivational support, limiting sustained engagement and diminishing their overall impact on user health. This paper presents *Synochi*, an intelligent coaching system that integrates mainstream wearable devices with established sports science principles. The system aims to deliver adaptive training programs, personalised feedback, and sustained motivational support—addressing key limitations of existing wearables. *Synochi* features an adaptive training engine grounded in sports science literature and implemented using fuzzy logic and regression techniques to personalise and optimise training progression. A pilot qualitative user evaluation demonstrated positive user perceptions regarding both the system's effectiveness and its ability to support motivation. These results suggest that the integration of sports science with wearable technologies offers significant potential to enhance long-term user engagement and maximise the health benefits of fitness wearables.

Keywords: Wearable computing · Fitness training · Sports science · Fuzzy logic · Gamification

1 Introduction

The rapid advancement of technology has spurred the development of cheaper, smaller, and more powerful devices that enrich the daily lives of consumers. One such area is fitness wearables—watches and bands that record biometric data to help individuals increase their physical activity. As discussed in the literature, many studies on the motivational aspects of wearables have found that while they are effective in promoting exercise, they often fail to sustain this effect in the long term for many users, due to a lack of engaging motivational features over extended periods [1]. Indeed, many commercial wearables focus on static daily goals, such as step counts, which can become repetitive and lose their motivational appeal over time [2]. Although these devices also collect valuable data—such as heart rate, sleep patterns, and exercise metrics—they often provide little in the way of motivational insight. This represents a lost opportunity to leverage such data to deliver more personalised insights and dynamic, long-term engagement features.

Reviewing current solutions [9–14] reveals that a few contemporary wearable systems have begun to offer more dynamic features, such as AI-based coaching and training insights. However, to the best of our knowledge, the theoretical or empirical foundations of such functionality are not made transparent, and these systems typically require specialised and expensive devices. This gap provided the primary motivation for designing the proposed system to address the long-term motivational challenges associated with fitness wearables. The system is firmly grounded in sports science, which informs all of its functionality. It achieves this by incorporating research findings transparently into both its metrics and coaching feedback. Drawing on the literature, a conceptual framework for a programmable training model was developed. The model employs High-Intensity Interval Training (HIIT), chosen for its highly parameterised structure and well-studied physiological effects and recommended values. Complementing this, Borg's Rating of Perceived Exertion (RPE) is used to provide a user-driven feedback mechanism for adapting training difficulty. Additionally, a novel index—the Heart Rate–Running Speed index (HRRS)—is introduced as a non-intrusive, personalised estimate of running performance. This model serves as the foundation for the entire software coaching system.

This paper presents *Synochi*, a wearable-based training system designed to address two key limitations of current commercial wearables: long-term motivational challenges and underutilisation of collected data. The novel contributions of Synochi, as outlined in the literature review, lie in its integration of evidence-based insights from exercise science with functionalities drawn from existing systems—delivered through a widely accessible wearable device. Key contributions of this work include the development of a fuzzy logic model that translates exercise research into personalised coaching feedback, and a promising set of evaluation results that highlight the system's potential to inform future advancements in wearable fitness technologies.

The remainder of the paper is structured as follows. Section 2 provides a review of related work and existing wearable systems, highlighting key limitations in motivation and adaptability. Section 3 presents the system design of *Synochi*, detailing the conceptual training model and the integration of sports science principles. Section 4 describes the system implementation, including its architecture and technical components. Section 5 reports the evaluation results based on user studies assessing motivation and usability. Finally, Sect. 6 concludes the paper and discusses potential directions for future work.

2 Literature Review

2.1 Effectiveness of Wearables

Many studies have investigated the effectiveness of wearables, with results obtained primarily through statistical analyses of wearable outputs and qualitative surveys such as the Healthcare Technology Self-Efficacy (HTSE) questionnaire. Research involving randomised or targeted participant demographics has quantitatively demonstrated increases in physical activity over periods of weeks or months [2–4,6,9]. Furthermore, Karapanos et al. found that users' perceptions of autonomy and competence were the most important psychological needs supported by these devices, and were strongly correlated with positive affect— i.e., the degree to which users attribute their positive behaviours to the system [6]. In the context of wearables, autonomy refers to the control users have over their physical activity, while competence relates to the sense of effectiveness in their actions. The importance of these psychological factors is a recurring theme throughout the literature and presents an ongoing challenge for the effectiveness and sustained adoption of wearable technology [1,4,5].

2.2 Existing Solutions

Table 1. Comparison of existing wearable solutions and the proposed system.

Feature/System	Eval metrics	Training approach	AI-based goal setting	AI-based training feedback	Co-operative gamification	Hardware
Runtastic	Pace & distance	Preset running programs	None	None	None	Mobile phone
C25K (Couch to 5K)	None	Preset running program	None	None	None	Mobile phone
Fitbit Coach	Steps, stairs climbed, active minutes	Video workout sessions	None	None	None	Fitbit wearable
FeetMe Sport	Running efficiency, propulsion force	Improve running technique	None	Feedback on running technique	None	Specialized shoe sole
Vi	Completed runs in a program	Preset running programs	Progression of difficulty towards goal	Voice-guided heart rate zone and pace feedback	None	Specialized in-ear headphones
Proposed (Synochi)	HRRS index, weekly consistency	Adaptive HIIT program	Regression model (based on past performance)	Fuzzy logic-based feedback	Weekly group challenges	Hardware independent (Fitbit-based prototype)

The wearable market is growing rapidly [9], and the supporting software for these devices is continuously improving, helping to mitigate some of the challenges discussed earlier. While literature on emerging novel solutions remains scarce, it is worth briefly describing several of these systems and their potential impact on the future of wearable software. As a speculative exploration of future trends in wearables, we conducted a brief analysis of features across a selection of established solutions, as well as newer systems such as *Vi*, summarised in Table 1. The systems we compared are roughly ordered from the least recent at the top to the most recent at the bottom, alongside our proposed solution.

Most mature, existing solutions focus on static performance metrics—such as step count, active minutes, and completed runs—and typically follow a preset training approach that lacks personalisation. They also largely omit AI-based or adaptive goal setting and training feedback, leaving this responsibility to the user. One example of a more novel system is Vi, a personal artificial intelligence (AI) trainer that employs custom hardware with voice feedback to provide coaching and fitness tracking [10]. By delivering quantitative feedback and motivation through voice, this approach has the potential to address the limitations of user perception by emulating a personal fitness coach. However, while this method offers a more personalised experience, it still relies on fixed, pre-defined running programs rather than dynamically adapting training plans to the user.

More recent solutions have begun incorporating increasingly sophisticated and meaningful features that go beyond simply tracking numerical targets. However, many of the novel solutions we surveyed still require specialised and often expensive hardware, limiting accessibility to such systems [10–14]. Our proposed system can be viewed as a natural progression of these prior solutions, aiming to further advance adaptability and personalisation.

2.3 Summarised Findings

The efficacy of wearables in increasing physical activity has been well studied and demonstrated across many demographic groups; however, significant challenges remain that lead many users to discontinue use and thus fail to realise the long-term benefits of using wearables to promote sustained physical exercise. These challenges are often rooted in the motivational aspects afforded by the technology, which are, in turn, shaped by the user's perception of such systems. Extensive literature has shown that factors such as trust, perceived competence, and social engagement are critical for users to successfully internalise and benefit from the motivational features offered—an essential process for achieving long-term adoption of wearable devices. Below, we summarise the key challenges identified in our review and highlight corresponding opportunities informed by current research:

- **Quantifying fitness.** Accurately quantifying fitness is a crucial prerequisite for delivering personalised and effective training programs. By leveraging indices from the literature—such as HRRS, HR_{max} estimators, and RPE—we can capture multiple aspects of an individual's fitness level, enabling programmable training, personalised goal setting, and automated coaching feedback.
- **Programmable training.** By employing HIIT, a well-studied and clearly parameterised training approach, we can implement evidence-based running prescriptions that adapt dynamically to the user, using HIIT parameters and self-reported RPE. These parameters also enable the generation of coaching feedback through fuzzy logic techniques.

- **Static numerical goals can demotivate.** Rather than relying on fixed numerical targets (such as step counts), we use the HRRS index to estimate the user's VO_2max, providing a more meaningful reflection of their evolving performance. When combined with regression-based predictions, this approach supports the delivery of personalised and adaptive goals that mitigate the demotivation commonly associated with static goal-setting strategies.
- **Lack of co-operative social gamification.** While competitive social gamification is widespread in wearable fitness systems, we propose a co-operative model in which friends and family join a group to work collaboratively toward achieving a weekly challenge—a model supported by research as being more motivating for certain user types.

3 System Design

3.1 Feature Overview

Building on insights from the literature, we designed a conceptual framework for a programmable, research-driven coaching system (see Fig. 1). At the core of this framework is a High-Intensity Interval Training (HIIT) model, which provides both structured exercise prescriptions and adaptive feedback grounded in sports science research. The system begins with an initial calibration run to establish user-specific baselines, including heart rate zones and peak running speeds. Based on these baselines, personalised HIIT parameters are prescribed, with recommended values and ratios drawn from empirical studies in the sports science literature. Key parameters—such as the durations of high- and low-intensity intervals—are dynamically updated after each completed run, informed by the user's self-reported Rating of Perceived Exertion (RPE). For example, if a user reports "low exertion", the system increases the duration of the high-intensity intervals in subsequent sessions to ensure progressive training adaptation.

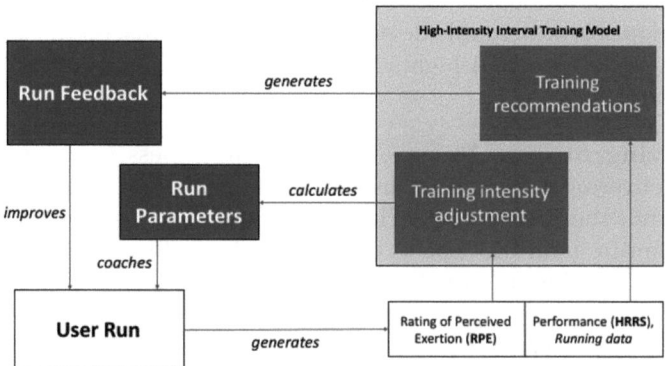

Fig. 1. Conceptual framework for training.

The primary control variable for personalising the workout is the high-intensity interval duration, while the rest and low-intensity phases are automatically adjusted to maintain optimal ratios based on best practices from running training studies. The system uses the Heart Rate–Running Speed (HRRS) index as its primary performance metric, enabling continuous tracking of progress and fine-tuning of the training program. In addition to adaptive training prescription, the system also analyses completed run data—including HRRS, RPE, heart rate, intensity, speed, and other metrics—and evaluates them against recommendations from sports science literature to generate constructive coaching advice. This advice focuses on helping the user improve the training effectiveness of their next run, specifically targeting physiological responses. The operation of each system component is described in more detail in the following sections.

High-Intensity Interval Training (HIIT). There are as many as nine parameters involved in prescribing a HIIT session, and the ratios and interactions between them are well-studied [18]. In addition to having well-defined parameters, the effectiveness of HIIT in eliciting a strong physiological response within a relatively short amount of exercise time—compared to other training methods—makes it a suitable, structured, and programmable approach for casual fitness wearable users. *Synochi* prescribes several key parameters using corresponding models, most of which are derived from extensive studies on HIIT. For example, high-intensity interval durations are based on findings from a meta-study on HIIT, which states that "preliminary data suggest that HIT performed somewhere between 50–60% of T_{max} may be optimal for improving endurance performance" [16], where T_{max} is defined as the time to exhaustion when running at maximum speed. By modelling these parameters, *Synochi* embeds training recommendations from the literature directly into its generated running prescriptions.

Borg's Rating of Perceived Exertion (RPE). As the main form of user feedback, the RPE scale facilitates adaptive training by adjusting the intensity of the next run based on the runner's perceived exertion. Although perception may differ from physiological response, clinical trials have shown that the RPE scale strongly correlates with objective exertion measures [19]. This makes RPE a valid and practical tool for estimating a runner's actual exertion and for dynamically adjusting the difficulty of training to match the user's fitness level. For example, if the user reports an exertion level of 17 ("very hard"), the system will reduce the high-intensity interval duration for the next run by a certain percentage—with a greater reduction applied if the user reports a level of 19 ("extremely hard"). In addition to the numeric RPE ratings (ranging from 6 to 20), Synochi provides corresponding descriptive cues (e.g., "light exertion") to help users more accurately gauge and interpret the scale.

Heart Rate–Running Speed (HRRS) Index. At the core of any training model lies some form of performance monitoring or evaluation. For a human running coach, this typically relies on expertise and experience to assess a trainee's running ability. More objective methods are used in many clinical trials, where participants are assessed in laboratory settings—an approach that is too expensive and intrusive for regular coaching. *Synochi* addresses this by using the *HRRS* index, a non-intrusive estimate of VO_2max—a well-researched physiological indicator of running performance that is typically measured in a laboratory. The HRRS estimate, developed by Vesterinen et al. [20], has been shown to strongly correlate with VO_2max and is suitable for use in training programs to track performance. The index is calculated as shown in Eq. 1. The constant k, described in Eq. 2, represents the slope of a user's heart rate–speed graph. In Synochi, the HRRS index is computed after each run, following the initial calibration run where the user's thresholds (resting, maximum, and peak heart rates) are established. This index serves as the system's primary performance measure, informing feedback generation, performance goal setting, and progress visualisation for the user.

$$\text{HRRS index} = S_{\text{avg}} - \frac{HR_{\text{avg}} - HR_{\text{standing}}}{k}, \tag{1}$$

$$k = \frac{HR_{\text{max}} - HR_{\text{standing}}}{S_{\text{peak}}}, \tag{2}$$

where S_{avg} and HR_{avg} are the average speed and heart rate during the run, HR_{standing} and HR_{max} are the user's standing (resting) and maximum heart rates (with S_{peak} the peak running speed reached).

3.2 Real-Time and AI-Based Coaching

Real-Time Coaching. The real-time component of the training model is primarily focused on helping the runner adhere to their HIIT prescription while collecting relevant running data, such as heart rate and speed. For a wearable-based coach, real-time coaching centres on leveraging the wearable device's display and haptic feedback during the run. The display shows the current training phase—warm-up, high intensity, low intensity, or rest—along with an associated timer. A vibration pattern signals transitions between phases, providing a physical cue so that the runner can switch phases without needing to look at the display. During high-intensity phases, the runner's speed and heart rate are recorded, and current heart rate is continuously displayed. At the end of the run, a simple RPE scale is presented on the watch, with descriptive labels for each level. After the user selects an exertion level, the collected running data is stored and sent for processing. Although this coaching approach gives the user more autonomy compared to that of a personal coach, *Synochi* is primarily focused on closing the feedback loop—generating actionable insights that help users refine their technique and improve future performance.

Personalised Coaching Feedback. In addition to providing guidance during a running session, a personal coach typically offers feedback on the trainee's technique and performance to help optimise future training. From a programming perspective, this type of reasoning cannot be effectively modelled using crisp conditional statements. For example, if one defines a speed above 5 m/s as "fast," then a speed of 4.9 m/s would not be classified as such—despite being virtually equivalent from a human perspective. Human reasoning, as used by coaches to generate feedback, is more akin to fuzzy logic, where classes and thresholds are represented as ranges rather than strict boundaries. For instance, a coach might advise a trainee to "run harder during the warm-up" to increase their heart rate if their sprinting during high-intensity intervals was "a bit low" relative to the target heart rate zone. If the runner remained within the correct heart rate zone for about 80% of the time, then in the coach's judgment, a result of 79% would not be meaningfully different from 82%—they would be considered roughly the same. Fuzzy logic supports this style of reasoning by allowing the use of sets with imprecise boundaries (e.g., "high heart rate zone"), which can be combined to form conclusions using fuzzy rules (e.g., IF low heart rate zone AND low exertion OR low performance THEN recommend higher intensity). Standard fuzzy logic operators—AND, OR, and NOT—can be implemented using Zadeh's operators [21], which map these logical functions to corresponding mathematical operations, as shown in Table 2.

Table 2. Zadeh's fuzzy operators

Logic operator	Mathematical operation
AND(x, y)	$\min(x, y)$
OR(x, y)	$\max(x, y)$
NOT(x)	$1 - x$

With this fuzzy logic approach, the process of integrating sports research into *Synochi* to generate coaching-like feedback was largely a knowledge engineering challenge, involving feature extraction, fuzzy membership definitions, and the translation of recommendations from the literature into fuzzy rules. In some cases, value ranges or broad guidelines were interpreted from the literature and would require iterative refinement and expert feedback to improve the validity of the generated feedback.

3.3 Additional Features

Feature Extraction. Specific characteristics of the available running data were extracted for use in fuzzy logic reasoning. The selected features were chosen based on their frequent appearance in HIIT recommendations throughout the literature. These features included HRRS, RPE, intensity, and cardiac drift—the latter describing the increasing divergence between a steady running pace and a rising heart rate. Intensity was defined as the percentage of high-intensity interval duration during which the user's heart rate exceeded 80% of their maximum heart rate, a threshold derived from recommendations in prior studies [16,17].

Membership Functions. As part of the knowledge engineering process, classes needed to be defined for the extracted features and modelled with functions that described the degree of membership to each class. For example, one class defined for intensity was HIGH_INTENSITY. Quoted values from HIIT studies were compiled, and approximate quartiles were used to define the bounds corresponding to a membership value of 1. Outside of these bounds, membership values decreased linearly to 0, forming a trapezoidal function. The zero-membership bounds were determined based on the maximum and minimum recommended values observed in the collected literature. For example, the resulting membership function for recommended heart rate had a value of 0 at 60% HR_{MAX}, increasing to 1 by 80%, and returning to 0 by 120%.

Fuzzy Rules and Feedback Generation. With the membership functions defined, fuzzy rules were created to infer conclusions based on the fuzzy memberships. Following a similar process to that used for defining the membership functions, recommendations were compiled from several meta-studies [16–18]. This process was relatively straightforward, as demonstrated by the following example from the literature: "... athletes achieve excellent results with a modest proportion of training performed at intensities between 85% and 100% of VO_2MAX." Combined with recommendations on exertion levels, this insight was translated into a fuzzy rule such as:

UNDER_EXERTION: intensity is NOT HIGH AND exertion is LOW OR OK.

This rule was added to the complete inference model. The outcome of each rule was assigned a key corresponding to a recommendation drawn from the literature. In the example above, the key UNDER_EXERTION would retrieve a recommendation discussing the physiological benefits of performing HIIT within the optimal intensity range. During feedback generation, all rules were evaluated; if a rule's output exceeded a defined threshold (e.g., 0.5), the associated recommendation was included in the feedback. In cases where rule pairs produced conflicting advice, thresholds and conditions were adjusted to ensure that they operated in a mutually exclusive manner.

Personalised Goals. Many goal-setting features in common wearable devices rely on static daily targets that must be manually updated by the user. As discussed in various studies, such goals often lack long-term motivational efficacy [23]. In contrast, Synochi generates personalised performance goals using a polynomial regression model trained on the user's past HRRS values. By learning from past performance, each new goal becomes dynamic and achievable, providing a more sustainable and personalised goal-setting mechanism over the long term.

Cooperative Gamification. In research on wearables and long-term motivational challenges, significant attention has been given to the role of social features. While many studies highlight the overall motivational benefits of social interaction, some observe that competitive gamification may not appeal to all user types [23]. To address this, Synochi incorporates a cooperative social feature, allowing users to create or join groups that participate in weekly challenges completed through the collective effort of all members. These challenges—including consistency, time, and distance goals—are based on group averages, with participants earning experience points and bonuses when goals are exceeded. While currently implemented as a proof-of-concept, this feature lays the groundwork for future enhancements aimed at fostering socially motivated engagement.

4 System Implementation

In this section, we explore the architecture of *Synochi*, detailing the communication between modules and highlighting key implementation details and design decisions.

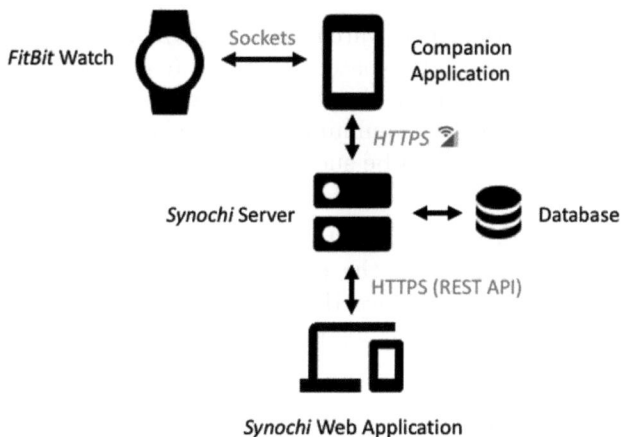

Fig. 2. System architecture of Synochi.

As illustrated in Fig. 2, Synochi comprises three main modules: 1) Fitbit watch and companion application 2) Node.js server 3) Web application. All modules were implemented in JavaScript (ES6/7), as required by the Fitbit SDK and to maintain consistency across the stack.

4.1 Watch and Companion Application

To support real-time coaching and collect running data, a watch component was developed for Fitbit Ionic and Versa models using the Fitbit SDK [22]. This module consists of two subcomponents: the watch application and the companion application. The watch application runs directly on the Fitbit device, while the companion application functions as a plugin within the Fitbit mobile app. Screens for the real-time coaching interface and the RPE scale are illustrated in Fig. 3. During the run, data is collected using a `setTimeout` loop that polls the device sensors. The data is stored in local arrays for later transfer.

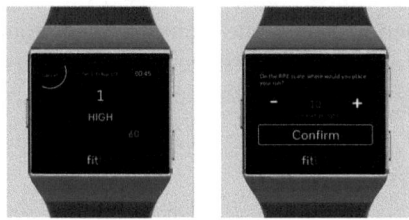

Fig. 3. Watch application user interface for real-time coaching.

Socket Communication. The watch and companion application communicate via socket messaging. JSON-based messages (e.g., `LOGIN`, `RUN_DATA`) are exchanged, with retry mechanisms implemented to ensure robustness given the asynchronous nature of socket communication. If connectivity is lost, the run data is stored locally until it can be successfully transmitted.

OAuth 2.0 Device Matching. OAuth login is implemented using Fitbit's web authentication, initiated through the companion application. Upon successful login, the Fitbit user ID and device ID are linked to the user's account. This enables the watch to automatically log in the user and retrieve HIIT parameters when the application starts.

4.2 Server Application

The Node.js server is responsible for processing run data, computing new HIIT parameters, and managing both feedback generation and personalised goal setting. It also exposes a REST API that is accessed by both the watch and web applications, facilitating seamless communication between components. The REST API is designed around a modular architecture, enabling features to be developed and tested independently. Key API routes include: **Auth**, which handles OAuth and session management; **Me**, for accessing user profile and personal data; **Social**, which supports cooperative challenge groups; **User**, for registration and data retrieval; **Watch**, which handles HIIT parameter delivery and run data uploads; and **Admin**, which provides diagnostics, logging, and maintenance capabilities. This modular approach not only improves code maintainability but also ensures flexibility for future feature extensions.

Personalised Goals. Goals are generated using a polynomial regression model trained on each user's historical HRRS data. Implemented with TensorFlow.js, the model is trained over 100 epochs with a learning rate of 0.001. Predictions are spaced according to each user's running frequency and capped (0–10 goals) to ensure reliability. The model uses a second-order polynomial to capture the non-linear nature of fitness progression:

$$f(n) = a_0 + a_1 n + a_2 n^2 \tag{3}$$

where $f(n)$ represents the predicted HRRS index goal for the n-th future run, a_0 is the baseline HRRS index value reflecting the user's current fitness level, a_1 captures the linear trend in performance improvement or decline, and a_2 models the quadratic component that accounts for acceleration or deceleration effects in fitness adaptation.

Feedback Generation. The fuzzy logic system is implemented from scratch. Rules are defined in a human-readable format and applied to features such as RPE and intensity. Rules with output values exceeding a defined threshold (e.g., 0.5) trigger the inclusion of corresponding feedback.

Asynchronous Job Queue. Tasks such as model training and feedback generation are offloaded to a database-backed job queue implemented using Agenda.js. This enables asynchronous execution and prevents blocking of client-facing requests.

4.3 Progressive Web Application

The front end is a responsive progressive web app (PWA), accessible via both mobile and desktop. The app uses adaptive layouts (e.g., collapsible menus on mobile) and displays training feedback and visualised goals, as shown in Fig. 4.

4.4 Data Privacy and Security

The system implements privacy-by-design principles to protect user data. All personal health information and data processing are performed locally on the user's device, where possible. User consent is obtained for all data collection and processing activities, and users retain control over their data with options to export or delete their information. The system adheres to relevant data protection regulations and follows established best practices for handling sensitive health data in wearable applications.

The server-side code follows a test-driven development (TDD) approach using Mocha and Chai. Each feature is tested by simulating complete user workflows through API requests. This ensures that functionality is validated from the perspective of full user interaction rather than through isolated unit tests.

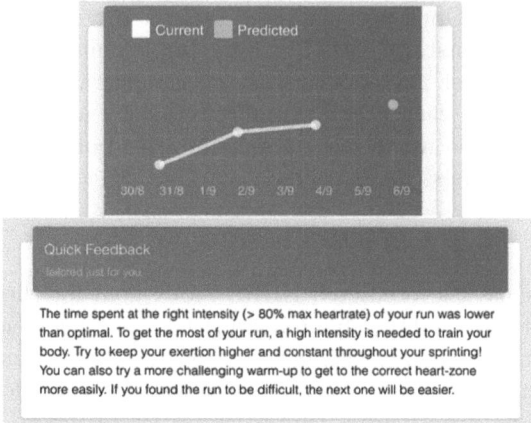

Fig. 4. Web application interface showing feedback and performance goals.

5 Evaluation

Due to the scope and limitations of the project, we were unable to conduct a controlled experiment to compare Synochi against real coaching programs. Instead, we adopted a two-part user evaluation. The first evaluation focused on assessing users' perceptions of motivational factors and the overall efficacy of the coaching system; the second evaluation was a user experience study of the web application interface.

5.1 User Studies

All participants were fourth-year engineering undergraduates. Nine participants were selected for the motivation and efficacy evaluation, and a different set of ten participants was chosen for the user experience evaluation. For the motivation and efficacy evaluation, participants were introduced to the watch and web application and instructed on how to use both in training. Each participant was briefed on the system workflow, from starting a run on the watch to viewing feedback and predictions on the web application. They were then asked to use the watch and try the system themselves, with initial HIIT parameters set to lower values. After completing a brief run, participants reviewed their results and feedback on the web application and subsequently completed a post-evaluation survey. The survey was based on the Technology Acceptance Model (TAM), with Likert-scale questions adapted from a previous Fitbit wearable study [23]. The results of these questions are presented in Fig. 5, which shows largely positive sentiment from the users—an encouraging outcome for the continued development and evaluation of the system.

The second part of the evaluation assessed the user experience of the web application. Participants in this part were not introduced to the full system,

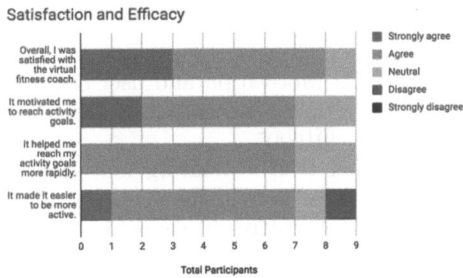
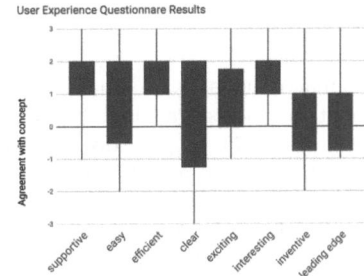

Fig. 5. TAM questionnaire results.   **Fig. 6.** UEQ results.

allowing the evaluation to focus solely on user interface interactions and navigation. These participants were provided with a test login and asked to freely explore the web application on both desktop and mobile views. They then completed a short Likert-scale questionnaire based on the User Experience Questionnaire (UEQ) [24]. The results are presented in Fig. 6 as a candlestick chart, where the lines represent the range and the boxes represent the upper and lower quartiles. While most of the responses were positive, participants provided some constructive feedback regarding the clarity and inventiveness of the user interface.

6 Conclusions

The key contributions of this work are twofold: first, the development of a systematic process for integrating sports science into a wearable coaching system; and second, the demonstration of promising results through user evaluation. The proposed system, *Synochi*, addresses the challenge of sustaining long-term motivation in wearable fitness devices by employing an adaptive training model grounded in sports science literature, supported by fuzzy logic and regression techniques. The system has been implemented and tested with Fitbit devices.

A user study demonstrated that participants responded positively to the system's motivational features and user experience. While these initial results are encouraging, a future clinical trial will be necessary to quantitatively assess the system's impact on fitness outcomes. In addition, ongoing work will focus on enhancing the training model through expert consultation and data mining, broadening device compatibility, and expanding social features such as group-based challenges and cooperative competitions. We hope that *Synochi* will serve as a foundation for the next generation of wearable systems—systems that are not only more adaptive and personalised, but also more effective in promoting long-term engagement and improving users' health outcomes.

References

1. Asimakopoulos, S., Asimakopoulos, G., Spillers, F.: Motivation and user engagement in fitness tracking: heuristics for mobile healthcare wearables. Informatics **4**(1), 5 (2017). https://doi.org/10.3390/informatics4010005
2. Strath, S.J., et al.: A pilot randomized controlled trial evaluating motivationally matched pedometer feedback to increase physical activity behavior in older adults. J. Phys. Act. Health **8**(Suppl. 2), S267 (2011). http://www.ncbi.nlm.nih.gov/pubmed/21918241
3. Butryn, M.L., et al.: Enhancing physical activity promotion in midlife women with technology-based self-monitoring and social connectivity: a pilot study. J. Health Psychol. **21**(8), 1548–1555 (2016). https://doi.org/10.1177/1359105314558895
4. Mercer, K., et al.: Behavior change techniques present in wearable activity trackers: a critical analysis. JMIR mHealth uHealth **4**(2), e40 (2016). http://www.ncbi.nlm.nih.gov/pubmed/27122452
5. Rupp, M.A., et al.: The role of individual differences on perceptions of wearable fitness device trust, usability, and motivational impact. Appl. Ergon. **70**, 77–87 (2018)
6. Karapanos, E., et al.: Wellbeing in the making: peoples' experiences with wearable activity trackers. Psychol. Well-Being **6**(1), 1 (2016). https://doi.org/10.1186/s13612-016-0042-6
7. Dyer, O.: Wearable fitness device does not help maintain weight loss, study finds. BMJ **354**, i5204 (2016). https://doi.org/10.1136/bmj.i5204
8. Etkin, J.: The hidden cost of personal quantification. J. Consum. Res. **42**(6), 967–984 (2016)
9. Shih, P., et al.: Use and adoption challenges of wearable activity trackers. In: iConference 2015 Proceedings (2015)
10. The First AI Personal Trainer & Running Headphones, Vi, product website. https://www.getvi.com/
11. Runtastic: Running, Cycling & Fitness GPS Tracker, product website. https://www.runtastic.com/
12. FeetMe Sport, your personal running coach, product website. https://feetmesport.com/
13. Fitbit Official Site for Activity Trackers & More, product website. https://www.fitbit.com/
14. C25K – 5K Trainer, mobile application website. http://www.c25kfree.com/
15. Bishop, D.: Warm up II. Sports Med. **33**(7), 483–498 (2003)
16. Laursen, P.B., Jenkins, D.G.: The scientific basis for high-intensity interval training. Sports Med. **32**(1), 53–73 (2002)
17. Seiler, S.: What is best practice for training intensity and duration distribution in endurance athletes? Int. J. Sports Physiol. Perform. **5**(3), 276–291 (2010)
18. Buchheit, M., Laursen, P.B.: High-intensity interval training, solutions to the programming puzzle. Sports Med. **43**(10), 927–954 (2013)
19. Scherr, J., et al.: Associations between borg's rating of perceived exertion and physiological measures of exercise intensity. Eur. J. Appl. Physiol. **113**(1), 147–155 (2013)
20. Vesterinen, V., et al.: Heart rate-running speed index may be an efficient method of monitoring endurance training adaptation. J. Strength Cond. Res. **28**(4), 902–908 (2014)

21. Zadeh, L.A.: Fuzzy logic = computing with words. IEEE Trans. Fuzzy Syst. **4**(2), 103–111 (1996)
22. Fitbit SDK, developer documentation. https://dev.fitbit.com
23. Mercer, K., et al.: Acceptance of commercially available wearable activity trackers among adults over 50 and with chronic illness: a mixed-methods evaluation. JMIR mHealth uHealth **4**(1) (2016)
24. Laugwitz, B., Held, T., Schrepp, M.: Construction and evaluation of a user experience questionnaire. In: Proceedings of USAB 2008: HCI and Usability for Education and Work. LNCS, vol. 5298, pp. 63–76 (2008)

Does Wellness Predict Performance? Player-Specific Insights from Daily Monitoring in College Men's Soccer

Leili Javadpour[✉], Ashwinth Reddy Kondapalli, Mehdi Khazaeli, and Adam Reeves

University of the Pacific, Stockton, CA 95219, USA
ljavadpour@pacific.edu

Abstract. This ongoing study investigates the potential effect of pre-training wellness metrics on daily training performance in NCAA Division I men's soccer players. A custom mobile application was developed to collect daily self-reported wellness data from players prior to their training sessions. The dataset includes 15 wellness indicators (e.g., sleep quality, muscle soreness, stress) paired with GPS-derived training outputs such as high-speed running, accelerations, and session-RPE. To assess whether wellness states influence session-specific performance, we conducted correlation analyses at both the team and individual levels. Preliminary findings indicate no significant correlations between the wellness metrics and training performance to date. As this research continues, the long-term goal is to develop a decision-support tool for coaches to personalize training regimens, reduce injury risk, and optimize athlete readiness.

Keywords: Performance prediction · Soccer analytics · Player wellness

1 Introduction

The increasing interest in sports analytics over the last two decades can be largely attributed to advances in technology, which have enabled teams and individuals to leverage data for gaining competitive advantages. The sheer volume of data collected during each game presents a big data challenge, as extracting meaningful insights from raw data is not readily feasible. As a result, data-driven decision-making is now being integrated across various aspects of sports, from gambling and fantasy leagues to improving team dynamics, enhancing performance, guiding tactical decisions, and preventing injuries.

Soccer is a high-intensity sport characterized by numerous complex factors that enable a team to compete effectively. At the highest levels, consistently optimized player performance is essential for team success. However, multiple factors influence a player's physical readiness to meet the demands of competition. Recent research highlights the central role of athlete wellness in shaping performance outcomes, particularly regarding recovery, exertion capacity, and injury risk. Wellness indicators such as sleep quality, fatigue, muscle soreness, stress, and perceived energy levels significantly impact on a player's ability to perform high-speed efforts, sustain workloads, and meet training demands. Systematic monitoring of these factors allows coaching staff to identify when athletes may be at risk of underperforming or require targeted recovery interventions.

In this study, we collected objective training data using Catapult GPS tracking systems, which provided detailed metrics on external load and physical performance during training sessions. To complement this, we developed a custom mobile application that enabled players to self-report their wellness status each morning. The app captured key wellness indicators such as sleep quality, fatigue, muscle soreness, stress, and perceived energy levels. By integrating these two data sources, our primary aim was to explore potential correlations between daily self-reported wellness and objective training metrics. This approach allows us to assess whether fluctuations in wellness are reflected in training performance, and ultimately, to determine the value of daily wellness monitoring for optimizing athlete readiness and performance in collegiate men's soccer.

2 Literature Review

Recent research demonstrates that athlete wellness, including sleep quality, fatigue, muscle soreness, and mood, plays a significant role in training performance and readiness. Studies in professional soccer have shown that self-reported wellness scores correlate with external training loads such as sprint distance and player load, with stronger associations for accumulated and variable training demands. The acute-to-chronic workload ratio (ACWR) is a key framework for managing training loads to reduce injury risk and maintain performance [1, 2]. Combining subjective wellness measures with objective data (e.g., GPS-derived metrics) enhances prediction of athlete readiness and performance outcomes, often outperforming physiological data alone [3, 4].

Psychological and social factors, such as emotional regulation, stress, and social support, are also critical for athletic performance. Psychological fatigue or unresolved stress can impair performance regardless of physical capability [5]. Mental toughness and resilience support consistent performance under [6, 7], while chronic stress without recovery increases injury risk [8, 9]. Negative mood states and lack of social support are linked to poorer performance and higher injury rates, whereas strong support systems enhance motivation and coping [10, 11]. These psychological and social markers are increasingly integrated into performance models, improving prediction accuracy.

Methodologically, both statistical and machine learning approaches are used to model athletic performance. Recent research demonstrates the effectiveness of using training session data and machine learning to predict soccer match outcomes, supporting the value of data-driven approaches for performance forecasting in soccer [12].

3 Data

This study utilized two primary data sources collected during Spring 2025 from NCAA Division I male soccer players: (1) device-based performance metrics and (2) self-reported wellness indicators via a custom-built mobile application. Due to academic and external responsibilities, data submission among student-athletes was not always consistent. A core group of 6 to 7 players submitted daily wellness data reliably throughout the one-month collection period. These participants were between 19 and 22 years old.

Although all athletes had access to the wellness app, our analysis focused on those with consistent entries to ensure data quality.

In addition, each player's profile details, such as height, weight, maximum heart rate, and maximum velocity, were recorded at the beginning of the season. These attributes were incorporated into the analysis. For instance, body mass index (BMI) was derived from height and weight, and velocity-related features were normalized where appropriate.

3.1 Catapult Data

Player physical performance metrics were recorded using Catapult GPS vests, a professional-grade wearable system designed to monitor external training load, high-speed movement, and impact forces. Each training session generated a comprehensive set of performance metrics, including maximum acceleration, deceleration, sprint distance, high-speed efforts, and player load, among others. Training sessions occurred three to four times per week (Tables 1 and 2).

Table 1. Sample of Catapult Feature Descriptions

Attribute	Description
Duration	Total session time (in minutes)
Max Acceleration	Highest acceleration recorded during session
Max Deceleration	Highest deceleration recorded during session
Accel + Decel Efforts	Total of both acceleration and deceleration efforts
Accel + Decel Efforts Per Minute	Rate of combined efforts per minute
Distance	Total distance covered (in meters)
Player Load	Composite score of total exertion during session
Max Velocity	Highest speed recorded (m/s)
Player Load Per Minute	Player load normalized by session duration
Energy	Estimated energy expenditure
Sprint Efforts	Number of sprint events
Sprint Dist Per Min	Sprint distance normalized per minute
High Speed Movement	Metrics for high-speed movement

Table 2. Wellness Feature Descriptions

Attribute	Description
Sleep Quality	Self-rated quality of sleep (1–5 scale)
Fatigue Level	Self-reported fatigue prior to training (1–5 scale)
Soreness Level	Perceived muscle soreness (1–5 scale)
Stress Level	Academic/personal stress level (1–5 scale)
Energy Level	Subjective energy level on training day (1–5 scale)

3.2 Wellness App Data

A custom mobile application was developed to allow players to track their wellness characteristics each day before practice. The app was designed to be quick and intuitive, minimizing the time required for daily submissions. Players typically completed their wellness assessments in the morning prior to each training session.

The app collected data on sleep quality, fatigue level, soreness level, stress level, and energy level, using slider-based Likert scales ranging from 1 to 5. Additionally, the app featured an interactive human body diagram, enabling players to select specific regions and indicate their level of pain or soreness with a color-coded system: green (no pain), yellow (mild pain), orange (moderate pain), and red (severe pain).

Figure 1 shows the user interface of the application. We recognize that the data collected through this app is entirely subjective, and individual pain tolerance varies among players. However, the primary objective is to investigate whether correlations exist between the collected wellness data and players' performance during training sessions.

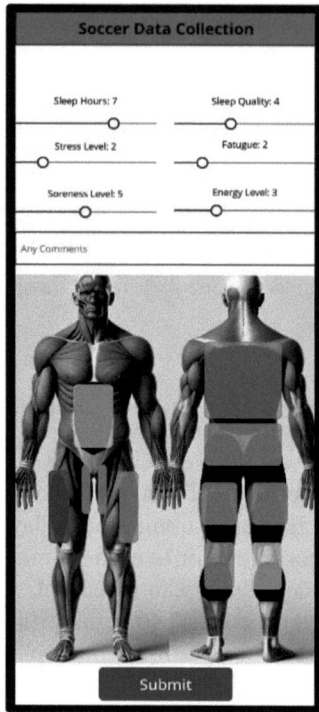

Fig. 1. Wellness Reporting App Interface

4 Correlation Analysis

A Spearman correlation matrix was computed across all numerical features to identify monotonic relationships, capturing both linear and nonlinear associations. This approach showed key interdependence between performance metrics. No strong correlations were observed between wellness variables, defined as Pain and Sleep-related metrics, and training load or movement outputs. However, one moderate negative correlation was found between Sleep Quality and Stress Level ($\rho \approx -0.42$), suggesting a possible link between poor sleep and elevated perceived stress. Additionally, weak but consistent negative associations were observed between Pain variables (e.g., glutes, hamstrings, back) and sprint-related metrics, implying a potential limiting effect of localized discomfort on explosive movement.

Due to the limited sample size and moderate variability, predictive models such as XGBoost yielded inconclusive performance. As an alternative, decision tree regressors were employed to identify interpretable patterns in how wellness inputs relate to training outcomes. Across multiple target variables (e.g., Max Velocity, Player Load, Accel + Decel Efforts), Energy Level consistently emerged as a top predictor, frequently appearing as the root node in trees. Additional split conditions included Fatigue, Right Groin Pain, and Sleep Hours, pointing to their conditional influence on workload and movement intensity.

While this study used Spearman correlation coefficients to explore monotonic associations, p-values were not emphasized due to the exploratory nature of the analysis and the limited sample size. However, we recognize the value of formal statistical reporting, and future iterations of this research will include p-values and confidence intervals for key associations to support stronger inference and transparency.

5 Conclusion and Future Work

This study investigated the extent to which self-reported wellness indicators can inform and potentially predict session-level training performance in NCAA Division I men's soccer athletes. Using a custom-built app, players submitted daily wellness entries each morning prior to training, reporting on factors such as sleep quality, fatigue, stress, and localized pain. These inputs were then aligned with GPS-derived performance data to explore how day-to-day fluctuations in wellness might impact physical output.

Despite the intent, our analysis did not reveal any strong correlations between wellness indicators and training outcomes. While some metrics, such as energy level and fatigue, showed weak to moderate associations with performance variables, these patterns were not consistent or robust across the dataset. Several factors may help explain the lack of stronger wellness-to-performance relationships:

- *Subjectivity of self-reported input:* Players' perceptions of fatigue, stress, or pain vary significantly. One athlete's "moderate pain" may be another's "high pain," leading to inconsistencies that reduce data reliability.
- *Timing of data collection*: Wellness responses were recorded in the morning, prior to any physical activity. It's likely that players' mood or readiness may change during warm-ups or once socially engaged with teammates, weakening the predictive power of pre-session inputs.
- *Limited sample size:* The relatively small and uniform dataset restricted the statistical power required to detect subtle or nonlinear trends.

Although the current findings did not uncover strong links between wellness and training performance, existing literature continues to support the influence of wellness on athletic output. This project contributes to the broader effort to integrate internal wellness monitoring with external load tracking in applied sport environments. The combination of correlation analysis and decision trees presents a viable exploratory framework for future studies.

Moving forward, we plan to enhance our data collection process by refining the app experience, seeking feedback from players and coaches, and potentially incorporating additional context-based features. Future research should aim to collect longitudinal data across competitive phases, expand sample sizes, and integrate richer contextual information such as match performance and psychological readiness.

This paper presents preliminary findings based on an initial dataset collected during the early phase of the study. The current analysis is exploratory and constrained by sample size, subjective reporting variability, and a limited collection window. At this stage, the focus is on methodological clarity and interpretability, rather than on presenting granular statistics. However, these early results lay the groundwork for a more comprehensive follow-up. Future phases of this research will involve extended data collection across multiple competitive periods, improved participant consistency, and the application of advanced statistical modeling techniques—such as mixed-effects and time-series analysis—to better capture temporal and individual-level variability. As data collection continues and the system matures, we anticipate uncovering more reliable patterns that will contribute to improved prediction accuracy and stronger insights into athlete readiness and performance.

References

1. Gabbett, T.J.: The training—injury prevention paradox: should athletes be training smarter and harder? Br. J. Sports Med. **50**(5), 273–280 (2016). https://doi.org/10.1136/bjsports-2015-095788
2. Windt, J., Gabbett, T.J.: How do training and competition workloads relate to injury? The workload—injury aetiology model. Br. J. Sports Med. **51**(5), 428–435 (2017). https://doi.org/10.1136/bjsports-2016-096040
3. Saw, A.E., Main, L.C., Gastin, P.B.: Monitoring the athlete training response: subjective self-reported measures trump commonly used objective measures: a systematic review. Br. J. Sports Med. **50**(5), 281–291 (2016). https://doi.org/10.1136/bjsports-2015-094758
4. Seshadri, D.R., et al.: Wearable technologies for personalized sports analytics: a review of current and emerging trends. NPJ Digital Med. **5**, 115 (2022). https://doi.org/10.1038/s41746-022-00651-w
5. Birrer, D., Morgan, G.: Psychological skills training as a way to enhance an athlete's performance in high-intensity sports. Scand. J. Med. Sci. Sports **20**(2), 78–87 (2010). https://doi.org/10.1111/j.1600-0838.2010.01188.x
6. Gould, D., Maynard, I.: Psychological preparation for the Olympic Games. J. Sports Sci. **27**(13), 1393–1408 (2009). https://doi.org/10.1080/02640410903081845
7. Jones, G., Hanton, S., Connaughton, D.: A framework of mental toughness in the world's best performers. Sport Psychol. **21**(2), 243–264 (2007). https://doi.org/10.1123/tsp.21.2.243
8. Kellmann, M.: Preventing overtraining in athletes in high-intensity sports and stress/recovery monitoring. Scand. J. Med. Sci. Sports **20**(S2), 95–102 (2010). https://doi.org/10.1111/j.1600-0838.2010.01192.x
9. Hanton, S., Fletcher, D., Coughlan, G.: Stress in elite sport performers: a comparative study of competitive and organizational stressors. J. Sports Sci. **23**(10), 1129–1141 (2005). https://doi.org/10.1080/02640410500131480
10. Rees, T., Hardy, L.: Matching social support with stressors: effects on factors underlying performance in tennis. Psychol. Sport Exerc. **5**(3), 319–337 (2004). https://doi.org/10.1016/S1469-0292(03)00021-7
11. Freeman, P., Rees, T.: Perceived social support from team-mates: direct and stress-buffering effects on self-confidence. Eur. J. Sport Sci. **10**(1), 59–67 (2010). https://doi.org/10.1080/17461390903049998
12. Javadpour, L., Khazaeli, M., Molenaar, R.: From practice to performance: predicting soccer match outcomes from training data. SN Comput. Sci. **6**, 324 (2025). https://doi.org/10.1007/s42979-025-03870-0

Locating Tennis Ball Impact on the Racket in Real Time Using an Event Camera

Yuto Kase[✉] [iD], Kai Ishibe [iD], Ryoma Yasuda [iD], Yudai Washida [iD], and Sakiko Hashimoto [iD]

Mizuno Corporation, 1-12-35 Nanko Kita, Suminoe-ku, Osaka, Japan
{ykase,kishibe,ryasuda,ywashida,skhashim}@mizuno.co.jp
https://corp.mizuno.com

Abstract. In racket sports, such as tennis, locating the ball impact on the racket is important in clarifying player and equipment characteristics, thereby aiding in personalized equipment design. High-speed cameras are used to measure the impact location; however, their excessive memory consumption limits prolonged scene capture, and manual digitization for location detection is time-consuming and prone to human error. These limitations make it difficult to effectively capture the entire playing scene, hindering the ability to analyze the player's performance. We propose a method for locating the tennis ball impact on the racket in real time using an event camera. Event cameras efficiently measure brightness changes (called 'events') with microsecond accuracy under high-speed motion while using lower memory consumption. These cameras enable users to continuously monitor their performance over extended periods. Our method consists of three identification steps: time range of swing, timing at impact, and contours of ball and racket. Conventional computer vision techniques are utilized along with an original event-based processing to detect the timing at impact (PATS: the amount of polarity asymmetry in time symmetry). The results of the experiments were within the permissible range for measuring tennis players' performance. Moreover, the computation time was sufficiently short for real-time applications.

Keywords: Computer Vision · Sports Analytics · Event-Based Vision Sensor

1 Introduction

In racket sports, such as tennis, increasing the ball velocity is a key factor. Naß et al. [16] confirmed that the velocity depends on the impact location because the different restitution coefficients of the racket surface vary depending on the area, and the impact locations are also quite different among players. Therefore, the measurement of the impact location for each player can aid in personalized equipment design.

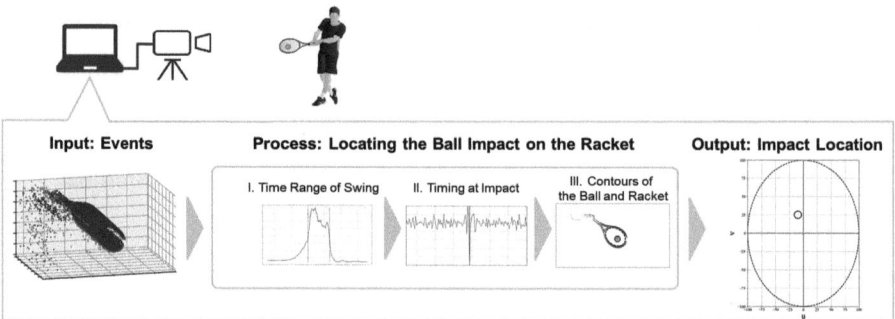

Fig. 1. Overview of the proposed method. A set of events captured by an event camera while playing tennis is input; + polarity events are represented in blue and − polarity events are represented in red. Locating the ball impact on the racket involves three steps. The output is the visualization of the impact location, which is the relative position of the ball on the racket. (Color figure online)

To measure the impact location, conventional frame-based cameras (high-speed cameras) or motion capturing systems [8] are utilized. High-speed cameras can accurately measure locations while preventing motion blur; however, they have a limitation regarding high memory consumption, consuming a large amount of memory as shutter speed increases, which limits the capturing period. This hinders capturing actual playing scenes, making it difficult to analyze players' performance efficiently. Motion capturing systems can accurately measure the marker positions, which need to be pre-attached to the ball and racket. As stated by Karditsas [10], to obtain the impact location from the data captured by these systems, manual digitization is a valid process; however, the digitization of multiple points across many images is a time-consuming process that is prone to human error. Furthermore, it is noted that automated processing offers an efficient and effective solution to this issue. To address these problems, we propose a method for locating the tennis ball impact on the racket in real time using an event camera.

Event cameras (also known as event-based cameras or dynamic vision sensors) [2,9,19] differ from conventional frame-based cameras in that instead of capturing full images at a fixed rate, they asynchronously measure per-pixel brightness changes and output a stream of events that encode the time, location, and sign of the brightness changes. This sign is referred to as polarity with '+' and '−' representing the brightness increase and decrease, respectively. These cameras offer attractive properties: high temporal resolution (on the order of microseconds) resulting in reduced motion blur, low data volume, low latency, low power consumption, and high dynamic range. In sports, these technologies enable users to easily measure and analyze their performance and equipment. In sports research, methods utilizing event cameras have been proposed to interpolate frame rates using a frame-based camera [4], estimate the ball's rotation rate [7,14], and detect the position of the ball's trajectory [15]. Regarding tennis impact locations, Yasuda et al. [29] conducted a precision evaluation of manually digitized impact images captured simultaneously by a high-speed camera and an event camera. The results fell within an error margin of 1/4 of the ball's diameter, demonstrating the usefulness of the event camera.

Our main contributions are summarized as follows:

- A method to locate the tennis ball impact on the racket in real time (Fig. 1).
- An original event-based processing method to detect the timing at impact, referred to as the amount of polarity asymmetry in time symmetry (PATS).

2 Related Work

2.1 Impact Location Methods in Sports

There are two automated approaches based on sensing systems for locating the ball impact on the racket:

The first is an inertial measurement unit (IMU) installed on a tennis racket [28]. It identifies the impact location using the vibration data resulting from the impact between the racket and the ball. The data are subjected to frequency analysis, using methods such as the Fast Fourier Transform (FFT), whereby the system matches the frequency characteristics with a database storing the characteristics of each racket to identify the impact location. This approach has the advantages of a less burdensome setup and prolonged scene capture. However, it requires the database to estimate the impact area of the racket.

The second requires high-speed cameras to be installed in a stationary device [11,12]. They identify the impact location for golf using object detection methods, such as the Hough transform, applied to the image at the moment of impact. These approaches have the advantage of precisely measuring the impact location. To detect the timing at impact, an image subtraction method is used, which subtracts the current image from the previous image pixel by pixel, and then compares it to a threshold. Additionally, there is a method that incorporates Doppler radars for triggering at golfing impact [26]. These methods reduce the memory consumption for high-speed image storage. Although useful in golf, where the impact location is fixed, it is less suitable for tennis, where the impact location may vary and the player's body may enter the frame.

These existing approaches make it difficult to accurately identify the timing and location of the ball's impact in tennis. Therefore, our method aims to overcome this challenge by using an event camera.

2.2 Event Camera Representation

In using an event camera, each event includes x, y, polarity, and time retrieved in chronological order. The events are transformed into various alternative representations to facilitate the extraction of meaningful information and solve a particular task, as shown in the survey [5]. The following representations are related to our method.

The first is an event packet [5]. This representation is a set of events within the specified accumulation time interval. The number of events in the packet per second is called the event rate [7]. This representation has the advantages of precise timestamp, the number of events, and low computational complexity. However, it lacks meaningful information, such as the coordinates and polarity.

The second is an event image [20]. This representation is an image where each pixel stores a polarity value related to the last event at that pixel. It has the advantage of being usable with conventional image-based computer vision algorithms. However, the time information can be lost.

The third is a time surface [13,24]. This representation is an image where each pixel stores a single time value related to the last event at that pixel, generated separately for each polarity value. The value of each pixel in the image is higher for more recent events, as in an intensity map. This has the advantages of being compatible with conventional image-based computer vision algorithms, similar to the event image while retaining the time information. However, this representation cannot simultaneously handle both polarities.

In this paper, we represent the event packet for input (Fig. 2(a)), event rate of the event packet to identify the time range of the swing (Fig. 2(b)), and event image to identify the contours of the ball and racket (Fig. 2(c)). Inspired by the event image and time surface, we propose an original processing method to detect the timing at impact (Fig. 5).

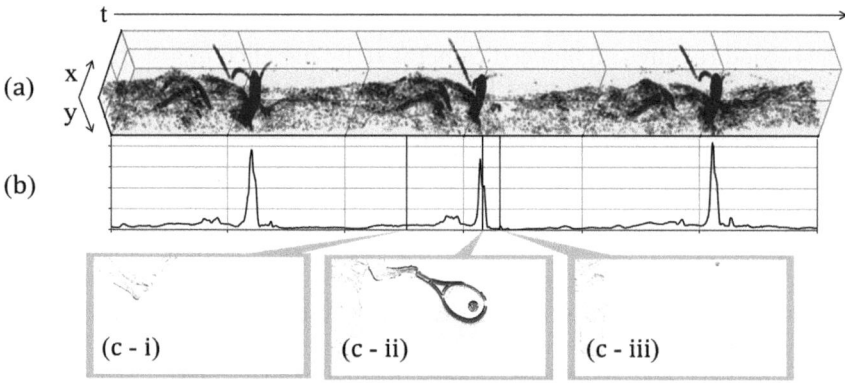

Fig. 2. Time series of event packets ε_t in tennis scenes. (a) event data e_k colored according to polarity (+ in blue, − in red); the areas with smaller x-coordinates show clusters of events caused by racket swings, whereas the areas with larger x-coordinates show clusters of events caused by ball bounces. (b) time series of event rates ($|\varepsilon_t|$ per second) which increases while a player is swinging the racket; this player swings three times in these scenes. (c) event image at a specific point in time. Image (c - i) is captured before swinging; image (c - ii) is captured at impact; and image (c - iii) is captured after swinging. (Color figure online)

3 Method

As shown in Fig. 1, the proposed method can be divided into three parts: input (Sec. 3.1), process (Sec. 3.2–3.4), and output (Sec. 3.5).

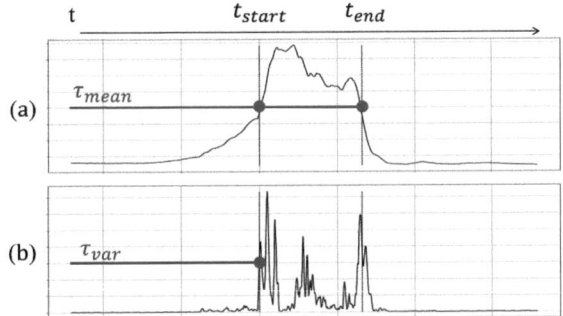

Fig. 3. Time series of the mean and variance for the consecutive n_ε event rates. (a) time series of the mean, where the red solid line represents the threshold τ_{mean}. (b) time series of the variance, where the solid red line represents the threshold τ_{var}. The black dashed lines represent t_{start} and t_{end}, respectively. τ_{mean} and τ_{var} are used to identify t_{start}, while τ_{mean} is used for t_{end}. (Color figure online)

3.1 Input

Individual events captured by a camera are expressed as $e_k = (x_k, y_k, p_k, t_k)$, where k is an index for each event retrieved in chronological order; x_k and y_k are pixel coordinates (px) with intervals of $[0, width - 1]$ and $[0, height - 1]$ based on the camera's pixel size; t_k is a timestamp in microseconds (μs); and p_k is the polarity ($p_k \in \{+, -\}$).

A set of events within the specified accumulation time interval is called an event packet. The event packet ε_t is defined as

$$\varepsilon_t = \left\{ e_k \left| t - \frac{t_{acc}}{2} < t_k \leq t + \frac{t_{acc}}{2} \right. \right\}, \tag{1}$$

where t is the reference time (μs), and t_{acc} is the accumulation time (μs). Therefore, ε_t is a $|\varepsilon_t| \times 4$ matrix, which represents the attributes (x_k, y_k, p_k, t_k) as columns, and the event packet ε_t is the input.

3.2 Time Range of Swing

As shown in Fig. 2, when a player starts to swing the racket, the number of events in the packet $|\varepsilon_t|$ increases. The racket moves relatively faster than other objects, including the ball, player's body, and background. To determine the swing, we utilize $|\varepsilon_t|$ per second, called event rates, as explained in Sect. 2.2. A time series of event rates identifies the time range of the swing as follows.

As the reference time t progresses at constant intervals with stride time t_{strd}, the event rate ($|\varepsilon_t|$ per second) is calculated for each packet. The mean and variance for the consecutive n_ε event rates are then determined.

The start time of the swing t_{start} is defined as the first moment that exceeds both the mean threshold τ_{mean} and the variance threshold τ_{var}. The end time of the swing t_{end} is defined as the first moment that occurs after a time interval τ_t and falls below τ_{mean}, as shown in Fig. 3.

Subsequently, we detect the impact timing t_{imp} in the time range of the swing $[t_{start}, t_{end}]$.

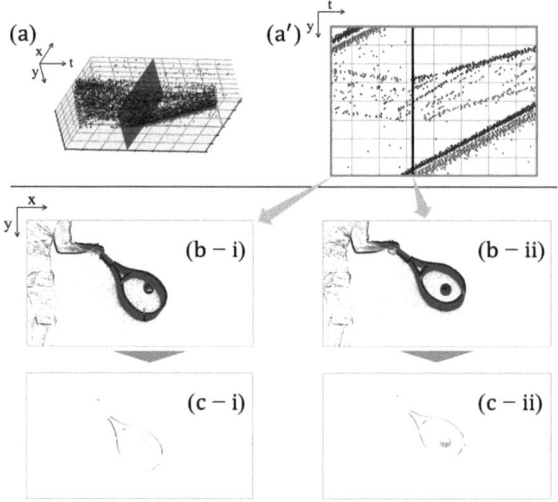

Fig. 4. Comparison of the visualization of PATS images before and at impact. (a) 3D plot of events e_k near the impact timing t_{imp}, where the blue points represent $+$; the red points represent $-$; the black plane represents t_{imp}; and the orange edge represents a plane parallel to the t-axis that contains the ball events. Plot (a') shows a 2D plot of events extracted along the orange edge of plot (a). The V-shaped events near the center are generated by the bouncing ball on the racket, whereas the upper left events are generated by the top of the swinging racket, and the lower right events are generated by the bottom of the swinging racket. (b) event images. Image (b - i) is captured at $t_{imp} - 6300$ μs (the time at the left edge of plot (a')), whereas image (b - ii) is captured at t_{imp} (the time at the black line of plot (a')). (c) PATS images as in Eq. 11, corresponding to images (b - i) and (b - ii). Image (c - ii) detected more ball events than image (c - i). (Color figure online)

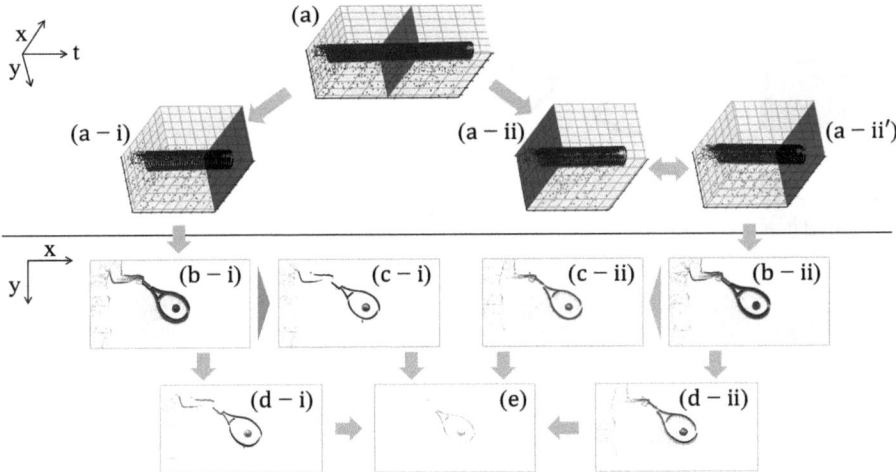

Fig. 5. Procedures for timing at impact. (a) 3D plot of events e_k, where blue points represent $+$; red points represent $-$; and the black plane represents reference time t. Plot (a) represents ε_t; plot (a - i) represents ε_{prev}; plot (a - ii) represents ε_{next}; and plot (a - ii') represents $\overleftarrow{\varepsilon_{next}}$. (b) (c) event images of i: ε_{prev} and ii: $\overleftarrow{\varepsilon_{next}}$, respectively, where blue pixels represent $+$; red pixels represent $-$; and white pixels represent *none*. Image (b - i) represents $\mathbf{F}_{prev}$; image (c - i) represents $\mathbf{F}^+_{prev}$; image (b - ii) represents $\mathbf{F}_{next}$; and image (c - ii) represents $\mathbf{F}^-_{next}$. (d) grayscale image as a focal time function, where black pixels represent 1.0, and white pixels represent 0.0 or *none*. Image (d - i) represents $\mathbf{G}_{prev}$, and image (d - ii) represents $\mathbf{G}_{next}$. (e) PATS image $|\mathbf{F}^+_{prev} * \mathbf{G}_{prev}| * |\mathbf{F}^-_{next} * \mathbf{G}_{next}|$, which is a pink grayscale image, where pink pixels represent 1.0 and white pixels represent 0.0. (Color figure online)

3.3 Timing at Impact

As shown in Fig. 4(a'), at the impact timing t_{imp}, the ball bounces on the racket. An event camera captures the bouncing location and shows the changes in the events from $+$ to $-$ in individual pixels. Assume that the ball is brighter than the background.

Therefore, we propose an original indicator (PATS: the amount of polarity asymmetry in time symmetry) that focuses on the changes to detect the impact. As shown in Fig. 5, the procedures for the indicator are as follows:

First, we split the event packet ε_t from reference time t into two packets: the packet before t: ε_{prev} and the packet after t: ε_{next}, defined as

$$\varepsilon_{prev} = \{e_k \in \varepsilon_t \,|\, t_k < t\}, \tag{2}$$

$$\varepsilon_{next} = \{e_k \in \varepsilon_t \,|\, t \leq t_k\}. \tag{3}$$

We reverse ε_{next} in chronological order to detect whether the events changed from $+$ to $-$ considering the symmetry about t. The reversed ε_{next} is defined as

$$\overleftarrow{\varepsilon_{next}} = \text{reverse}(\varepsilon_{next}), \tag{4}$$

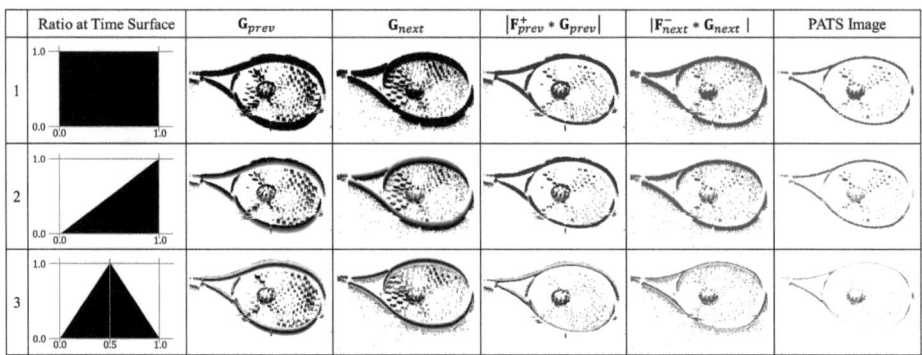

Fig. 6. Patterns of various focal time functions. Pattern 1 shows a function with uniform ratio (1.0), which is the same as an event image. Pattern 2 shows a function where the ratio increases linearly, which is equivalent to a time surface. Pattern 3 shows a function where the ratio increases linearly up to 0.5 time and then decreases linearly up to 1.0 time. The first column is the pattern index. The second column shows the ratio at the value of the time surface. $\mathbf{G}_{prev}$ and $\mathbf{G}_{next}$ columns show the enlarged grayscale images of the area near the racket, where black pixels represent 1.0, and white pixels represent 0.0 or *none*. $|\mathbf{F}^{+}_{prev} * \mathbf{G}_{prev}|$ and $|\mathbf{F}^{-}_{next} * \mathbf{G}_{next}|$ columns show the convolved images. The last column shows the PATS image, which refers to $|\mathbf{F}^{+}_{prev} * \mathbf{G}_{prev}| * |\mathbf{F}^{-}_{next} * \mathbf{G}_{next}|$. Pattern 3 most effectively reduces the string flickering.

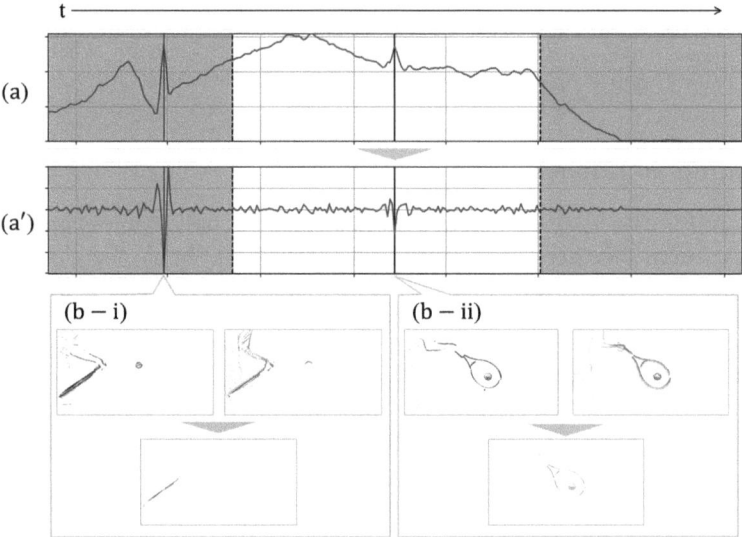

Fig. 7. Peaks in the time series of ρ_t values (*i.e.* PATS). Graph (a) time series of ρ_t. Graph (a') shows the Laplacian filtered time series, where the black solid line represents the estimated impact timing t_{imp} within the estimated time range of the swing (black dashed line). The red solid line represents false detection if the time range is overestimated. (b) results in the PATS image, respectively. (Color figure online)

where reverse(·) is a function that sorts rows in descending chronological order.

Second, the event images [20] of ε_{prev} and $\overleftarrow{\varepsilon_{next}}$ are generated. They are defined as

$$\mathbf{F}_{prev} = \text{image}(\varepsilon_{prev}), \tag{5}$$

$$\mathbf{F}_{next} = \text{image}(\overleftarrow{\varepsilon_{next}}), \tag{6}$$

where image(·) is a function that generates a $height \times width$ (px) event image and consists of $\{1, -1, 0\}$ assigned as '+' to 1, '−' to −1, and 'none' to 0, respectively. Note that the pixels of $\mathbf{F}_{next}$ are overwritten by the older events because $\overleftarrow{\varepsilon_{next}}$ is sorted in descending chronological order.

Third, the time surfaces of ε_{prev} and $\overleftarrow{\varepsilon_{next}}$ are generated. They are defined as

$$\mathbf{T}_{prev} = \text{timeSurface}(\varepsilon_{prev}), \tag{7}$$

$$\mathbf{T}_{next} = \text{timeSurface}(\overleftarrow{\varepsilon_{next}}), \tag{8}$$

where timeSurface(·) is a function that generates a $height \times width$ (px) time surface and linearly transforms the oldest time to 0.0 and the most recent time to 1.0. Similarly to $\mathbf{F}_{next}$, note that $\overleftarrow{\varepsilon_{next}}$ is sorted in descending chronological order. The obtained time surfaces are transformed using focalTime(·), which focuses on the selected time. They are defined as

$$\mathbf{G}_{prev} = \text{focalTime}(\mathbf{T}_{prev}), \tag{9}$$

$$\mathbf{G}_{next} = \text{focalTime}(\mathbf{T}_{next}). \tag{10}$$

The focal time images $\mathbf{G}_{prev}$ and $\mathbf{G}_{next}$ are such that the pixel values of the time surface at the most focused time are set to 1, and the values that are the least focused are set to 0, as shown in Fig. 6.

Fourth, we convolve the event images with the focal time images, and calculate the sum of the convolved image. This is defined as

$$\rho_t = \sum (|\mathbf{F}_{prev}^+ * \mathbf{G}_{prev}| * |\mathbf{F}_{next}^- * \mathbf{G}_{next}|), \tag{11}$$

The indicator ρ_t is referred to as PATS, and the convolved image is the PATS image. The reason for convolving the focal time images is to enhance robustness under direct sunlight. This is because the flickering of the tennis strings due to sunlight causes false impact timing. Considering that this flicker (100 μs) takes less time than the impact (4000 μs, as demonstrated by [3]), we designed a focal time function to exclude the flicker time, as shown in Fig. 6.

Finally, the peak in the time series of ρ_t values is detected to identify t_{imp}, as shown in Fig. 7. The Laplacian filter is utilized for peak detection, and the time corresponding to the smallest value in the filtered time series is the estimated impact timing t_{imp}.

Alternatively, n_c candidates for t_{imp} can be obtained in ascending order of peak size, and the one with the centroid position in the PATS image closest to the center can be selected. This prevents false detection of times when the racket frame overlaps near the start of the swing due to a broad estimation of the swing interval, as shown in Fig. 7(b - i).

In the next step, contours of the ball and racket are identified at t_{imp}.

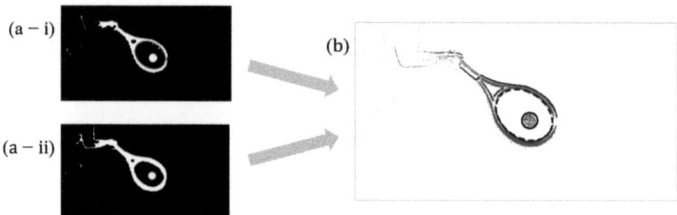

Fig. 8. Procedures for contours of the ball and racket. (a) binary images at the impact timing t_{imp} with different accumulation times t_{acc}. Image (a - i): $\mathbf{B}^r_{t_{imp}}$ detects the racket with $t_{acc} = 500$, and image (a - ii): $\mathbf{B}^b_{t_{imp}}$ detects the ball with $t_{acc} = 2000$, where the red ellipse represents the detected objects. (b) event image at t_{imp} with $t_{acc} = 500$ and ellipse detection overlaid, where the black solid line represents the ball's ellipse and the dashed line represents the racket's ellipse.

3.4 Contours of Ball and Racket

To locate the ball on the racket, we identified the approximate contours of the ball and racket as ellipses in this step, as shown in Fig. 8.

First, the appropriate accumulation times for the ball and racket are set because the number of events differs depending on the difference in speed as captured by an event camera. We generated the event images for the ball $\mathbf{F}^b_{t_{imp}}$ and racket $\mathbf{F}^r_{t_{imp}}$ at t_{imp} using the image($\cdot$) function, as shown in Eq. (5). In the above generation, the activity noise filter [21] was applied to remove noise from these images.

Subsequently, $\mathbf{F}^b_{t_{imp}}$ and $\mathbf{F}^r_{t_{imp}}$ are converted to the binary images $\mathbf{B}^b_{t_{imp}}$ and $\mathbf{B}^r_{t_{imp}}$, respectively. In these binary images, each pixel event is assigned a value of 1 for $\{+, -\}$ or 0 otherwise (*none*), as shown in Fig. 8. We applied a morphological transformation [17] that performs a closing operation on these binary images to improve the accuracy of ellipse detection. To prevent false detections, we can optionally crop the images to the region with a large number of events (*i.e.* ROI) in advance.

Finally, ellipse detection [18] is applied to $\mathbf{B}^b_{t_{imp}}$ and $\mathbf{B}^r_{t_{imp}}$. This detection retrieves the properties of the ellipses, which include the center coordinates, semi-major axis, semi-minor axis, and rotation angle. In $\mathbf{B}^r_{t_{imp}}$, the second largest detected ellipse is estimated to be inside the racket frame, whereas in $\mathbf{B}^b_{t_{imp}}$, the largest detected ellipse inside the frame is estimated to be the ball.

3.5 Output

As output, the impact location is expressed as a percentage of the relative position of the ball on the racket. The center of the ball relative to the racket is located using the properties of the ellipses. In particular, the center coordinates of the ball's ellipse are mapped to the axes of the racket's ellipse in the uv coordinate system, and the ratios of the axes are calculated. Note that the $+u$

direction is defined as upward along the racket, and the $+v$ direction is defined as toward the tip of the racket.

4 Experiment

To verify the proposed method, we conducted experiments to capture tennis scenes using an event camera.

4.1 Setup

For the event camera, SilkyEvCam HD [2] was used for these experiments. The laptop used for processing event data was equipped with an Intel Core i7-12700H CPU (14 cores, 20 threads, 2.3 GHz), an NVIDIA GeForce RTX 3080 Ti Laptop GPU (16 GB GDDR6 memory), and 32 GB of LPDDR5 memory (5200 MHz). This method was implemented in Python.

Fig. 9. Play scene and equipment for our experiments. The event camera captured the scene from behind the player, and the ball was tossed from the machine positioned in the front right, as shown in the left figure. In the case of left-handed players, the position of the machine was reversed. A white racket frame and black polyester strings were used, as shown in the right figure.

4.2 Play Scene

As shown in Fig. 9, the impact location was measured on an outdoor tennis court. The camera was positioned at a height of 0.7 m to match the impact location and then placed 2.7 m behind it. A zoom lens was used for the measurement. The data were captured both under direct sunlight and in the absence of sunlight.

These tennis scenes were designed as rallies, where a player consecutively hit back 12 balls from a tossing machine, and then each ball bounced once after being tossed. Three players were right-handed, whereas two players were left-handed. The equipment consisted of a white racket and black polyester strings.

4.3 Metrics

The visual definitions of the timing and location at impact using an event camera are based on [29]. The impact timing is defined as the moment when the area of the ball for + events is at its maximum. We visually inspected the timing and location at impact, using the following definitions.

Each of the three identification steps was verified.

I. **Time Range of Swing:** We verified whether the visually inspected timing was included in the identified swing range $[t_{start}, t_{end}]$. If it was included, the next step was proceeded to as a success; if it was not included, the next step was not proceeded to, considering it a failure.
II. **Timing at Impact:** The absolute difference in time (μs) between the identified timing t_{imp} and the visually inspected timing was verified. If the difference was 2000 μs or less, the next step was proceeded to as a success; if it was greater, the next step was not proceeded to, considering it a failure. Note that Cross [3] demonstrated that the contact time of the ball during the impact in tennis is approximately 4000 μs.
III. **Contours of Ball and Racket (Output):** The relative absolute difference in percentage points (%pt) between the identified and visually inspected location was verified. If the contours were not detected, the step was considered a failure. Note that this location was utilized because it is easier to evaluate than the contour.

Table 1. Variables of our proposed method set using the training data.

Section	Symbol	Value
Time Range of Swing	t_{acc}	500
	t_{strd}	500
	n_ε	10
	τ_{mean}	1×10^7
	τ_{var}	6×10^{11}
	τ_t	100000
Timing at Impact	t_{acc}	4000
	t_{strd}	500
	n_c	3
Contours of Ball	t_{acc}	2000
Contours of Racket	t_{acc}	500

4.4 Variables

For the 60 data obtained from 12 consecutive balls hit by 5 players, the first 2 balls were used as training data (10 data), whereas the subsequent 10 balls were used as test data (50 data). Using the training data, the variables were set in Table 1. For the focal time function, we selected Pattern 3 in Fig. 6.

Table 2. Results of our experiments. The ratio at each step represents the number of successful instances/total number of instances that passed the previous step. For players No. 4 and 5, the time range of the swing was overestimated by one instance each. Output indicates that the red plots represent the visually inspected impact locations. The crosses (×) represent instances that failed in the time range of swing or timing at impact, while the triangles (△) indicate failures in the contours of the ball and racket. The circles (○) indicate successful estimation instances, with black circles representing the estimated impact locations, and their diameter corresponds to 1/4 the size of a tennis ball. The black solid lines connect the corresponding visually inspected and estimated impact locations.

Player No.	1	2	3	4	5
Under Direct Sunlight				✓	✓
Left-Handed			✓		✓
Time Range of Swing	9/10	10/10	10/10	11/10	11/10
Timing at Impact (μs)	322 ± 244	350 ± 240	322 ± 79	460 ± 254	430 ± 224
	9/9	8/10	9/10	10/10	10/10
Output (%)					
	8/9	8/8	8/9	0/10	3/10
Computation Time (s)	1.83 ± 0.34	1.86 ± 0.25	1.41 ± 0.11	1.75 ± 0.23	1.97 ± 0.19

4.5 Result

Verification was conducted using 10 test data from each of the 5 players, totaling 50 data (instances) in Table 2. No. 1 – 3 were measured in the absence of sunlight, whereas No. 4 and 5 were measured under direct sunlight. No. 3 and 5 are left-handed. The mean computation time per instance was less than 2 s.

I. **Time Range of Swing:** 49/50 instances were correctly estimated for $[t_{start}, t_{end}]$. For player No. 1, one instance of range estimation did not include the visually inspected impact timing. For players No. 4 and 5, one instance each of overestimating the swing range occurred.

II. **Timing at Impact:** 46/49 instances were correctly detected for t_{imp}. The three instances that were not correctly detected involved misestimations of the timing: one instance was estimated to be 26900 μs (No. 2) after the visually inspected impact timing, and two instances were estimated to be 8300 μs (No. 2) and 13500 μs (No. 3) before it.

III. **Contours of Ball and Racket (Output):** In the absence of direct sunlight, 24/26 instances were successfully detected for contour detection. Whereas under direct sunlight, only 3/20 instances were detected. For all instances where contour detection was successful, the relative absolute difference was less than 12.1 %pt for u and 9.1 %pt for v. This means the difference was kept below 15 mm, corresponding to less than 1/4 of the diameter of a tennis ball.

5 Discussion

5.1 Consideration

Utility of Our Method: Our proposed PATS proved effective even under direct sunlight. This method eliminates the need for manually searching for impact timing and allows for the combined use of event and frame-based cameras. By triggering the shutter of the frame-based camera at the impact timing identified by the event camera, we can determine the impact location by the frame-based camera, leveraging the advantages of both camera types.

For all successful contour detection instances, the relative absolute difference in impact location was within the permissible range (15 mm) for measuring tennis players' performance, as indicated by [29]. We found that impact locations varied for each player, suggesting that impact location is useful for analyzing player and equipment characteristics.

The computation time was generally less than 2 seconds, shorter than the mean rally time of 2.44 to 2.68 seconds reported for three Grand Slam tournaments [1]. Thus, our method can be considered a real-time analysis system.

Effects of Variable Changes: For the variables in Table 1, the following effects are expected when the values change.

In time range of the swing, reducing τ_{mean} may result in a later t_{end}, extending the range $[t_{start}, t_{end}]$. This extension increases the search range for impact timing detection, potentially increasing computation time. Conversely, decreasing τ_{var} may lead to an earlier t_{start} and a longer range, which may also raise the likelihood of false detections, as illustrated in Fig. 7. Furthermore, shortening τ_t may cause the impact timing to fall outside the designated range.

In timing at impact, decreasing t_{acc} reduces computation time but may result in a relatively lower peak at the impact timing t_{imp} detected by PATS, increasing the likelihood of false detections. Increasing t_{acc} slows down computation time with little effect on detection accuracy. Considering both factors, a value of 4000 μs, corresponding to the impact time range in tennis [3], appears appropriate.

Regarding contours of the ball and racket, increasing t_{acc} may cause motion blur from events, affecting positional accuracy. Decreasing t_{acc} may result in a lower detection rate of ellipses due to insufficient events.

5.2 Limitation

Solid-Colored Constraints in Equipment: A solid-colored racket and strings were used to clarify the contours of both the racket and the ball. The racket was colored white for contrast with the background, whereas the strings were black to contrast with the ball, increasing the number of events. Due to the characteristics of event cameras, appropriate solid-colored equipment should be used based on the capturing environment. Additionally, using multi-colored equipment may result in event loss at color boundaries, complicating contour detection.

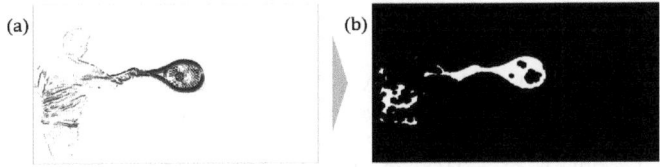

Fig. 10. Failure of contour estimation due to string flickering under direct sunlight. (a) event image at the impact timing. (b) binary image converted from image (a).

Requirement for a Noise-Free Background: The motivation for this research was to develop equipment, leading to methods that assume a noise-free background. However, in a match scenario, the movements of opponents and spectators can induce noise into the background. In time range of the swing, this noise may result in false detections due to variations in event rates. In timing at impact, if more events change from + to − than those of the ball, false positive timing may be detected. These limitations could be improved by combining existing image recognition methods [22, 27] to narrow the ROI.

Influence of Direct Sunlight on Detection Rates: Under direct sunlight, the detection rate for estimating the impact timing did not decrease; however, the detection rate for contour estimation significantly decreased. As shown in Fig. 10, this is due to the simultaneous occurrence of events related to both string flickering caused by sunlight reflection and the ball, which makes it difficult to detect the contours of the ball in the binary image.

5.3 Future Work

To improve the detection rate of contours under direct sunlight, the utilization of the proposed PATS image is considered. Additionally, implementing existing motion compensation methods [6, 23, 25] to clarify the contours in the event images may also be a viable approach.

Extensions to 3D measurements for estimating the tilt of the racket at impact and adaptations to other sports are also considered.

6 Conclusion

We propose a method for locating the tennis ball impact on the racket in real time using an event camera. The process comprises three identification steps: the timing range of the swing, the timing at impact, and detecting the contours of the ball and racket. Particularly, the impact timing estimation method PATS is suggested to eliminate the effort of manually searching for the impact timing.

The experimental results were within the permissible range for measuring tennis players' performance, and the computation time was sufficiently short for real-time applications. Thus, the proposed method is a valuable real-time analysis system.

References

1. Carboch, J., Siman, J., Sklenarik, M., Blau, M.: Match characteristics and rally pace of male tennis matches in three grand slam tournaments. Phys. Act. Rev. **7**, 49–56 (2019)
2. CenturyArks: SilkyEvCam HD - CenturyArks Co., Ltd. https://centuryarks.com/en/silkyevcam-hd/. Accessed 09 June 2025
3. Cross: Dynamic properties of tennis balls. Sports Eng. **2**(1), 23–33 (1999)
4. Deckyvere, A., Cioppa, A., Giancola, S., Ghanem, B., Droogenbroeck, M.V.: Investigating event-based cameras for video frame interpolation in sports. In: 2024 IEEE International Workshop on Sport, Technology and Research (STAR), pp. 138–143 (2024)
5. Gallego, G., et al.: Event-based vision: a survey. IEEE Trans. Pattern Anal. Mach. Intell. **44**(1), 154–180 (2022)
6. Gallego, G., Rebecq, H., Scaramuzza, D.: A unifying contrast maximization framework for event cameras, with applications to motion, depth, and optical flow estimation. In: IEEE/CVF Conference on Computer Vision and Pattern Recognition (CVPR), pp. 3867–3876 (2018)
7. Gossard, T., Krismer, J., Ziegler, A., Tebbe, J., Zell, A.: Table tennis ball spin estimation with an event camera. In: IEEE/CVF Conference on Computer Vision and Pattern Recognition (CVPR) Workshops, pp. 3347–3356 (2024)
8. Ikenaga, M., et al.: Influence of ball impact location on racquet kinematics, forearm muscle activation and shot accuracy during the forehand groundstrokes in tennis. In: 13th Conference of the International Sports Engineering Association (2020)
9. iniVation: iniVation – Neuromorphic vision systems (2023). https://inivation.com. Accessed 09 June 2025
10. Karditsas, H.E.: Large-Scale Method for Identifying the Relationships between Racket Properties and Playing Characteristics. Sheffield Hallam University (United Kingdom) (2020)
11. Kiraly, C., Merloti, P.: Golf club head measurement system, United States Patent US8951138B2 (2015)
12. Kiraly, C., Wintriss, V.: Flight parameter measurement system, European Patent EP1509781B1 (2015)
13. Lagorce, X., Orchard, G., Galluppi, F., Shi, B.E., Benosman, R.B.: Hots: a hierarchy of event-based time-surfaces for pattern recognition. IEEE Trans. Pattern Anal. Mach. Intell. **2017**(7), 1346–1359 (2017)
14. Nakabayashi, T., Higa, K., Yamaguchi, M., Fujiwara, R., Saito, H.: Event-based ball spin estimation in sports. In: IEEE/CVF Conference on Computer Vision and Pattern Recognition (CVPR) Workshops, pp. 3367–3375 (2024)
15. Nakabayashi, T., Kondo, A., Higa, K., Girbau, A., Satoh, S., Saito, H.: Event-based high-speed ball detection in sports video. In: 6th International Workshop on Multimedia Content Analysis in Sports, pp. 55–62 (2023)
16. Naß, D., Hennig, E.M., Schnabel, G.: Ball impact location on a tennis racket head and its influence on ball speed, arm shock and vibration. In: 16 International Symposium on Biomechanics in Sports (1998)
17. OpenCV: OpenCV Image Filtering #morphologyEx (2025). https://docs.opencv.org/4.x/d4/d86/group__imgproc__filter.html. Accessed 09 June 2025
18. OpenCV: OpenCV Structural Analysis and Shape Descriptors #fitEllipse (2025). https://docs.opencv.org/4.x/d3/dc0/group__imgproc__shape.html. Accessed 09 June 2025

19. Prophesee: Prophesee | Metavision Technologies (2025). https://www.prophesee.ai. Accessed 09 June 2025
20. Prophesee: SDK Core Algorithms – Metavision SDK Docs 5.1.0 documentation #BaseFrameGenerationAlgorithm (2025). https://docs.prophesee.ai/stable/api/python/core/bindings.html. Accessed 09 June 2025
21. Prophesee: SDK CV Python bindings API — Metavision SDK Docs 5.1.0 documentation #ActivityNoiseFilterAlgorithm (2025). https://docs.prophesee.ai/stable/api/python/cv/bindings.html. Accessed 09 June 2025
22. Sandler, M., Howard, A., Zhu, M., Zhmoginov, A., Chen, L.C.: Mobilenetv2: inverted residuals and linear bottlenecks. In: Proceedings of the IEEE Conference on Computer Vision and Pattern Recognition, pp. 4510–4520 (2018)
23. Shiba, S., Aoki, Y., Gallego, G.: Secrets of event-based optical flow. In: European Conference on Computer Vision (ECCV), pp. 628–645 (2022)
24. Sironi, A., Brambilla, M., Bourdis, N., Lagorce, X., Benosman, R.: Hats: histograms of averaged time surfaces for robust event-based object classification. In: IEEE/CVF Conference on Computer Vision and Pattern Recognition (CVPR) (2018)
25. Stoffregen, T., Gallego, G., Drummond, T., Kleeman, L., Scaramuzza, D.: Event-based motion segmentation by motion compensation. In: IEEE/CVF International Conference on Computer Vision (ICCV), pp. 7244–7253 (2019)
26. Tuxen, F.: System and method for determining impact characteristics of sports ball striking element, United States Patent US10953303B2 (2021)
27. Wang, A., Chen, H., Liu, L., Chen, K., Lin, Z., Han, J., et al.: Yolov10: real-time end-to-end object detection. Adv. Neural. Inf. Process. Syst. **37**, 107984–108011 (2024)
28. Yamashita, K., Matsunaga, H.: Sensor device, analyzing device, and recording medium for detecting the position at which an object touches another object, United States Patent US9551572B2 (2017)
29. Yasuda, R., Kase, Y., Ishibe, K., Washida, Y., Hashimoto, S.: [a proposal for tennis impact location measurement method with an event camera] event camera wo mochiita koushiki tennis no daten ichi sokutei shuhou no teian (in Japanese). In: Sports Informatics (SI), vol. 2024, pp. 1–6 (2024)

An Analysis of Differences in Golf Performance Between Age Groups for the Development of an XR Metaverse Platform and Content for Inclusive Digital Leisure

Yun-hwan Lee[1], Yeong-hun Kwon[2], Jin-i Hong[2], Jongsung Kim[3], and Jongbae Kim[4(✉)]

[1] Rehabilitation Science Technology Laboratory, Yonsei University, Wonju, South Korea
[2] Department of Occupational Therapy, Graduate School, Yonsei University, Wonju, South Korea
[3] Creative Contents Research Division ETRI, Daejeon, South Korea
[4] Department of Occupational Therapy, College of S/W Digital Healthcare Convergence, Yonsei University, Wonju, South Korea
jongbae@yonsei.ac.kr

Abstract. This study aimed to analyze age-related differences in golf swing performance within a virtual reality to provide foundational data for the development of XR metaverse platforms and content for inclusive digital leisure. A total of 80 participants were divided into four age groups: Teenager (TA), Youth (YU), Middle-aged (MA), and Old-aged (OA). Participants performed at least three driver shots in a virtual golf content while wearing a head-mounted display (HMD) and motion trackers at the head, wrists, ankles, and chest. Three-dimensional position and quaternion coordinates were collected from each device. The collected data were preprocessed, and the magnitudes of the linear velocity of both hand controllers and the angular velocity of the chest tracker were extracted as kinematic features. The analysis focused on the downswing (DS) phase. Since the assumption of normality is not met, nonparametric statistical analyses were conducted using the Kruskal–Wallis H test and Dunn's test.

The results revealed significant age-related differences in both the linear velocity of the controllers and the angular velocity of the chest tracker during the downswing. Specifically, the YU group exhibited significantly higher values, while the OA group showed the lowest values among all age groups. These findings suggest that physical factors such as muscle strength, neuromuscular control, and trunk flexibility may influence swing performance in virtual reality.

The results of this study may serve as foundational data for functional adaptation and augmentation based on users' physical abilities, contributing to the development of XR metaverse-based digital leisure content.

Keywords: Golf · Inclusive digital leisure · XR metaverse

1 Introduction

Leisure is defined as non-obligatory activities that are intrinsically motivated and freely pursued based on individual interests and preferences, such as sports or the arts [1]. Participation in leisure is essential, as it contributes to personal health management, enhancing quality of life and social relationships [2, 3]. In South Korea, public participation in leisure has been increasing annually, with participation in sports showing a particularly notable rise [4].

Golf has become a popular sport in South Korea. In the past, it was perceived as expensive and business-related. However, it has become more familiar to the public through various forms of media and has attracted a younger age group in recent years [5]. In particular, with the emergence of screen golf based on virtual simulation technology, golf has come to be a leisure culture that can be enjoyed by anyone, both indoors and outdoors [6].

Recently, leisure content using the metaverse has attracted growing attention. The metaverse allows users to participate in leisure activities through avatars in virtual spaces, without being limited by time or location [7]. This improves access to leisure activities like exercise, games, and performance viewing. In particular, XR technology, which supports the metaverse, offers users greater immersion and realism, making the experience more similar to the real world [8].

In South Korea, metaverse users are primarily in their teens and twenties, a group that is generally familiar with digital content [9]. In contrast, middle-aged and older adults tend to be less accustomed to digital content and may experience difficulty in accepting and using the metaverse platform and content [10]. South Korea is facing a declining birth rate and a rapidly aging population. It highlights the need for strategies to promote leisure participation among middle-aged and older adults [11]. Moreover, physical ability and digital literacy vary across life stages; it is essential to consider age-specific characteristics in developing the metaverse platform and content [12]. In particular, golf requires various physical abilities, such as muscle strength and coordination [13]. It is important to consider differences in performance related to physical ability when designing a metaverse platform and content.

Therefore, this study aims to analyze differences in performance between age groups by acquiring motion data during golf activities using XR devices, and provide foundational data to support compensation for differences in physical ability in the development of XR-based metaverse golf platforms and content.

2 Method

2.1 Participants

Participants were categorized into four age groups: Teenager (TA, 12–19), Youth (YU, 20–39), Middle-Aged (MA, 40–59), and Old-Aged (OA, 60 and above). 80 participants were recruited, with 20 participants in each group (Table 1).

Participants were excluded if they (1) had musculoskeletal disorders such as lower back pain, shoulder pain, or arthritis; (2) had vestibular impairments that could interfere with the use of XR devices; (3) had communication difficulties or cognitive impairments; or (4) were professional golf players.

Table 1. General characteristics of participants

		TA(N = 20)	YU(N = 20)	MA(N = 20)	OA(N = 20)
AGE*		14.0 ± 2.3	28.2 ± 3.8	49.4 ± 5.8	64.2 ± 2.0
Gender	Male (%)	11(55.0)	10(50.0)	8(40.0)	9(45.0)
	Female (%)	9(45.0)	10(50.0)	12(60.0)	11(55.0)

*: Mean ± SD

2.2 Device and Content

The HTC VIVE Pro Eye head-mounted display (HMD) along with dual controllers and HTC VIVE Tracker 3.0 are used in this study.

Both devices support six degrees of freedom (6-DoF) tracking and provide real-time three-dimensional position and quaternion coordinates using integrated IMU sensors and optical tracking via base stations. Five trackers were attached to the wrists, ankles, and chest. The content used a golf game available on the SteamVR platform (Fig. 1).

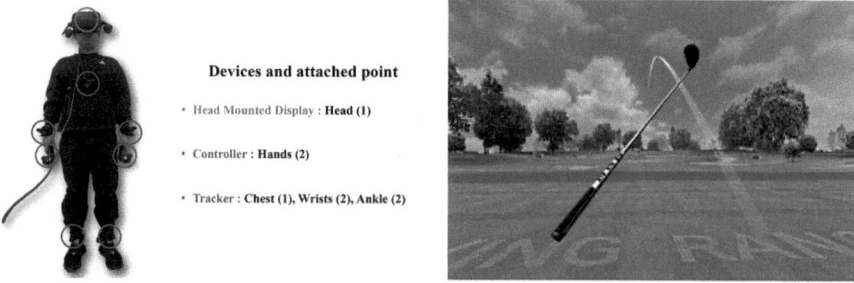

Fig. 1. XR Device and golf game content

2.3 Motion Data Acquisition Platform

In this study, a Unity-based platform was developed to acquire tracked three-dimensional position and quaternion coordinates through XR Devices.

The platform integrates XR devices using the SteamVR Plugin and OpenVR SDK. Each device was assigned a virtual object, and a skeleton structure was created by connecting them around the chest tracker using the line renderer function. It can provide real-time visualization of the swing motion and monitor tracked position and quaternion coordinates.

In addition, it was configured to acquire three-dimensional positional and quaternion coordinates of each XR device at 90 fps and save them in .csv format.

2.4 Procedure

2.4.1 Motion Data Acquisition

The experiment for motion data acquisition was conducted in a 5.5 m × 5.5 m space to minimize sensor detection errors and allow smooth movement. Base stations were installed at all four corners of the space to enable full-body tracking.

Before wearing the devices, participants were given instructions and practice time to become familiar with the golf swing. Subsequently, they wore the HMD and trackers and performed at least three driver shots in a virtual golf content. Participants were instructed to conduct additional trials if any sensor detection or improper swing motions.

The motion data acquisition platform was activated to acquire each device's three-dimensional position and quaternion coordinates at 90 fps during each swing. The data were saved in .csv format. Simultaneously, the swing motion and content screens were recorded using OBS (Fig. 2).

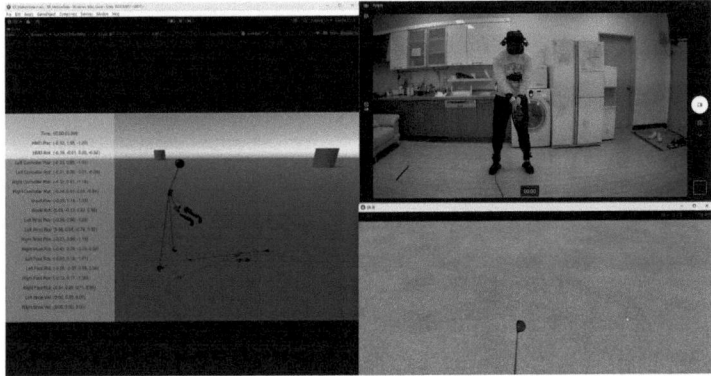

Fig. 2. Acquiring coordinates using the platform while playing golf in the virtual game content

2.4.2 Labeling and Preprocessing

The collected motion data were labeled according to key phases of the golf swing (Fig. 3). The recorded swing videos were reviewed frame by frame using Adobe Premiere Pro, and labels were assigned to the corresponding segments of the time-series position and quaternion coordinates in the .csv file according to the swing phases (Fig. 4).

The labeled data were preprocessed in three steps. First, the acquired quaternion coordinates were converted into rotation vector data to derive kinematic features. Second, since most participants were right-handed, the data obtained from left-handed participants' controllers and wrist trackers were switched to align with right-handed. Third, the X, Y, and Z components of the position and rotation vector data were interpolated to compensate for missing values and outliers and ensure the continuity and stability of the time-series data.

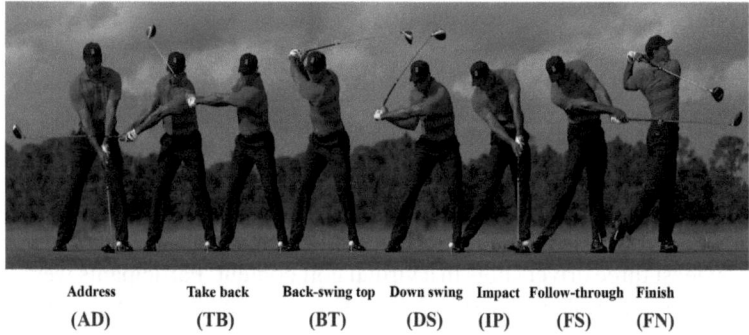

Address
(AD)
Take back
(TB)
Back-swing top
(BT)
Down swing
(DS)
Impact
(IP)
Follow-through
(FS)
Finish
(FN)

Fig. 3. The key phases of the golf swing

Labeling	TimeStam	HMDPosX	HMDPosY	HMDPosZ	HMDQutX	HMDQutY	HMDQutZ	HMDQutW	LeftContlPc	LeftContlPos	LeftContlPos	LeftContlQut	LeftContlQut	LeftContlQut	LeftContlQut	RightContlPosX	RightContlPosY	RightContl
AR	00:08.0	-0.44	1.46	-1.14	-0.45	0.04	-0.01	-0.89	-0.42	0.75	-1.01	-0.34	-0.01	0.01	-0.94	-0.43	0.71	-1.09
AR	00:08.0	-0.44	1.46	-1.14	-0.45	0.04	-0.01	-0.89	-0.42	0.75	-1.01	-0.34	-0.01	0.01	-0.94	-0.42	0.71	-1.09
AR	00:08.0	-0.43	1.46	-1.14	-0.45	0.04	-0.01	-0.89	-0.42	0.75	-1.01	-0.34	-0.02	0.01	-0.94	-0.42	0.71	-1.09
AR	00:08.0	-0.43	1.46	-1.14	-0.45	0.04	-0.01	-0.89	-0.42	0.75	-1.01	-0.34	-0.02	0.01	-0.94	-0.42	0.71	-1.09
AR	00:08.0	-0.43	1.46	-1.14	-0.44	0.04	-0.01	-0.89	-0.41	0.75	-1.01	-0.33	-0.02	0.01	-0.94	-0.42	0.71	-1.09
AR	00:08.0	-0.43	1.46	-1.14	-0.44	0.04	-0.01	-0.90	-0.41	0.75	-1.01	-0.33	-0.02	0.01	-0.94	-0.42	0.71	-1.09
AR	00:08.0	-0.43	1.46	-1.14	-0.44	0.04	-0.01	-0.90	-0.41	0.75	-1.01	-0.33	-0.02	0.02	-0.94	-0.41	0.71	-1.09
AR	00:08.0	-0.42	1.45	-1.14	-0.44	0.05	-0.01	-0.90	-0.41	0.75	-1.01	-0.33	-0.02	0.02	-0.94	-0.41	0.71	-1.09
AR	00:08.0	-0.42	1.45	-1.14	-0.44	0.05	-0.01	-0.90	-0.40	0.75	-1.01	-0.33	-0.01	0.02	-0.94	-0.41	0.71	-1.08
AR	00:08.1	-0.42	1.45	-1.14	-0.44	0.05	-0.01	-0.90	-0.40	0.75	-1.01	-0.33	-0.02	0.02	-0.94	-0.41	0.71	-1.08
AR	00:08.1	-0.41	1.45	-1.14	-0.44	0.05	-0.01	-0.90	-0.40	0.75	-1.01	-0.33	-0.02	0.02	-0.94	-0.40	0.71	-1.08
AR	00:08.1	-0.41	1.45	-1.14	-0.44	0.05	-0.01	-0.90	-0.39	0.75	-1.01	-0.33	-0.02	0.02	-0.94	-0.40	0.71	-1.08
AR	00:08.1	-0.41	1.45	-1.14	-0.44	0.05	-0.01	-0.90	-0.39	0.75	-1.01	-0.33	-0.02	0.02	-0.94	-0.40	0.71	-1.08
AR	00:08.1	-0.40	1.45	-1.14	-0.44	0.05	-0.01	-0.90	-0.38	0.75	-1.01	-0.33	-0.02	0.02	-0.94	-0.39	0.71	-1.08
AR	00:08.1	-0.40	1.45	-1.14	-0.44	0.05	-0.01	-0.90	-0.37	0.75	-1.02	-0.33	-0.03	0.02	-0.95	-0.38	0.71	-1.08
AR	00:08.1	-0.39	1.45	-1.15	-0.43	0.06	-0.01	-0.90	-0.37	0.75	-1.02	-0.32	-0.03	0.01	-0.95	-0.38	0.71	-1.08
AR	00:08.1	-0.39	1.45	-1.15	-0.43	0.06	-0.01	-0.90	-0.36	0.75	-1.02	-0.32	-0.03	0.01	-0.95	-0.37	0.71	-1.08
AR	00:08.1	-0.38	1.45	-1.15	-0.43	0.06	-0.01	-0.90	-0.35	0.75	-1.02	-0.32	-0.03	0.01	-0.95	-0.36	0.71	-1.08
AR	00:08.2	-0.37	1.45	-1.15	-0.43	0.06	-0.01	-0.90	-0.34	0.75	-1.02	-0.32	-0.04	0.01	-0.95	-0.35	0.71	-1.08
AR	00:08.2	-0.37	1.45	-1.15	-0.43	0.06	-0.01	-0.90	-0.33	0.75	-1.02	-0.32	-0.05	0.00	-0.95	-0.34	0.71	-1.08
AR	00:08.2	-0.36	1.45	-1.15	-0.43	0.06	-0.01	-0.90	-0.32	0.75	-1.02	-0.32	-0.05	0.00	-0.95	-0.33	0.71	-1.08
AR	00:08.2	-0.36	1.44	-1.16	-0.43	0.06	-0.01	-0.90	-0.31	0.75	-1.02	-0.32	-0.06	0.00	-0.95	-0.32	0.71	-1.08
TB	00:08.2	-0.35	1.44	-1.16	-0.42	0.06	-0.01	-0.90	-0.29	0.75	-1.02	-0.32	-0.06	0.00	-0.95	-0.30	0.71	-1.08
TB	00:08.2	-0.34	1.44	-1.16	-0.42	0.07	-0.01	-0.90	-0.28	0.75	-1.02	-0.31	-0.07	-0.01	-0.95	-0.29	0.71	-1.08
TB	00:08.2	-0.34	1.44	-1.16	-0.42	0.07	-0.01	-0.90	-0.26	0.75	-1.02	-0.31	-0.08	-0.01	-0.95	-0.27	0.71	-1.08
TB	00:08.2	-0.33	1.44	-1.17	-0.42	0.07	-0.01	-0.91	-0.24	0.75	-1.02	-0.31	-0.09	-0.02	-0.95	-0.25	0.71	-1.08
TB	00:08.2	-0.32	1.44	-1.17	-0.42	0.07	-0.01	-0.91	-0.23	0.75	-1.02	-0.31	-0.09	-0.02	-0.95	-0.24	0.71	-1.08
TB	00:08.3	-0.32	1.44	-1.17	-0.41	0.07	-0.01	-0.91	-0.20	0.76	-1.02	-0.30	-0.10	-0.03	-0.95	-0.21	0.71	-1.08
TB	00:08.3	-0.31	1.44	-1.18	-0.41	0.07	-0.01	-0.91	-0.18	0.76	-1.02	-0.30	-0.11	-0.03	-0.95	-0.20	0.71	-1.08
TB	00:08.3	-0.30	1.44	-1.18	-0.41	0.07	-0.01	-0.91	-0.16	0.76	-1.03	-0.30	-0.12	-0.04	-0.95	-0.17	0.72	-1.09
TB	00:08.3	-0.30	1.44	-1.18	-0.41	0.07	-0.01	-0.91	-0.14	0.77	-1.03	-0.30	-0.12	-0.05	-0.95	-0.15	0.72	-1.09
TB	00:08.3	-0.29	1.44	-1.18	-0.40	0.07	-0.01	-0.91	-0.11	0.77	-1.03	-0.29	-0.13	-0.05	-0.95	-0.12	0.72	-1.09
BT	00:08.3	-0.28	1.44	-1.19	-0.40	0.07	-0.01	-0.91	-0.09	0.78	-1.03	-0.29	-0.14	-0.06	-0.95	-0.10	0.73	-1.09
BT	00:08.3	-0.28	1.44	-1.19	-0.40	0.07	-0.01	-0.91	-0.06	0.78	-1.04	-0.28	-0.15	-0.07	-0.94	-0.07	0.74	-1.10
BT	00:08.3	-0.27	1.44	-1.19	-0.40	0.07	-0.01	-0.91	-0.04	0.79	-1.04	-0.28	-0.17	-0.08	-0.94	-0.04	0.74	-1.10
BT	00:08.3	-0.27	1.43	-1.19	-0.40	0.07	-0.02	-0.91	-0.01	0.80	-1.05	-0.27	-0.18	-0.09	-0.94	-0.01	0.75	-1.10
BT	00:08.4	-0.26	1.43	-1.20	-0.40	0.07	-0.02	-0.92	0.02	0.81	-1.05	-0.27	-0.20	-0.09	-0.94	0.01	0.76	-1.11
BT	00:08.4	-0.25	1.43	-1.20	-0.40	0.07	-0.02	-0.92	0.06	0.83	-1.06	-0.26	-0.23	-0.10	-0.93	0.05	0.77	-1.12
BT	00:08.4	-0.25	1.43	-1.20	-0.40	0.07	-0.02	-0.92	0.09	0.84	-1.07	-0.26	-0.24	-0.10	-0.93	0.08	0.78	-1.12
BT	00:08.4	-0.24	1.43	-1.20	-0.39	0.07	-0.02	-0.92	0.12	0.85	-1.08	-0.25	-0.26	-0.11	-0.92	0.11	0.80	-1.13
BT	00:08.4	-0.24	1.44	-1.21	-0.39	0.07	-0.02	-0.92	0.16	0.87	-1.09	-0.24	-0.29	-0.12	-0.92	0.14	0.82	-1.14
BT	00:08.4	-0.23	1.44	-1.21	-0.39	0.06	-0.02	-0.92	0.19	0.89	-1.10	-0.24	-0.32	-0.13	-0.91	0.17	0.84	-1.15
BT	00:08.4	-0.23	1.44	-1.21	-0.39	0.06	-0.02	-0.92	0.22	0.91	-1.12	-0.23	-0.35	-0.13	-0.90	0.20	0.86	-1.16
BT	00:08.4	-0.22	1.44	-1.21	-0.39	0.06	-0.02	-0.92	0.25	0.93	-1.13	-0.22	-0.37	-0.14	-0.89	0.23	0.88	-1.18
BT	00:08.4	-0.22	1.44	-1.21	-0.39	0.06	-0.02	-0.92	0.28	0.95	-1.15	-0.21	-0.40	-0.15	-0.88	0.26	0.90	-1.19
FT	00:08.5	-0.22	1.44	-1.22	-0.39	0.06	-0.02	-0.92	0.31	0.98	-1.16	-0.19	-0.43	-0.16	-0.87	0.28	0.92	-1.20
FT	00:08.5	-0.21	1.44	-1.22	-0.39	0.05	-0.02	-0.92	0.33	1.01	-1.18	-0.18	-0.45	-0.17	-0.86	0.31	0.95	-1.22
FT	00:08.5	-0.21	1.44	-1.22	-0.38	0.05	-0.02	-0.92	0.36	1.04	-1.20	-0.16	-0.47	-0.19	-0.85	0.33	0.98	-1.23

Fig. 4. The labeled raw data according to the swing phase

2.4.3 Deriving Kinematic Feature

Linear and angular velocities were calculated by differentiating the X, Y, and Z components of each device's preprocessed position and rotation vector data, which were defined as kinematic features.

The magnitude of each feature was computed as the Euclidean norm of the corresponding three-dimensional vector, as shown in Eqs. (1) and (2). Subsequently, the maximum value for each swing phase was extracted and used as the final feature.

$$\textit{Magnitude of linear velocity}: |v| = \sqrt{v_x^2 + v_y^2 + v_z^2} \quad (1)$$

$$\textit{Magnitude of angular velocity}: |\omega| = \sqrt{\omega_x^2 + \omega_y^2 + \omega_z^2} \quad (2)$$

2.4.4 Comparative Analysis Between Age Group

This study focused on the downswing (DS) phase. These are the points where the energy accumulated during the backswing is transferred to the ball through the club, directly influencing the distance, direction, and accuracy [14, 15]. The club speed and the rotational velocity of the trunk during the swing have been widely used as key indicators for evaluating swing efficiency and performance [16, 17]. Accordingly, a comparative analysis focused on the controllers' linear velocity and the chest tracker's angular velocity.

Statistical analysis was performed using Python 3.12.0. The Shapiro-Wilk test was used to assess the normality of kinematic features for each group in the DS phase. If the assumption of normality was met, Levene's test was used to assess the homogeneity of variances. When equal variances were assumed, one-way ANOVA and Bonferroni post hoc tests were used; otherwise, Welch's ANOVA and the Games–Howell post hoc test were used. If the normality assumption was not met, the Kruskal–Wallis H test and Dunn's test were used. The significance level was set at .05.

3 Result

3.1 Normality Assumption for Kinematic Features

The Shapiro–Wilk test indicated that the normality assumption was not met for several kinematic features by age group in the DS phase (Table 2). Accordingly, the Kruskal–Wallis H test and Dunn's test were used to analyze differences.

Table 2. A Result of the normality test (Shapiro-Wilk test)

Phase	Feature	Group	Statistics	Df	p
DS	Left controller linear velocity	TA	.941	107	.000***
		YU	.914	70	.000***
		MA	.975	100	.059
		OA	.992	106	.824
	Right controller linear velocity	TA	.971	107	.020*
		YU	.873	70	.000***
		MA	.979	100	.119
		OA	.974	106	.038*
	Chest tracker angular velocity	TA	.924	107	.000***
		YU	.983	70	.505
		MA	.992	100	.828
		OA	.953	106	.001**

*$p < .05$, **$p < .01$, ***$p < .001$

3.2 Differences in Linear Velocity of the Controller and Angular Velocity of the Chest Tracker Between Age Groups in the DS Phase

The linear velocity of the left controller showed a significant difference among age groups ($H = 100.54$, $p < 0.001$). Post hoc results indicated that the YU (6.556) had significantly higher values than the TA (5.752) and MA (6.403), while the OA (4.664) had significantly lower values than the TA (5.752), YU (6.556), and MA (6.403).

The linear velocity of the right controller also significantly differed among age groups in the DS phase ($H = 100.54$, $p < 0.001$). The OA (4.949) had significantly lower values than the TA (7.283), YU (7.158), and MA (7.375) (Fig. 5).

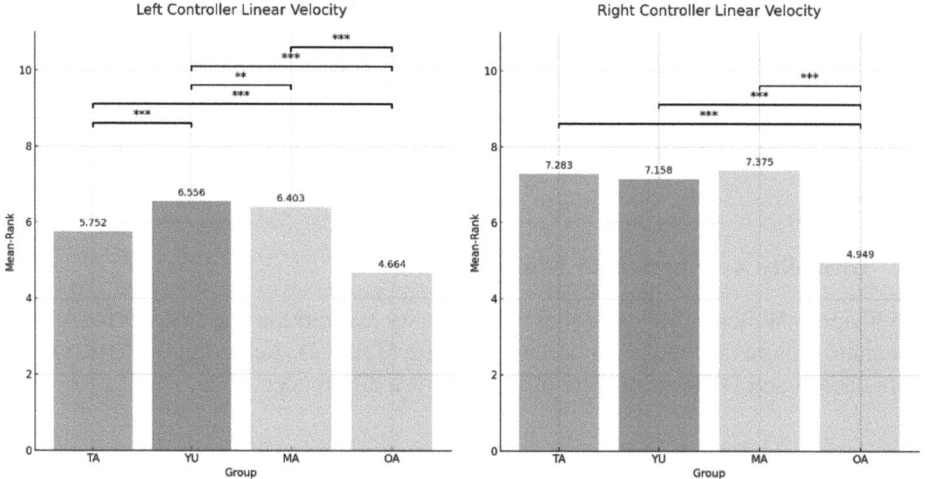

Fig. 5. Differences in linear velocity of the controllers between age groups in the DS phase

The angular velocity of the chest tracker showed a significant difference between age groups ($H = 46.63$, $p < 0.001$). Post hoc results indicated that the YU (8.150) had significantly higher values than the TA (7.029) and MA (7.191), while the OA (5.428) had significantly lower values than the TA (7.029), YU (6.556), and MA (6.403) (Fig. 6).

These results are associated with age-related declines in neuromuscular control, muscle strength, and trunk mobility. The lower linear velocity of controllers in OA may reflect diminished muscle strength and upper limb coordination [18, 19]. Likewise, the lower angular velocity of the chest tracker in OA may reflect diminished core muscle strength and trunk flexibility, which are essential for effective energy transfer and direction during the downswing [20].

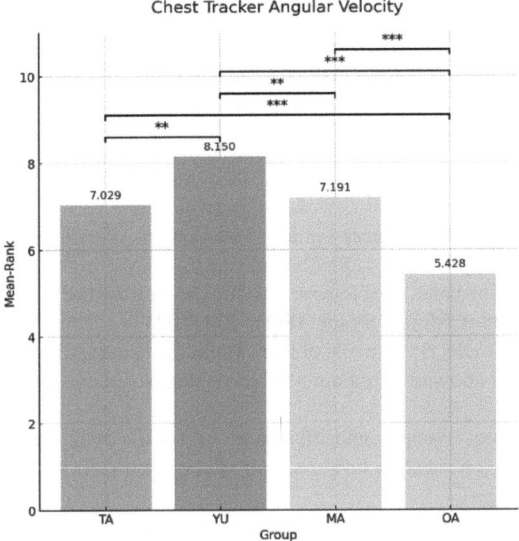

Fig. 6. Differences in angular velocity of the chest tracker between age groups in the DS phase

4 Conclusion

This study analyzed differences in golf performance between age groups within a virtual environment. Eighty participants were categorized into four age groups: Teenager (TA, 12–19), Youth (YU, 20–39), Middle-age (MA, 49–59), and Old-age (OA, 60 and above). Participants performed golf swings in a virtual golf content while wearing a head-mounted display (HMD) and trackers attached to the head, wrists, ankles, and chest. 3-dimensional position and quaternion data were acquired from each device and preprocessed to extract kinematic features. The linear velocity of both hand controllers and the angular velocity of the chest tracker were selected as the final features, and the analysis focused on the downswing phase.

The results revealed significant differences in the linear velocity of both hand controllers and the angular velocity of the chest tracker between age groups in the downswing phase. In particular, the YU exhibited relatively higher linear and angular velocities compared to the other groups, whereas the OA showed the lowest values. These findings suggest that age-related differences in physical capabilities, such as muscle strength, neuromuscular control, and trunk flexibility, may have influenced the results.

This finding can serve as foundational data for implementing functional adaptation and augmentation based on differences in users' physical abilities in the development of inclusive XR metaverse platforms and golf content for digital leisure. Future research should consider a comprehensive analysis of various kinematic features, including acceleration and angular acceleration across the entire swing phase, as expanding both the range of features and phases could provide more robust and in-depth evidence and enhance adaptability to diverse user profiles.

Acknowledgments. This research was supported by Culture, Sports and Tourism R&D Program through the Korea Creative Content Agency grant funded by the ministry of Culture, Sports and

Tourism in 2023(Project Name: Development of universal XR platform technology to build a metaverse supporting digital cultural inclusion, Project Number: RS-2023–00270006, Contribution Rate: 100%).

References

1. Boop, C., et al.: Occupational therapy practice framework: domain and process fourth edition. Am. J. Occup. Therapy **74**(S2), 1–85 (2020)
2. Huang, I.W., et al.: The benefits of leisure activities on healthy life expectancy for older people with diabetes. Diabetol. Metab. Syndr. **16**(1), 100 (2024)
3. Yu, J., Mock, S.E., Smale, B.: The role of health beliefs in moderating the relationship between leisure participation and wellbeing among older Chinese adults. Leis. Stud. **40**(6), 764–778 (2021)
4. Ministry of Culture, Sports and Tourism.: (2023). 2022 National Leisure Activity Survey. https://www.mcst.go.kr/kor/s_policy/dept/deptView.jsp?pSeq=1898&pDataCD=0406000000&pType=
5. Park, C.M., Bang, S.H.: Covid-19 era, changes in the golf market environment due to the popularization of golf. Interdisc. Res. Arts Cult. **2**(2), 53–61 (2021)
6. Kang, Y.S., Kwon, O.R., Kim, J.H.: Golf as sportainment and its spirit. J. Golf Study **16**(2), 167–176 (2022)
7. Jung, J.E., Son, N.Y., Kim, H.J.: Case studies of cultural contents using Metaverse. J. Cult. Ind. **22**(1), 201–213 (2022)
8. Özkan, A., Özkan, H.: Meta: XR-AR-MR and mirror world technologies business impact of Metaverse. J. Metaverse **4**(1), 21–32 (2024)
9. Korea Creative Content Agency: Research on content usage trends in the digital transformation (2021). https://welcon.kocca.kr/ko/info/trend/1951143
10. Lee, H.S., Kim, H.C.: Metaverse and the future of education for the elderly: focusing on the liberal arts education for the elderly in senior welfare centers. Korean J. Educ. Gerontol. **8**(1), 1–20 (2022)
11. Korea Statistics: Korea population dashboard (2025). https://kosis.kr/visual/populationKorea/PopulationDashBoardMain.do
12. Lee, Y.H., et al.: Physical human factor for the development of universal XR platform to build a metaverse supporting digital inclusive leisure & culture. Augmented, Virtual Mixed Reality Simul. **118**(118), 80–87 (2023)
13. Cole, M.H., Grimshaw, P.N.: The biomechanics of the modern golf swing: implications for lower back injuries. Sports Med. **46**, 339–351 (2016)
14. Bourgain, M., Rouch, P., Rouillon, O., Thoreux, P., Sauret, C.: Golf swing biomechanics: a systematic review and methodological recommendations for kinematics. Sports **10**(6), 91 (2022)
15. Kraśna, S., Čoh, M., Prebil, I., Mackala, K.: Comparative analysis of golf clubhead motion at impact. Facta Universitatis, Ser.: Phys. Educ. Sport, 437–452 (2020)
16. Okuda, I., Gribble, P., Armstrong, C.: Trunk rotation and weight transfer patterns between skilled and low skilled golfers. J. Sports Sci. Med. **9**(1), 127–133 (2010)
17. Torres-Ronda, L., Sánchez-Medina, L., González-Badillo, J.J.: Muscle strength and golf performance: a critical review. J. Sports Sci. Med. **10**(1), 9–18 (2011)
18. Severin, A.C., Tackett, S.A., Barnes, C.L., Mannen, E.M.: Three-dimensional kinematics in healthy older adult males during golf swings. Sports Biomech. **21**(2), 165–178 (2022)
19. Hunter, S.K., Pereira, H.M., Keenan, K.G.: The aging neuromuscular system and motor performance. J. Appl. Physiol. **121**, 982–995 (2016)

20. Zemková, E., Jeleň, M., Zapletalová, L.: Trunk rotational velocity in young and older adults: a role of trunk angular displacement. In: 3rd International Scientific Conference, 223–228 (2018)

Scalable Tactical Tennis Insights: Hybridizing Automated Reports and LLM-Powered Analytics

Zizhen Li(✉), Zhaoyu Liu(✉), and Kan Jiang(✉)

School of Computing, National University of Singapore, Singapore, Singapore
lizizhen@u.nus.edu, liuzy@nus.edu.sg, jiangkan@comp.nus.edu.sg

Abstract. Personalized, data-driven tactical analysis is crucial in modern tennis, yet automated systems that deliver player-specific tactical insights across matches remain scarce. We present a hybrid solution consisting of two complementary components: TennisTact and ChatTennis. TennisTact is an automated reporting system that continuously collects and processes extensive match data to deliver personalized tactical reports for tennis singles players. To overcome the rigidity of static reports, we pair TennisTact with ChatTennis, a multi-agent generative AI chatbot that draws on both shot-level and player-level databases to deliver on-demand, context-sensitive analyses. By combining the consistency and scalability of scheduled reporting with the flexibility of real-time dialogue, the system offers both commonly used player evaluation metrics and more customized evaluations while updating automatically after each day's matches. This work bridges the gap between abundant online tennis data and actionable, player-specific tactical insights.

Keywords: Tennis Analytics · Data Mining · Natural Language Processing · Retrieval-Augmented Generation

1 Introduction

In the modern tennis landscape, personalized and data-driven tactical analysis has become crucial for athletes and coaches seeking a competitive advantage [9,15]. A key responsibility of coaches is to analyze players' tactical strengths and weaknesses. Traditionally, this involves watching past tennis matches, identifying significant strokes, and quantifying the player's patterns of play to determine areas for improvement [4,8,26]. However, this manual process is time-consuming; for example, a coach may need to watch approximately 25 matches to detect a reliable breakpoint pattern of a top ATP player [12].

To mitigate these challenges, systems such as Hawk-Eye [13] have been developed to automatically index important strokes. Additionally, advanced data analytics and machine learning algorithms have facilitated the extraction of both general [20,30,32,37] and player-specific [6,11,21,22,36,38] tactical recommendations from extensive match data. Despite these advancements, there remains a gap in automated and widely accessible player-specific tactical analysis systems. Such systems are essential for providing customized insights that can enhance a player's tactical skills and improve performance against specific opponents. Specifically, player-specific information can be compared with general tactical suggestions to aid a player's development. Additionally, understanding an opponent's tactical patterns can offer the player a competitive edge in upcoming matches.

To bridge this gap, the first component of this research introduces an automated, player-specific reporting system for tennis singles. By continuously collecting and analyzing match data, the system delivers timely and actionable tactical insights. However, it is impractical to present all possible metrics within static reports. Thus, there is a need for a flexible analysis platform capable of generating detailed insights on demand.

Recent advances in large language models (LLMs) have demonstrated their effectiveness in providing natural-language interactions tailored to user inquiries. Mainstream models, such as ChatGPT and DeepSeek, though broadly useful, often lack specialized tennis expertise and real-time accuracy. Consequently, the second component of this study presents a specialized tennis-focused LLM chatbot that utilizes our collected data and the statistical outputs of the automated system to provide precise, clear, and targeted responses to user queries.

This research has two primary objectives: (1) to develop a comprehensive, automated, and accessible player-specific tennis reporting system that provides continuous tactical updates; and (2) to integrate an LLM-driven chatbot capable of supplementing automated reports through interactive, on-demand analytical conversations. Together, these components enable players and coaches to benefit from timely tactical summaries and detailed strategic explorations. Players can utilize these insights to identify their own strengths and weaknesses, understand their opponents better, and prepare effectively for upcoming matches.

2 Related Works

2.1 Sports Analytics

Recent works in sports analytics have explored a wide range of challenges, including strategy modeling using probabilistic reasoning and deep learning [8,14,22,24–26,29], injury prediction [27], fine-grained event detection in sports videos [28], and specific tasks such as court detection and ball tracking in broadcast footage [15–17]. These efforts demonstrate the growing integration of AI, data mining, and formal methods in advancing sports performance analysis and tactical understanding [23].

2.2 Tennis Tactical Analysis

Tactical analysis in tennis typically falls into two categories: general tactics analysis, broadly applicable across players, and player-specific tactics analysis, tailored to individual playing styles. Example general tactical studies include identification of optimal serve angles that increase the likelihood of aces [37], and the strategic use of the inside-out forehand [30].

Player-specific tactical analyses employ various data-driven techniques. For instance, subgroup discovery methods reveal conditions under which a specific player performs optimally, such as avoiding certain strokes or favoring short rallies [6]. Minimum Description Length algorithms identify recurring shot sequences that reflect a player's tactical tendencies across matches [38]. Additionally, probabilistic simulations using Markov Decision Processes help evaluate the impact of specific tactics offering actionable insights against particular opponents [22].

Emerging smart sensor technologies also contribute significantly by capturing high-precision and richly detailed metrics such as ball spin, impact location, and movement patterns, enabling personalized training recommendations [11,21].

2.3 Automated Reporting Systems

Most of the aforementioned player-specific tactical analysis studies rely heavily on manual data collection, which significantly limits their scalability. For instance, works such as [6] and [38] used manually labeled datasets and acknowledged the need for automated systems as a future direction. Even studies that leveraged Hawk-Eye data [36] fail to implement scalable automation. While smart sensors have been employed in certain cases [11,21], limited access to such technologies and datasets restricts broader analysis and comparison.

In addition, existing tools such as CourtTime [33] and TenniVis [34] enable tactical pattern visualization but require manual data input. Similarly, platforms like TennisAnalytics[1] depend on human annotators, thereby constraining scalability. GIS-based modeling tools [7] have also been explored for performance analysis, yet they too lack automation.

Moreover, these tools are typically designed for single-match analysis [7,33,34]. Analyzing only individual matches fails to capture consistent playing patterns across multiple matches, which are crucial for understanding a player's recurrent strategies and tendencies.

Overall, current systems are limited by their reliance on manual processes and match-specific focus, ultimately hindering the discovery of consistent, player-specific tactical patterns

2.4 LLMs and Chatbots in Sports

Large language models (LLMs) have begun to reshape sports technology. Systems such as PanelGPT provide round-the-clock athlete support [31], while LLM commentators generate near-real-time football play-by-play [5] or fuse vision and ontologies to produce dynamic baseball commentary [19]. Beyond live commentary, LLM pipelines now turn raw statistics into readable stories—e.g. SNIL for basketball statistics [2]—and even auto-extract video highlights from multimodal cues [18].

---

[1] https://www.tennisanalytics.net.

Conversational systems remain rarer. Early rule- or retrieval-based bots such as GameBot answer NBA queries by template filling [40], but their fixed responses limit analytical depth. Recent works combine generative models with retrieval to personalize feedback, as in a swimming coach that layers RAG over GPT-4o to cut hallucinations [3]. Hybrid vision–language agents go further: ChatMatch orchestrates four specialized agents to interpret badminton video and generate statistics on demand [39]. For tennis, however, efforts are confined to rule lookup (MEGAN) rather than tactical analytics [35].

In short, while LLMs can already narrate games and offer generic advice, there is still a lack of a data-rich chatbot that ingests longitudinal match records and delivers tennis insights—the gap our ChatTennis aims to fill.

3 Methodology

3.1 Framework

Our system adopts an architecture that combines (i) an automated, large-scale data-processing pipeline that generates daily reports (TennisTact) and (ii) an on-demand, multi-agent chatbot (ChatTennis). TennisTact delivers commonly used performance metrics, whereas ChatTennis enables flexible queries to generate analyses on-demand. Figure 1 summarizes the overall structure of our hybrid system.

3.2 Data Acquisition and Pre-processing

To ensure up-to-date data coverage, we developed a web crawler and parser to automatically extract point-by-point textual descriptions from https://www.tennisabstract.com, a comprehensive tennis statistics website maintained by volunteers. The system runs on a daily schedule using cron jobs to automate data retrieval. The extracted raw text is then parsed into structured CSV files, which serve as the basis for subsequent analysis.

3.3 TennisTact: Automated Reporting System

TennisTact visualizes data and generates new reports upon receiving daily updates. This data analysis and visualization component is engineered to extract and present tactical insights tailored for player-specific performance assessments. Utilizing Python libraries such as Pandas for data processing and Matplotlib for visualization, the system dissects various match elements to uncover actionable insights that inform strategic preparation.

Because a player's tactics often differ when facing left- or right-handed opponents, the system produces two separate reports for each player—one summarizing performance against left-handers, the other against right-handers. It can also generate a head-to-head comparison between two players directly.

Each report contains several key metrics:

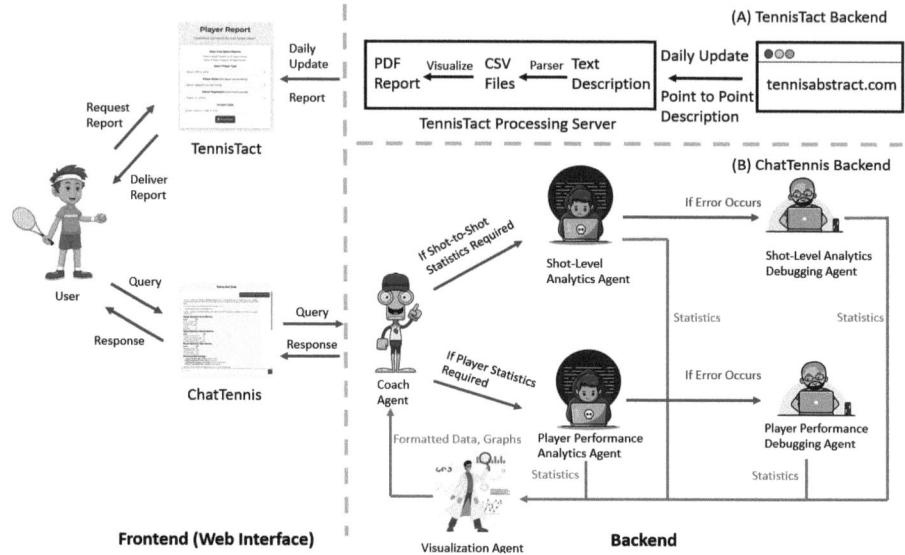

Fig. 1. System architecture of the proposed hybrid platform. It includes a frontend web interface for user interaction and a backend comprising: (A) TennisTact for automated daily reporting, and (B) ChatTennis, a multi-agent chatbot for interactive analysis. Icon credits: User, Visualization Agent, Analytics Agent © Freepik. Debugging Agent © Dribbble. Coach Agent © iStock.

1. Serve Efficiency and Error Management: Evaluates a player's first and second serve effectiveness against opponents, along with their management of winners and errors.
2. Serve and Return Patterns: Provides a detailed breakdown of serve and return directions by analyzing the frequency and point win percentage of each option (e.g., wide, T, body) across first and second serves, court sides (Ad and Deuce), and under pressure situations such as break points.
3. Rally Patterns: Examines stroke direction from both Ad and Deuce courts, analyzing the frequency of shots hit to Ad, Deuce, or central zones, along with their associated point win rates.
4. Rally Analysis: Computes win probabilities and classifies point outcomes (e.g., winner or error) based on rally length.
5. Momentum Analysis: Tracks the net point differential (points won minus points lost) over time to assess performance consistency and mental readiness toward upcoming matches.
6. Shot Type Analysis: Highlights effectiveness across various strokes, including forehands, backhands, volleys, and others.
7. On the Run Return Direction: Analyzes situations where the player must chase down aggressive shots, revealing defensive tendencies and strategic responses under pressure.

8. Smash Direction: Reports the distribution of smashes to the Ad side, middle, or the Deuce side.

Figures 2 and 3 show a sample report covering the above metrics.

Knowing these metrics can guide a player's self-improvement, while studying an opponent's report can help anticipate their habits and uncover potential weaknesses. For instance, recognizing that an opponent historically serves wide on more than 60% of first serves and tends to smash toward the Ad court can inform proactive tactical preparation.

A web platform has been developed to improve the accessibility of our system. Sample reports can also be found on our website https://depintel.pythonanywhere.com.

3.4 ChatTennis: An Interactive Complement

Due to the data-intensive nature of tennis, including every possible metric within static reports is impractical. This limitation highlights the need for a flexible platform capable of generating on-demand insights. However, directly querying Large Language Models (LLMs) is often unreliable, as they are prone to hallucinations when addressing domain-specific queries or those beyond their training data [10]. Additionally, LLMs lack access to up-to-date tennis match data.

To overcome these limitations, we present ChatTennis, a Retrieval-Augmented Generation (RAG) system built on a modular, multi-agent architecture. As illustrated in Fig. 1 (B), the backend comprises six specialized agents that work collaboratively to process user queries.

When a user submits a query, the Coach Agent (CA) serves as the primary interface, initially evaluating whether further statistical computation is necessary. If not, the CA responds directly, prefixing the reply with F| to indicate that no additional assistance is required. Otherwise, it initiates a request for further analysis by starting the response with T|, followed by S or A to specify the appropriate agent. A prefix of T|S delegates the task to the Shot-Level Analytics Agent (SLA), while T|A directs it to the Player Performance Analytics Agent (PPA) (see Fig. 4A for an example).

The SLA generates and executes Python code to access detailed shot-level data stored in the CSV files described in Sect. 3.2 (Fig. 4B). Concurrently, the PPA processes relevant player statistics from CSV files containing historical performance metrics in reports generated by TennisTact. In case of execution errors, both the SLA and PPA relay their generated code and error logs to their respective Debugging Agents (DA), which correct the errors and rerun the analysis (example in Fig. 5).

Once the required statistics are successfully computed, the Visualization Agent (VA) converts them into formatted tables and visualizations (Fig. 4C). These are returned to the CA, which then constructs and delivers the final response to the user (Fig. 4D).

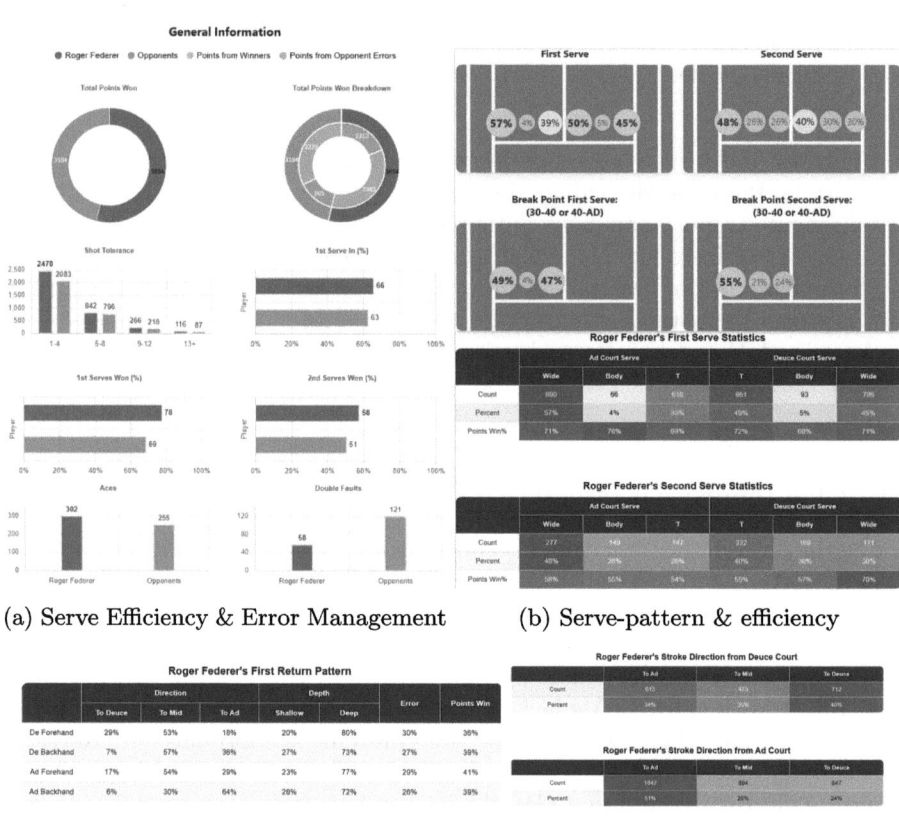

Fig. 2. First part of a sample report generated by TennisTact.

(a) Rally & Momentum Analysis

(b) Shot Type Analysis

(c) On the run return & smash analysis

Fig. 3. Second part of a sample generated by TennisTact.

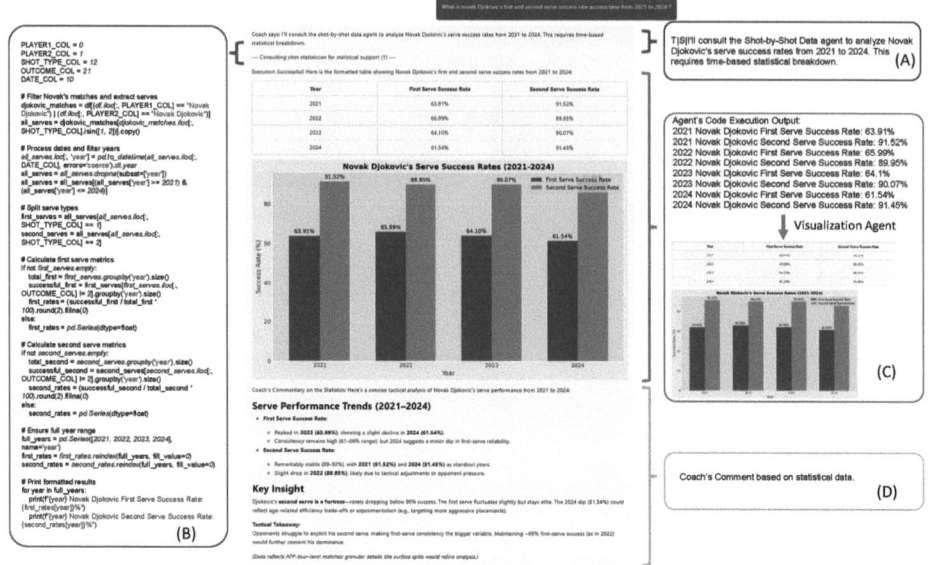

Fig. 4. Sample ChatTennis input and response workflow.

The SLA, PPA, and their respective DAs use DeepSeek-R1 for enhanced coding capabilities [1], while the remaining agents use DeepSeek-V3 for faster and more fluent responses. DeepSeek is chosen over ChatGPT due to cost efficiency, as ChatGPT's equivalent model (e.g., OpenAI o1) is over 25 times more expensive [1]. All agents are guided by carefully crafted, task-specific prompts.

Through this targeted prompt engineering, the CA decides whether analysis is needed and then turns the results into user-friendly narration. The SLA and PPA complement each other: SLA mines raw, point-by-point data for fine-grained questions, whereas PPA taps TennisTact's aggregated reports to answer higher-level queries efficiently. The Debugging Agents (DAs) automatically fix code errors to boost reliability, and the Visualization Agent (VA) converts otherwise hard-to-interpret numbers into clear tables and charts.

A detailed overview of each agent's responsibilities and corresponding prompt-engineering strategies is provided in Table 1.

4 Discussion and Future Work

This hybrid system combining automated reporting system (TennisTact) and an interactive chatbot (ChatTennis) can deliver players an abundance of useful metrics. TennisTact can provide users with the most commonly used metrics, and if additional metrics are needed, they can turn to ChatTennis for additional information. ChatTennis harnesses the flexibility of generative AI. By drawing on a comprehensive database, ChatTennis can deliver a wide range of insights—from tournament breakdown to in-depth player evaluations and general tactical

Table 1. Agent Roles and Prompt Directives in the ChatTennis System

Agent	Primary Responsibilities	Prompt-Engineering Directives
CA	• Conversational front-end • Decide agent invocation • Comment on retrieved data	• Respond concisely, stats-focused • Understand each agent's role • Prepend control flags: T|S|, T|A|, F| • Mimic professional tennis coach
SLA	• Generate & run Python code on shot-by-shot database • Deliver low-level stats	• Output *raw* runnable Python code • Follow CSV schema • Print label–value pairs only
PPA	• Generate & run Python code on performance-metric CSV • Provide higher-level player stats	• Same rules as SLA with different data instruction
DA	• Fix SLA/PPA code errors • Retry execution up to preset attempts	• Receive faulty code & traceback • Follow CSV schema • Return corrected code only • Stop after success retries
VA	• Convert stats to Markdown tables • Generate plots for visual insight • Rectify/discard unusual data	• Output well-formatted Markdown tables • Follow plotting code rules • Check values against pro-tennis norms

analysis—as demonstrated in Figs. 5, 6 and 7. Notably, many of these metrics are not included in our static reports.

In the future, more specialized agents can be added to our system for tasks like winning chance prediction, real-time data scraping, and rule clarification, alongside enhanced inter-agent coordination to produce richer insights.

Furthermore, user studies involving professional coaches, amateur players, and casual fans are needed to systematically evaluate the system's effectiveness, user satisfaction, and impact on player performance and strategic understanding. To facilitate this, we can design a set of evaluation questions to assess system performance and implement a feature on our website that displays pure LLM responses alongside those of ChatTennis, enabling users to directly indicate their preference. Additionally, we can distribute our reports to players and track whether their performance improves as a result.

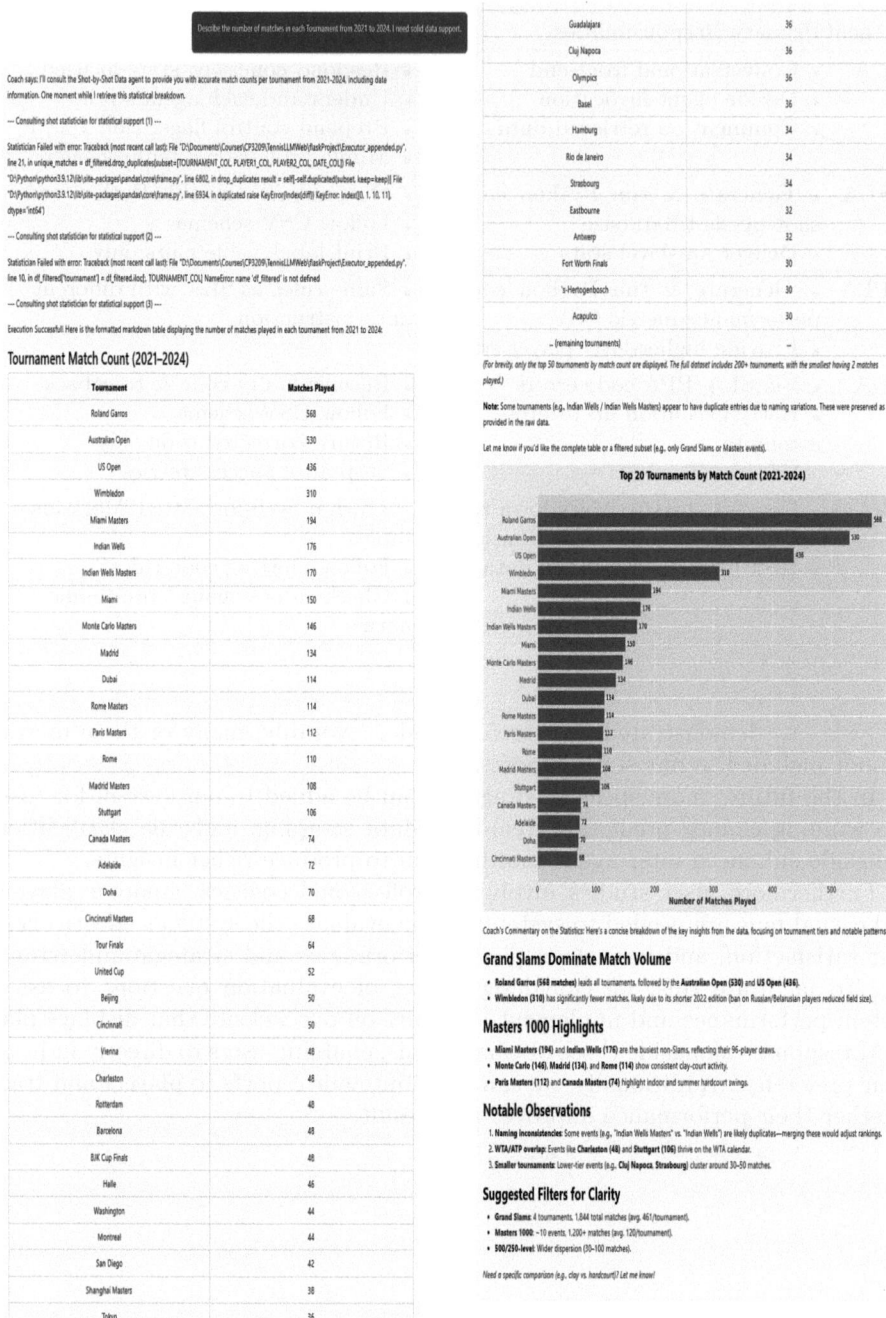

Fig. 5. Illustration of the debugging process and resulting tournament analysis.

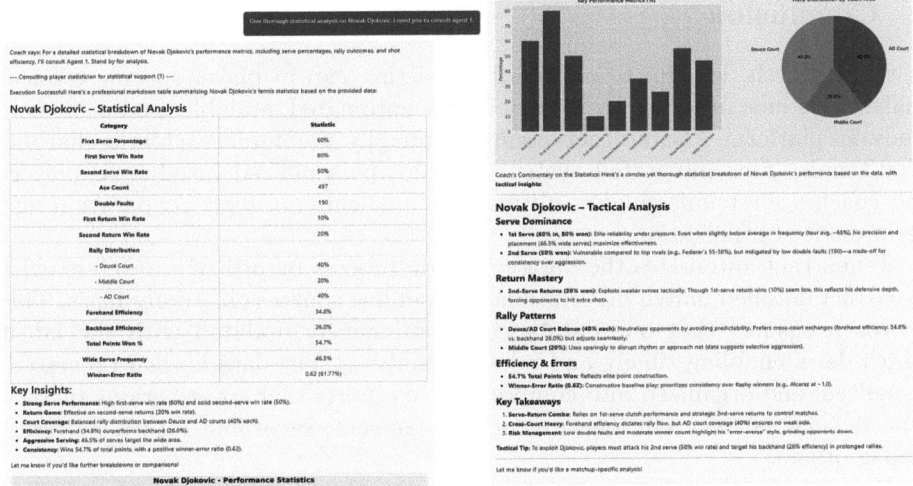

Fig. 6. Example of player specific tactical analysis.

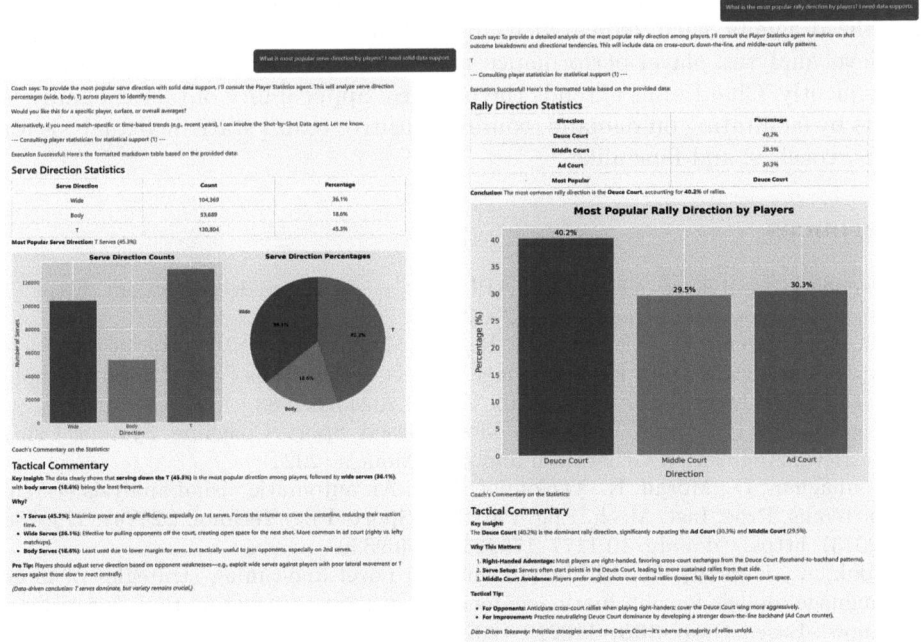

Fig. 7. Example of general tactical analysis.

5 Conclusion

In this research, we aim to address the existing gap in player-specific tactical analysis systems in tennis by developing an automated, scalable, and accessible analytics platform—TennisTact- complemented by an interactive tennis chatbot, ChatTennis. This integrated system provides personalized insights to players and coaches for tennis singles, ultimately enhancing strategic preparation and performance optimization.

TennisTact automates the data collection process by utilizing a web crawler to extract detailed match descriptions from online sources on a daily basis. This approach ensures that our database is continuously updated with the latest match data, enabling timely and relevant analyses. The data is then processed, visualized, and organized into comprehensive reports that cover various tactical aspects such as serve efficiency, error management, serve and return patterns, among others.

Recognizing that static reports alone cannot encompass every analytical angle, we have introduced ChatTennis—a multi-agent, generative AI framework designed to expand on TennisTact's core analytics through interactive, natural language conversations. ChatTennis utilizes a modular structure comprising specialized agents, each dedicated to distinct analytical tasks, such as detailed shot-level analyses, player performance summaries, debugging, and visualization. Consequently, ChatTennis extends the practical applicability of tactical analysis reports by facilitating on-demand, context-sensitive tennis statistical analysis for players, coaches, and fans alike.

References

1. DeepSeek-R1 Release | DeepSeek API Docs — API-docs.deepseek.com. https://api-docs.deepseek.com/news/news250120
2. Cheng, L., Deng, D., Xie, X., Qiu, R., Xu, M., Wu, Y.: SNIL: generating sports news from insights with large language models. IEEE Trans. Vis. Comput. Graph. 1–14 (2024). https://doi.org/10.1109/TVCG.2024.3392683
3. Comendant, C.: Large Language Model-Based Sport Coaching System Using Retrieval-Augmented Generation and User Models (2024)
4. Connaghan, D., Moran, K., O'Connor, N.E.: An automatic visual analysis system for tennis. Proc. Inst. Mech. Eng. Part P J. Sports Eng. Technol. **227**(4), 273–288 (2013). https://doi.org/10.1177/1754337112469330
5. Cook, A., Karakuş, O.: LLM-commentator: novel fine-tuning strategies of large language models for automatic commentary generation using football event data. Knowl.-Based Syst. **300**, 112219 (2024). https://doi.org/10.1016/j.knosys.2024.112219
6. de Leeuw, A.-W., Hoekstra, A., Meerhoff, L., Knobbe, A.: Tactical analyses in professional tennis. In: Cellier, P., Driessens, K. (eds.) ECML PKDD 2019. CCIS, vol. 1168, pp. 258–269. Springer, Cham (2020). https://doi.org/10.1007/978-3-030-43887-6_20
7. Demaj, D.: Geovisualizing spatio-temporal patterns in tennis: an alternative approach to post-match analysis. In: Proceedings of the 26th International Cartographic Conference, vol. 9, p. 16 (2013)

8. Dong, J.S., et al.: Sports analytics using probabilistic model checking and deep learning. In: 2023 27th International Conference on Engineering of Complex Computer Systems (ICECCS), pp. 7–11. IEEE (2023)
9. Dong, J.S., Shi, L., Jiang, K., Sun, J., et al.: Sports strategy analytics using probabilistic reasoning. In: 2015 20th International Conference on Engineering of Complex Computer Systems (ICECCS), pp. 182–185. IEEE (2015)
10. Gao, Y., et al.: Retrieval-Augmented Generation for Large Language Models: A Survey. https://doi.org/10.48550/arXiv.2312.10997. http://arxiv.org/abs/2312.10997
11. Giménez-Egido, J.M., Ortega, E., Verdu-Conesa, I., Cejudo, A., Torres-Luque, G.: Using smart sensors to monitor physical activity and technical-tactical actions in junior tennis players. Int. J. Environ. Res. Public Health **17**(3), 1068 (2020). https://doi.org/10.3390/ijerph17031068
12. Golden Set Analytics: WHY USE ANALYTICS – Golden Set Analytics. goldensetanalytics.com (2017). https://goldensetanalytics.com/why-use-analytics/
13. Hawk-Eye Innovations: Hawk-eye. hawkeyeinnovations.com (2019). https://www.hawkeyeinnovations.com/
14. Hundal, R.S., Liu, Z., Wadhwa, B., Hou, Z., Jiang, K., Dong, J.S.: Soccer strategy analytics using probabilistic model checkers. In: International Sports Analytics Conference and Exhibition, pp. 249–264. Springer (2024)
15. Jiang, K., Izadi, M., Liu, Z., Dong, J.S.: Deep learning application in broadcast tennis video annotation. In: 2020 25th International Conference on Engineering of Complex Computer Systems (ICECCS), pp. 53–62. IEEE (2020)
16. Jiang, K., Li, J., Liu, Z., Dong, C.: Court detection using masked perspective fields network. In: 2023 IEEE 28th Pacific Rim International Symposium on Dependable Computing (PRDC), pp. 342–345. IEEE (2023)
17. Jiang, K., Liu, Z., Wu, Q., Ma, M., Dong, J.S.: Tracking small and fast moving ball in broadcast videos using transfer learning and the enhanced interactive multimotion model. In: International Sports Analytics Conference and Exhibition, pp. 81–96. Springer (2024)
18. Khan, A.A., Shao, J., Ali, W., Tumrani, S.: Content-aware summarization of broadcast sports videos: an audio–visual feature extraction approach. Neural Process. Lett. **52**(3), 1945–1968 (2020). https://doi.org/10.1007/s11063-020-10200-3
19. Kim, B.J., Choi, Y.S.: Automatic baseball commentary generation using deep learning. In: Proceedings of the 35th Annual ACM Symposium on Applied Computing, Brno Czech Republic, pp. 1056–1065. ACM (2020). https://doi.org/10.1145/3341105.3374063
20. Klaus, A., Bradshaw, R., Young, W., O'Brien, B., Zois, J.: Success in national level junior tennis: tactical perspectives. Int. J. Sports Sci. Coach. **12**(5), 618–622 (2017). https://doi.org/10.1177/1747954117727792
21. Larson, A., Smith, A.: Sensors and data retention in grand slam tennis. In: 2018 IEEE Sensors Applications Symposium (SAS), Seoul, Korea (South), pp. 1–6. IEEE (2018). https://doi.org/10.1109/SAS.2018.8336712
22. Liu, Z., Jiang, K., Hou, Z., Lin, Y., Dong, J.S.: Insight analysis for tennis strategy and tactics. In: 2023 IEEE International Conference on Data Mining (ICDM), Shanghai, China, pp. 1169–1174. IEEE (2023). https://doi.org/10.1109/ICDM58522.2023.00143
23. Liu, Z., et al.: Analyzing the formation strategy in tennis doubles game. SN Comput. Sci. **6**(2), 100 (2025)

24. Liu, Z., Dong, C., Wang, C., Dong, T.Y., Jiang, K.: Exploring team strategy dynamics in tennis doubles matches. In: International Sports Analytics Conference and Exhibition, pp. 104–115. Springer (2024)
25. Liu, Z., Durrani, M., Xuan, L.Y., Simon, J.F., Deon, T.Y.F.: Strategy analysis in NFL using probabilistic reasoning. In: International Sports Analytics Conference and Exhibition, pp. 116–128. Springer (2024)
26. Liu, Z., Guo, J., Wang, M., Wang, R., Jiang, K., Dong, J.S.: Recognizing a sequence of events from tennis video clips: addressing timestep identification and subtle class differences. In: 2023 IEEE 28th Pacific Rim International Symposium on Dependable Computing (PRDC), pp. 337–341. IEEE (2023)
27. Liu, Z., Jiang, K., Dong, J.S.: Sports injury prediction in professional tennis. In: 2023 IEEE 28th Pacific Rim International Symposium on Dependable Computing (PRDC), pp. 304–308. IEEE (2023)
28. Liu, Z., Jiang, K., Ma, M., Hou, Z., Lin, Y., Dong, J.S.: F^3Set: towards analyzing fast, frequent, and fine-grained events from videos. arXiv preprint arXiv:2504.08222 (2025)
29. Liu, Z., Ma, M., Jiang, K., Hou, Z., Shi, L., Dong, J.S.: PCSP# denotational semantics with an application in sports analytics. In: The Application of Formal Methods: Essays Dedicated to Jim Woodcock on the Occasion of His Retirement, pp. 71–102. Springer (2024)
30. Martin-Lorente, E., Campos, J., Crespo, M.: The inside out forehand as a tactical pattern in men's professional tennis. Int. J. Perform. Anal. Sport **17**(4), 429–441 (2017). https://doi.org/10.1080/24748668.2017.1349528
31. McBee, J.C., et al.: Interdisciplinary inquiry via panelgpt: application to explore chatbot application in sports. Rehabilitation (2023). https://doi.org/10.1101/2023.07.23.23292452
32. Mergheş, P.E., Simion, B., Nagel, A.: Comparative analysis of return of serve as counter-attack in modern tennis. Timisoara Phys. Educ. Rehabil. J. **6**(12), 18–22 (2014). https://doi.org/10.2478/tperj-2014-0023
33. Polk, T., Jackle, D., Hausler, J., Yang, J.: Courttime: generating actionable insights into tennis matches using visual analytics. IEEE Trans. Vis. Comput. Graph. (2019). https://doi.org/10.1109/TVCG.2019.2934243
34. Polk, T., Yang, J., Hu, Y., Zhao, Y.: Tennivis: visualization for tennis match analysis. IEEE Trans. Visual Comput. Graphics **20**(12), 2339–2348 (2014). https://doi.org/10.1109/TVCG.2014.2346445
35. Priya, M.Y., Kamble, S.S., Shendre, S.P., Sridhar, S.: Megan - A sports chatbot using OpenAI APIs and Django framework with python. In: 2024 IEEE 9th International Conference for Convergence in Technology (I2CT), Pune, India, pp. 1–8. IEEE (2024). https://doi.org/10.1109/I2CT61223.2024.10543499
36. Wei, X., Lucey, P., Morgan, S., Carr, P., Reid, M., Sridharan, S.: Predicting serves in tennis using style priors. In: Proceedings of the 21th ACM SIGKDD International Conference on Knowledge Discovery and Data Mining, Sydney, NSW, Australia, pp. 2207–2215. ACM (2015). https://doi.org/10.1145/2783258.2788598
37. Whiteside, D., Reid, M.: Spatial characteristics of professional tennis serves with implications for serving aces: a machine learning approach. J. Sports Sci. **35**(7), 648–654 (2017). https://doi.org/10.1080/02640414.2016.1183805
38. Wu, J., Guo, Z., Wang, Z., Xu, Q., Wu, Y.: Visual analytics of multivariate event sequence data in racquet sports. In: 2020 IEEE Conference on Visual Analytics Science and Technology (VAST), Salt Lake City, UT, USA, pp. 36–47. IEEE (2020). https://doi.org/10.1109/VAST50239.2020.00009

39. Zhang, J., Han, D., Han, S., Li, H., Lam, W.K., Zhang, M.: Chatmatch: exploring the potential of hybrid vision-language deep learning approach for the intelligent analysis and inference of racket sports. Comput. Speech Lang. **89**, 101694 (2025). https://doi.org/10.1016/j.csl.2024.101694
40. Zhi, Q., Metoyer, R.: GameBot: a visualization-augmented chatbot for sports game. In: Extended Abstracts of the 2020 CHI Conference on Human Factors in Computing Systems, Honolulu, HI, USA, pp. 1–7. ACM (2020). https://doi.org/10.1145/3334480.3382794

Analyzing Basketball Lineups with MDP Using NBA Statistics and Player Tracking Data

Zhaoyu Liu[✉] and Shenyi Su

National University of Singapore, Singapore, Singapore
liuzy@nus.edu.sg

Abstract. Predicting basketball game outcomes is a challenging task due to the sport's dynamic nature, requiring rapid decision-making and strategic execution. Previous research has applied machine learning and reinforcement learning methods to predict game results, evaluate lineup performance, and determine optimal strategies, such as shot selection. While these studies have achieved significant predictive accuracy, they often lack interpretability. This work uses a Markov Decision Process (MDP) framework to model NBA games between fixed lineups, integrating historical NBA statistics and player-tracking data. By analyzing player interactions and decision outcomes, our model simulates half-court possessions, starting once the offensive team crosses half-court. The model estimates the probabilities of critical decisions, including passing, shooting, and rebounding. By simulating 100 possessions per team—typical for an NBA game—the model predicts game outcomes by comparing projected team scores. Model predictions are compared against actual NBA results and betting odds to evaluate predictive accuracy and profitability, applying betting strategies such as the Kelly Criterion. Our approach, leveraging data from the NBA SportVU dataset and NBA API, offers valuable insights into lineup performance.

Keywords: Basketball Analytics · Game Outcome Prediction · Lineup Evaluation · Markov Decision Process

1 Introduction

Basketball is a dynamic sport requiring rapid decision-making and strategic execution. Predicting game outcomes necessitates a detailed understanding of the numerous decisions players make throughout a game. Markov Decision Processes (MDP) provide a robust framework for modeling these decisions and their probabilistic outcomes. By leveraging historical data, we estimate the probabilities of key actions and their impact on the game.

Previous studies have employed machine learning techniques, such as logistic regression and reinforcement learning, to predict lineup performance [1]. While these methods achieve notable predictive accuracy, they often lack interpretability. The MDP framework addresses this limitation by structuring the game as a sequence of state-dependent decisions, capturing player interactions and strategic choices—such as shooting, passing, and dribbling—and their influence on scoring outcomes. By integrating MDP modeling, we aim to improve predictive accuracy while maintaining a transparent model structure that simulates and visualizes player decisions, providing deeper insights into lineup dynamics in a fixed setting.

The primary challenge is accurately modeling complex decision-making processes, including passing, shooting, and defensive maneuvers, to capture their probabilistic outcomes and predict overall game results.

2 Related Work

2.1 Sports Analytics

Recent works in sports analytics have explored a wide range of challenges, including strategy modeling using probabilistic reasoning and deep learning [3,4,9–11,13,15], injury prediction [12], fine-grained event detection in sports videos [14], and specific tasks such as court detection and ball tracking in broadcast footage [5–7]. These efforts demonstrate the growing integration of AI, data mining, and probabilistic reasoning in advancing sports performance analysis and tactical understanding [8].

2.2 Basketball Strategy Analysis

Recent advancements in basketball analytics have centered on predictive modeling for game outcomes, player evaluation, and strategic optimization, employing methods such as Markov Decision Processes (MDPs), network analysis, and machine learning.

Sandholtz and Bornn [17] proposed a non-stationary MDP using high-resolution tracking data from the 2015–2016 NBA season, modeling game states by ball carrier identity, court region, and defensive pressure, with a binary action space (shoot vs. not shoot). They later extended this model with a Bayesian hierarchical framework to capture non-stationary transition dynamics across eight segments of the shot clock [18], using tensor-structured transition probabilities and data pooling across players, teams, and regions to improve robustness with limited samples.

Complementary approaches include Huang and Chen's [16] use of XGBoost and SHAP to identify key predictors of NBA outcomes (e.g., FG%, defensive rebounds, turnovers), offering interpretable, real-time decision support. Ahmadalinezhad et al. [1,2] applied edge-centric multi-view network analysis to historical NBA data (2007–2019), achieving 80% accuracy in outcome prediction via logistic regression on network-derived features. Additionally, Loeffelholz et al. [21] leveraged feed-forward neural networks and Bayesian models, identifying FG%, 3P%, and steals as top predictors, achieving 74.33% accuracy with only four input features.

3 Data Source and Processing

This study leverages two primary data sources—player box score statistics and detailed player tracking data—to develop and validate our basketball simulation model. Box score data provide comprehensive player performance metrics, while tracking data enable a deeper analysis of player movements, ball trajectories, and interaction patterns on the court. Together, these datasets enable a thorough characterization of in-game decision-making and player dynamics.

3.1 Box Score for Individual Players

We collected player-level box scores directly from NBA.com (see Fig. 1 for an example illustrating LeBron James' box score). These data include individual statistics such as field goals made (FGM), field goals attempted (FGA), three-point field goals made (3PM), three-point field goals attempted (3PA), steals (STL), blocks (BLK), frequency of passes to teammates, minutes played (MIN), points scored (PTS), field goal percentage (FG%), three-point field goal percentage (3P%), free throws made (FTM), free throws attempted (FTA), free throw percentage (FT%), offensive rebounds (OREB), defensive rebounds (DREB), total rebounds (REB), assists (AST), and turnovers (TOV). Using these comprehensive metrics, the model can simulate an NBA game between two fixed lineups, even when using data from different seasons.

GP – Games Played					3P% – 3 Point Field Goal Percentage							TOV – Turnovers									
MIN – Minutes Played					FTM – Free Throws Made							STL – Steals									
PTS – Points					FTA – Free Throws Attempted							BLK – Blocks									
FGM – Field Goals Made					FT% – Free Throw Percentage							PF – Personal Fouls									
FGA – Field Goals Attempted					OREB – Offensive Rebounds							FP – Fantasy Points									
FG% – Field Goal Percentage					DREB – Defensive Rebounds							DD2 – Double Doubles									
3PM – 3 Point Field Goals Made					REB – Rebounds							TD3 – Triple Doubles									
3PA – 3 Point Field Goals Attempted					AST – Assists							+/- – Plus-Minus									

BY YEAR	TEAM	GP	MIN	PTS	FGM	FGA	FG%	3PM	3PA	3P%	FTM	FTA	FT%	OREB	DREB	REB	AST	TOV	STL	BLK	PF	FP	DD2
2024-25	LAL	58	34.9	25.0	9.6	18.5	51.7	2.3	5.9	38.4	3.6	4.7	77.0	1.0	7.2	8.2	8.5	3.9	0.9	0.6	1.5	48.3	30
2023-24	LAL	71	35.3	25.7	9.6	17.9	54.0	2.1	5.1	41.0	4.3	5.7	75.0	0.9	6.4	7.3	8.3	3.5	1.3	0.5	1.1	48.8	27
2022-23	LAL	55	35.5	28.9	11.1	22.2	50.0	2.2	6.9	32.1	4.6	5.9	76.8	1.2	7.1	8.3	6.8	3.2	0.9	0.6	1.6	50.3	18
2021-22	LAL	56	37.2	30.3	11.4	21.8	52.4	2.9	8.0	35.9	4.5	6.0	75.6	1.1	7.1	8.2	6.2	3.5	1.3	1.1	2.2	53.0	21
2020-21	LAL	45	33.4	25.0	9.4	18.3	51.3	2.3	6.3	36.5	4.0	5.7	69.8	0.6	7.0	7.7	7.8	3.7	1.1	0.6	1.6	47.0	18
2019-20	LAL	67	34.6	25.3	9.6	19.4	49.3	2.2	6.3	34.8	3.9	5.7	69.3	1.0	6.9	7.8	10.2	3.9	1.2	0.5	1.8	51.3	46
2018-19	LAL	55	35.2	27.4	10.1	19.9	51.0	2.0	5.9	33.9	5.1	7.6	66.5	1.0	7.4	8.5	8.3	3.6	1.3	0.6	1.7	52.0	32
2017-18	CLE	82	36.9	27.5	10.5	19.3	54.2	1.8	5.0	36.7	4.7	6.5	73.1	1.2	7.5	8.6	9.1	4.2	1.4	0.9	1.7	54.1	52
2016-17	CLE	74	37.8	26.4	9.9	18.2	54.8	1.7	4.6	36.3	4.8	7.2	67.4	1.3	7.3	8.6	8.7	4.1	1.2	0.6	1.8	51.3	42
2015-16	CLE	76	35.6	25.3	9.7	18.6	52.0	1.1	3.7	30.9	4.7	6.5	73.1	1.5	6.0	7.4	6.8	3.3	1.4	0.6	1.9	47.1	28

Fig. 1. Box score for LeBron James.

3.2 Video Tracking Data from NBA SportVU

We also utilize detailed video tracking data from the NBA SportVU Dataset[1] for the 2015–2016 NBA season. These data (see Fig. 2 for an example visualization) capture trajectories of all players and the basketball during games, represented in an x, y, z coordinate system scaled to the court dimensions. By analyzing the trajectory, velocity, and relative positioning of the ball and players, we can classify specific in-game actions, such as passes, dribbles, and shots. Furthermore, passing trajectories derived from these data enable analysis of each player's passing tendencies across distinct court zones. Identifying such zone-specific passing patterns provides additional granularity for our simulation model, helping us capture realistic in-game decision-making and interactions.

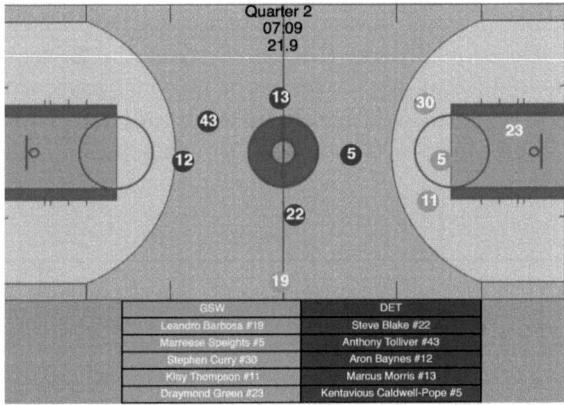

Fig. 2. Sample video tracking display.

Previous research has leveraged similar tracking data to represent complex basketball dynamics within a non-stationary Markov Decision Process (MDP) framework, adjusting transition probabilities dynamically according to intervals of the 24-second shot clock [17]. This prior approach employed transition and shot policy tensors structured in short intervals (2-second segments), effectively capturing time-dependent shifts in player behavior. However, we employ static transition matrices rather than non-stationary ones to ensure our initial model remains manageable. Although dynamic modeling better reflects realistic gameplay, the use of static matrices allows us to establish a foundational framework first.

4 The Proposed Approach

This section introduces our proposed approach to modeling NBA basketball games, detailing key assumptions, the state and action spaces, transition and

---

[1] https://paperswithcode.com/dataset/nba-sportvu.

reward functions, policy definitions, possession simulation methodology, and the implementation of dynamic lineups.

4.1 Assumptions

To effectively model a game of basketball, we establish several assumptions to streamline implementation. Our model assumes a fixed lineup throughout the game, with no substitutions. Additionally, fouls are not considered; however, to account for players skilled in drawing fouls, we incorporate each player's average Free Throws Made (FTM) into individual and team scoring statistics. Fatigue effects, which could diminish a player's shooting or passing efficiency over time, are also disregarded.

To simplify gameplay representation, the model focuses solely on half-court play. Since advancing the ball beyond half-court typically takes about four seconds, the Markov Decision Process (MDP) starts at the 20-second mark on the shot clock. Players' field goal percentages vary by court zone, with the court segmented into specific regions (Fig. 3), each linked by adjacency relationships. This zoning approach simulates player positioning, zone-specific shooting efficiency, and passing tendencies.

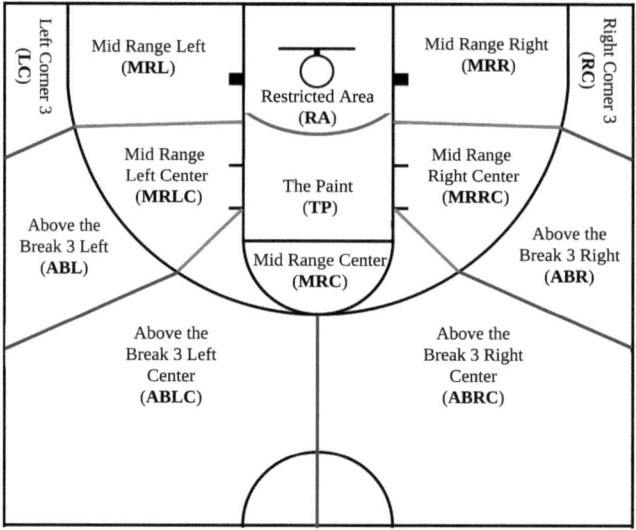

Fig. 3. Half court area decomposition.

Possession begins with a player based on their assist statistics, reflecting their ball-handling frequency. Defensive matchups are position-based, ensuring that each offensive player is guarded by their counterpart on the opposing team, preventing uncontested shots from skewing outcomes. Players can choose from three actions during dribbling, with decisions occurring at one-second intervals, resulting in a maximum of 20 MDP steps per possession. These actions include continuing to dribble or moving to an adjacent court zone, passing to a teammate, or attempting a shot. The probabilities of these choices sum to one. Dribbling frequency is estimated using assist statistics, with players who record more assists assumed to spend more time handling the ball. Passing frequencies between teammates are derived from NBA statistics to reflect real-world team dynamics. A player's shooting frequency is determined by their relative Field Goals Attempted (FGA) compared to teammates, ensuring that high-volume shooters take more attempts. Rebounding probabilities are modeled based on the ratio of a team's offensive rebounds to the opposing team's defensive rebounds. The overall game simulation evaluates each team's action tendencies and success probabilities, iterating across multiple simulations to project scores and estimate the likelihood of victory for each team.

4.2 MDP Model Architecture

We model each team's offensive possession as a sequence of decisions and outcomes within a Markov Decision Process (MDP). Possession begins with a player selected based on preprocessed statistical frequencies, who can choose among dribbling, passing, or shooting. The decision-making process is represented as an MDP, where each possession follows a structured sequence of states, actions, transitions, and rewards, structured as an MDP tuple (S, A, T, R, π).

State Space. Each state $s \in S$ represents the ball carrier's identity and location on the court, which is divided into specific zones, including the Restricted Area (RA), Paint (TP), Mid-Range zones (e.g., MRL, MRLC, MRC, MRRC, MRR), Left Corner 3 (LC), Right Corner 3 (RC), and Above the Break 3 zones (e.g., ABL, ABR, ABLC, ABRC). Each state captures the current ball handler's position, enabling the modeling of player movements and decisions based on the ball's location and handler.

Action Space. The ball handler in each state s can choose among three actions: dribbling, passing, or shooting. Dribbling allows the player to either continue dribbling at the current location or move to an adjacent court area, with success determined by the ball handler's turnover rate and the defender's steal rate. Passing transitions possession to a teammate, where the success of the pass depends on both the passer's and receiver's turnover rates. If successful, the receiving player becomes the new ball handler; otherwise, a turnover occurs. Shooting attempts a field goal, where the probability of scoring is based on the player's shooting percentage from that specific zone. A successful shot ends the possession with a reward, while a missed attempt leads to a rebound scenario.

Transition Function. The transition function $T(s, a, s') = P[S_{t+1} = s'|S_t = s, A_t = a]$ defines the probability of moving to a new state s' after taking action a in state s. When passing, a successful pass updates the state to reflect the new ball handler's location, while an unsuccessful pass results in a turnover, terminating the possession. Shooting transitions the state based on the outcome: a made shot ends the possession with an updated score, a missed shot with an offensive rebound allows the team to retain possession with the shot clock reset to 24 s, and a missed shot with a defensive rebound transfers possession to the opposing team. Dribbling transitions either to an adjacent court zone or remains within the same area. If successful, the new state reflects the ball handler's updated position; if a steal occurs, possession is lost, terminating the possession.

Reward Function. The reward function $R(s, a) = E[R_{t+1}|S_t = s, A_t = a]$ provides the expected points from taking action a in state s. A successful shot yields a reward of 2 or 3 points, depending on the shooting zone. If a shot is missed and the defensive team secures the rebound, the reward is 0, as the possession ends without scoring. Similarly, a turnover, whether from a failed pass or a steal during dribbling, results in a reward of 0 due to the loss of possession. An offensive rebound does not immediately generate points but temporarily yields a reward of 0, allowing the possession to continue and creating additional scoring opportunities.

Policy. The policy $\pi(s, a) = P[A_t = a|S_t = s]$ represents the probability that a player will choose action a (dribble, pass, or shoot) in a given state s. Shooting probability is determined by the player's historical shooting frequency and field goal attempts, with players who take more shots being more likely to attempt another. Passing likelihood is influenced by the player's passing tendencies and historical passing frequencies to specific teammates. Dribbling generally has a lower probability but varies based on the ball handler's tendencies and court location. Since our model employs a static transition matrix, $\pi(s, a)$ remains independent of time and is unaffected by the shot clock, relying solely on the ball handler's statistical tendencies.

Possession. In a typical NBA game, each team has approximately 100–105 possessions over 48 min, according to data from Team Ranking NBA[2]. For simplicity, our model assumes exactly 100 possessions per team per game. To enhance the robustness of our results, we simulate 10,000 possessions and compute the average outcome over each set of 100 possessions. This approach improves the approximation of expected results, effectively capturing the essential decision-making dynamics and variability in basketball plays. By leveraging this MDP framework with well-defined transition, reward, and policy functions, we can accurately predict game outcomes based on player tendencies and defensive interactions.

4.3 Dynamic Transition Matrix

In the prototype model, the transition matrix is invariant to the shot clock time, meaning that, for a fixed player, the probability of selecting an action at any time within a shot-clock interval is constant. However, this assumption is unrealistic because the shot clock resets at the beginning of each possession—each

---

[2] https://www.teamrankings.com/nba/stat/possessions-per-game.

new shot clock in basketball initiates a new episode rather than continuously progressing across episodes. Consequently, our model emphasizes within-episode temporal dynamics instead of global time-dependent factors commonly discussed in Markov Decision Process (MDP) literature. Moreover, NBA teams often utilize the initial portion of the shot clock for executing strategic passes to disrupt defensive formations, reserving shot attempts for the final seconds. Thus, incorporating a dynamic transition matrix for the policy $\pi(s,a)$ is justified, leading to the introduction of:

$$T(c_n) = \begin{cases} 1, c_n \in (0,2], \\ \vdots \\ 12, c_n \in (22,24], \end{cases}$$

where T is a transition matrix indexed by the current shot-clock interval c_n. Using SportVU data, we identified key actions (shot, pass, dribble) and calculated their frequencies within each defined interval across all episodes. An example of the resulting transition matrix for the shot-clock interval $(0,2]$ is shown in Table 1.

Table 1. Example transition matrix (shot clock interval (0,2]).

Position	Pass_Freq	Dribble_Freq	Shoot_Freq
Guard	0.0746	0.8754	0.0500
Forward	0.0779	0.8617	0.0603
Centre	0.0493	0.9041	0.0466

4.4 Dynamic Lineups

In the early stages of the research, we attempted to use a fixed lineup for each team by selecting the five players with the most minutes played over the season. However, this approach yielded mediocre model performance. In high-intensity leagues like the NBA, player rotations and injuries often lead to lineup changes, with key players sometimes missing games. The presence or absence of star players significantly influences game outcomes, making it unrealistic to rely on a single fixed lineup for simulation.

To address this, we implemented a dynamic lineup strategy. Using the NBA API, we retrieved the starting five players for each game based on its unique GameId. By simulating each game with its actual starting lineup, the model could better capture the impact of lineup variations and improve its predictive accuracy. This approach accounts for lineup changes and better reflects real-world game conditions, leading to a more accurate and reliable model performance.

4.5 Model Simulation Outcome

For each matchup between an offensive and defensive team, we simulate 10,000 possessions. To approximate a full game's outcome, we average these results over 100 possessions per team—the typical number in an NBA game. This calculation yields projected team scores, from which we determine win-loss outcomes by comparing scores.

The MDP model generates a comprehensive record of all in-game events, including timestamps for each action, player box scores, and overall team scores.

```
Sample event records from simulation:
- Pass from PG (area ABLC) to PF (area TP) at time 20
- Pass from PF (area TP) to SG (area ABR) at time 19
- Dribble by SG (area ABR) at time 18
- Dribble by SG (area RC) at time 17
- Dribble by SG (area MRRC) at time 16
- Dribble by SG (area MRRC) at time 15
- Pass from SG (area TP) to C (area RA) at time 14
- Shot by C (area RA) at time 13: MISS
- Offensive rebound by SG (area TP) at time 20
- Shot by SG (area TP) at time 20: GOAL
```

We simulated the entire 2023–24 NBA regular season to verify model accuracy, comparing the model's win-loss predictions to actual game outcomes (Table 2). This approach allows us to validate the model's performance over a full season and assess its predictive reliability.

Table 2. Model results for selected NBA regular season games (2023–24).

Team1	Team2	Score1	Score2	Odds1	Odds2	Pred_Score1	Pred_Score2	Winner
NOP	SAC	105	98	1.83	2.02	105.24	101.53	NOP
MIA	CHI	112	91	1.75	2.12	96.13	98.23	CHI
CHI	ATL	131	116	1.66	2.27	103.69	95.21	CHI
PHI	MIA	105	104	2.45	1.79	97.44	100.21	MIA
SAC	GSW	118	94	2.26	1.66	101.57	91.23	SAC

4.6 Substitution Effect

In basketball games, bench players significantly influence team performance by providing crucial contributions and allowing starters to rest. Therefore, ignoring their impact when modeling game outcomes is unrealistic. However, accurately modeling bench players presents considerable challenges, as introducing each bench player increases lineup combinations by $O(n)$. Many potential lineup

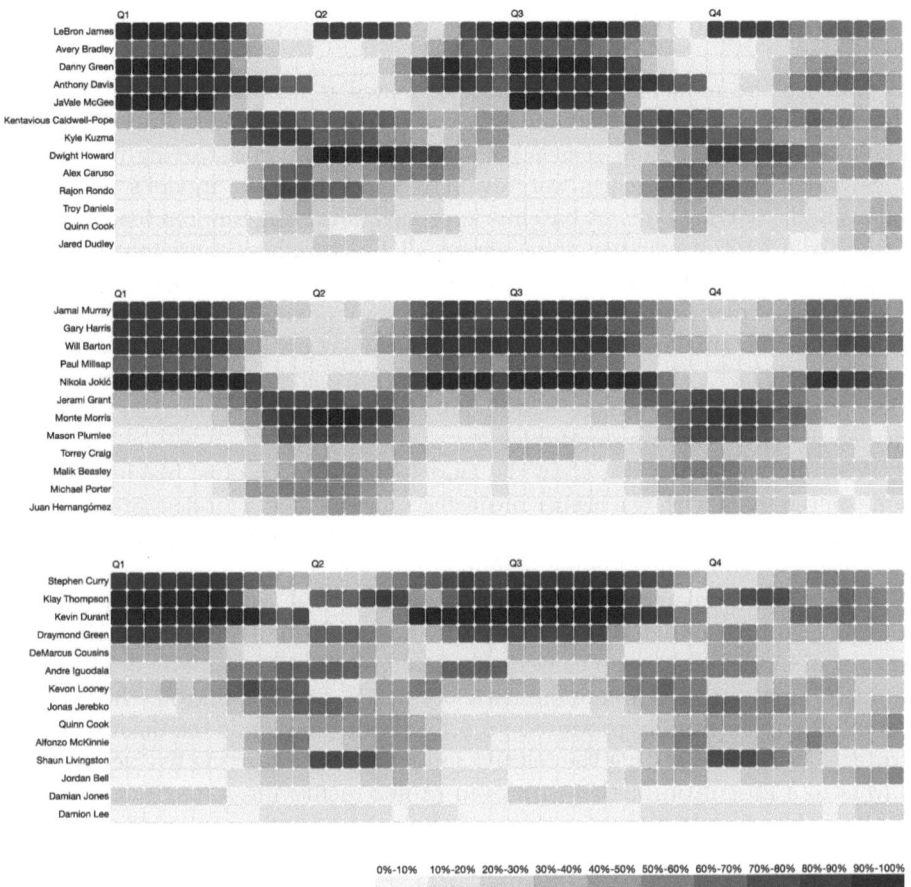

Fig. 4. Examples of NBA substitution patterns.

combinations have rarely or never appeared in actual games, resulting in data scarcity that can degrade model performance.

Nevertheless, NBA substitution patterns reveal a tendency for starting lineups and bench units to function as largely separate groups. Teams commonly substitute entire five-player lineups simultaneously, rather than using staggered substitutions. As shown in Fig. 4, this observation is supported by NBA substitution data visualizations provided by [20], where each row represents a player, each column corresponds to a game minute, and the cell color indicates the percentage of games in which that player participated during that minute.

We propose a weighted matchup approach based on these insights. Starting lineups from both teams are simulated against each other, and bench lineups are similarly matched. Weights are assigned to these matchups according to their average playing time, thereby improving the accuracy of our simulations.

5 Experimental Results

To evaluate the model's effectiveness, we tested it against NBA betting odds from the 2023–24 season. Betting data was sourced from Odds Portal[3], and we simulated all 1,231 regular-season games using each team's starting five players.

Several betting strategies were evaluated based on our model's predictions and available betting odds. A baseline approach, involving random team selection with a fixed $100 wager per game across all 1,231 games, consistently resulted in approximately a 10% loss of the initial bankroll. Conversely, a strategy trusting the odds—placing bets on teams favored by bookmakers (teams with lower odds)—led to a smaller loss of around 2%, achieving a high win-loss prediction accuracy of 69.2%. Betting against the odds, on teams with higher bookmaker odds, incurred substantial losses of about 11%, emphasizing the risks associated with consistently choosing underdogs.

When betting solely based on the model's predicted scores, placing $100 per game on the team with the higher projected score yielded a modest profit of 1.3% and a prediction accuracy of 59.2%. While this demonstrates some predictive effectiveness, the financial gains remained limited.

Finally, we implemented the Kelly Criterion [19], which adjusts bet sizes according to the predicted probability of winning. Initially, using the Kelly Criterion without a betting cap caused significant early losses, rapidly depleting the bankroll. However, when capping each bet at 1% of the bankroll, this strategy stabilized and eventually produced an 8.1% profit, making it the most profitable strategy. Besides, adjusting the relative influence of the starting five versus bench players to a 20:1 ratio further improved profits, achieving a maximum net gain of 14.8%. Conversely, reducing this ratio below 4:1 eliminated profitability.

Overall, the model achieved a prediction accuracy of 59.2%. While betting exclusively based on model predictions yielded only a minor gain of 1.3% over the season, integrating the Kelly Criterion with a capped betting strategy substantially enhanced profitability, underscoring the importance of optimized betting strategies in leveraging the model's predictive capabilities.

6 Conclusion

In this work, we developed a Markov Decision Process (MDP) model leveraging historical NBA statistics and player-tracking data to simulate basketball games between fixed lineups. Our model successfully captures essential dynamics of half-court basketball possessions, including passing, shooting, and rebounding decisions, enabling the prediction of game outcomes with 59.2% accuracy. Evaluations against actual NBA game results and betting odds demonstrated that the proposed MDP approach provides meaningful predictive insights, particularly when combined with optimized betting strategies such as the capped Kelly Criterion. However, our current model has limitations, including reliance on historical lineups and static transition probabilities, which do not fully reflect

---

[3] https://www.oddsportal.com/basketball/usa/nba-2023-2024/results/.

real-game complexities such as player substitutions and dynamic tactical adjustments. Future research incorporating dynamic player rotations, non-stationary transitions, and more nuanced tactical modeling could further enhance predictive performance and applicability.

References

1. Ahmadalinezhad, M., Makrehchi, M.: Basketball lineup performance prediction using edge-centric multi-view network analysis. Soc. Netw. Anal. Min. **10**(1), 1–11 (2020). https://doi.org/10.1007/s13278-020-00677-0
2. Ahmadalinezhad, M., Makrehchi, M., Seward, N.: Basketball lineup performance prediction using network analysis. In: Proceedings of the 2019 IEEE/ACM International Conference on Advances in Social Networks Analysis and Mining, pp. 519–524 (2019)
3. Dong, J.S., et al.: Sports analytics using probabilistic model checking and deep learning. In: 2023 27th International Conference on Engineering of Complex Computer Systems (ICECCS), pp. 7–11. IEEE (2023)
4. Hundal, R.S., Liu, Z., Wadhwa, B., Hou, Z., Jiang, K., Dong, J.S.: Soccer strategy analytics using probabilistic model checkers. In: International Sports Analytics Conference and Exhibition, pp. 249–264. Springer, Cham (2024)
5. Jiang, K., Izadi, M., Liu, Z., Dong, J.S.: Deep learning application in broadcast tennis video annotation. In: 2020 25th International Conference on Engineering of Complex Computer Systems (ICECCS), pp. 53–62. IEEE (2020)
6. Jiang, K., Li, J., Liu, Z., Dong, C.: Court detection using masked perspective fields network. In: 2023 IEEE 28th Pacific Rim International Symposium on Dependable Computing (PRDC), pp. 342–345. IEEE (2023)
7. Jiang, K., Liu, Z., Wu, Q., Ma, M., Dong, J.S.: Tracking small and fast moving ball in broadcast videos using transfer learning and the enhanced interactive multi-motion model. In: International Sports Analytics Conference and Exhibition, pp. 81–96. Springer, Cham (2024)
8. Liu, Z., et al.: Analyzing the formation strategy in tennis doubles game. SN Comput. Sci. **6**(2), 100 (2025)
9. Liu, Z., Dong, C., Wang, C., Dong, T.Y., Jiang, K.: Exploring team strategy dynamics in tennis doubles matches. In: International Sports Analytics Conference and Exhibition, pp. 104–115. Springer, Cham (2024)
10. Liu, Z., Durrani, M., Xuan, L.Y., Simon, J.F., Deon, T.Y.F.: Strategy analysis in NFL using probabilistic reasoning. In: International Sports Analytics Conference and Exhibition, pp. 116–128. Springer, Cham (2024)
11. Liu, Z., Guo, J., Wang, M., Wang, R., Jiang, K., Dong, J.S.: Recognizing a sequence of events from tennis video clips: addressing timestep identification and subtle class differences. In: 2023 IEEE 28th Pacific Rim International Symposium on Dependable Computing (PRDC), pp. 337–341. IEEE (2023)
12. Liu, Z., Jiang, K., Dong, J.S.: Sports injury prediction in professional tennis. In: 2023 IEEE 28th Pacific Rim International Symposium on Dependable Computing (PRDC), pp. 304–308. IEEE (2023)
13. Liu, Z., Jiang, K., Hou, Z., Lin, Y., Dong, J.S.: Insight analysis for tennis strategy and tactics. In: 2023 IEEE International Conference on Data Mining (ICDM), pp. 1169–1174. IEEE (2023)

14. Liu, Z., Jiang, K., Ma, M., Hou, Z., Lin, Y., Dong, J.S.: F^{3}set: towards analyzing fast, frequent, and fine-grained events from videos. arXiv preprint arXiv:2504.08222 (2025)
15. Liu, Z., Ma, M., Jiang, K., Hou, Z., Shi, L., Dong, J.S.: Pcsp# denotational semantics with an application in sports analytics. In: The Application of Formal Methods: Essays Dedicated to Jim Woodcock on the Occasion of His Retirement, pp. 71–102. Springer, Cham (2024)
16. Ouyang, Y., et al.: Integration of machine learning XGBoost and shap models for NBA game outcome prediction and quantitative analysis methodology. PLoS ONE **19**(7), e0307478 (2024)
17. Sandholtz, N., Bornn, L.: Replaying the NBA. In: The 12th Annual MIT Sloan Sports Analytics Conference (2018)
18. Sandholtz, N., Bornn, L.: Markov decision processes with dynamic transition probabilities: an analysis of shooting strategies in basketball (2020)
19. Thorp, E.O.: Portfolio choice and the Kelly criterion. In: Stochastic Optimization Models in Finance, pp. 599–619. Elsevier (1975)
20. Wainger, A.: https://alexwainger.github.io/NBASubstitutionPatterns/. Accessed 05 June 2025
21. Wang, J.: Predictive analysis of NBA game outcomes through machine learning. In: Proceedings of the 6th International Conference on Machine Learning and Machine Intelligence, pp. 46–55 (2023)

AI and Data Science in Sports Education

Ruchika Malhotra[1], Bimlesh Wadhwa[2], Shweta Meena[1(✉)], and Subodh Mor[3]

[1] Department of Software Engineering, Delhi Technological University, Delhi, India
shwetameena@dtu.ac.in
[2] National University of Singapore, Singapore, Singapore
[3] The Sports Rehabilitation Association of India, Delhi, India

Abstract. The integration of Artificial Intelligence (AI) and data science in sports is transforming - from athlete performance optimization to injury prevention, fan engagement, wearable devices, business analytics, and performance movements. It focuses on integrating interdisciplinary research in education at undergraduate and postgraduate level. Thus, Delhi Technological University has introduced a Minor Specialization in Sports Analytics as a part of its M.Tech. Data Science and Software Engineering curriculum. In this study, the structure of the specialization is discussed with integration of real-world tools, empirical learnings, and sports applications. Furthermore, the outline of the curriculum such as wearable technology, performance measurement techniques, training program, business analytics, data analytics is focused on improving athlete performance in upcoming years. It also includes implementation challenges, ethical considerations, and potential collaboration between academic, research institute, and sports industry such as Nike, Adidas and Puma. However, the outcome of this specialization is focused on producing more sports analysts and improving education in sports.

Keywords: Artificial Intelligence · Data Analytics · Interdisciplinary Research · Sports Education · Sports Industry

1 Introduction

Sports, coaches, and athletes observed that sports education has been transformative in the past few decades. Education in sports is recognizing contributions to individual development, national identity, public health, and economic growth. Thus, the growing modernization of sports curriculum and educational policy or framework reflects importance of sports as a multidimensional field that requires scientific, pedagogical, and technological input at every point. It is not restricted to physical activity or recreation of a sportsperson. Academicians and coaches now consider it as strategic component of holistic education. It is observed that government, educational institutions, and policy makers recognize that sports is important for every person with respect to their physical well-being, mental health, discipline, leadership, social skills, and entrepreneurs. Sports education mainly focused on sports tournament, sports business, sports measurement, sports infrastructure and development.

In some countries, it is emphasized to provide inclusive and quality education through sports, it is a fundamental human right as per UNESCO Kazan Action Plan (2017) and International Charter of Physical Education, Physical Activity, and Sport [1]. Digitization in sports education is crucial for urban energy systems and environmental sustainability [2]. Moreover, the sports sector is considered a vast industry, with costs expected to increase from $500 billion to $700 billion by 2026 [3]. Sports education mainly focused on various sports activities conducted to improve resource allocation, venue structure, funding channels, and program design [4]. Some of the Universities also offering various courses such as BS in Data Science – Sports Analytics Concentration at The University of Tennessee Knoxville, B. S. in Sports Analytics at Syracuse University, and Major Sports Analytics Degree at Rice University in the Department of Sports Management. However, existing courses are mostly at undergraduate level. DTU planned it to first implement this for post-graduate students. Educational activity also includes external cooperation and exchange, the specific content of the activity programs, the teaching strategies and methods during the activities and the response to the challenges of the epidemic. The output of such sport tournaments includes the number of audience and population distribution, evaluation and feedback of the activities, ways and means of promoting the activities and difficulties and challenges faced.

Sports are used as tools and enablers for a specified set of students. In relation to the sports business in higher education, it is an underdeveloped research area, particularly in the UK and many other countries. Thus, authors have conducted study to explore sport business in depth [5]. A web-based tool is developed to conduct test for sport measurement physical training of FKIP Unsri physical education students. The tool utilized the research methodology that is used to produce such products in the form of applications for measuring physical tests for various sports. The results provided by the tools help coaches, athletes, and students to be able to measure the process and results of training with validation [6].

Sport surface and infrastructure play a major role in athlete training and its performance at various levels of tournaments. Sports surfaces and infrastructure refer to the facilities and resources need for sports and recreational activities. It includes every basic component required for athlete training. It is always advisable to provide and develop will-equipped infrastructure for fostering athlete talent and promoting public health in view of today's lifestyle. Further, authors have conducted study to consider various environmental factors that affect athlete performance and training such as climatic conditions, ground surface, effective infrastructure, and maintained tools [8].

Delhi Technological University (DTU) plans to offer Minor Specialization in Sports Analytics for M.Tech. Software Engineering (with an intake of 25 students total) and Data Science (with an intake of 30 students total) from August 2025 session to remove gaps in sports industry and technology. In order to improve the gaps in sports field, tournaments, sports technology this course will be helpful. The course is mainly focused on integrating practical skills. Thus, a capstone project is offered as an elective subject. With the help of capstone project, students can collaborate with various companies working on athlete performance, coaches, academies, visualization and sports surfaces such as KheloMore, and FanCode. Further, this course will provide internship opportunities to the students with support from the DTU in various sports tech companies. The sports which are working in analytics and biomechanics will also collaborate with DTU to offer internships to such students. The detailed outcome of this course is expected to be available by Dec 2026 (mid of placement season at DTU) for longitudinal evaluation.

The organization of this paper is as follows: Sect. 2 discussed curriculum development, Sect. 3 presented and summarized structure of the curriculum, Sect. 4 discussed curriculum courses, Sect. 5 presented various modes of collaboration with Industry, Sect. 6 summarized the study in conclusion.

2 Curriculum Development

The curriculum of minor specialization in sports analytics was developed in discussion with various reputed industry experts, sports experts, and academicians. The discussion mainly focused on improving the upcoming athlete performance, sport business, sport measurement, and sport infrastructure and development. In recent years, the performance of athletes is not outstanding at various levels of tournaments. It is not only focused on athletes but also emphasized to integrate sports into students daily life. Further, the integration of wearable devices for predicting athlete health is also considered, helps in providing required training programs and injury prevention. Students will learn the techniques and methodology required for organizing various tournaments, and events at national and international levels.

3 Structure of the Curriculum

The curriculum of Sports in Minor Specialization consists of a total of seven courses. In order to be awarded a minor specialization degree, a candidate must study a minimum of 4 subjects from the basket designed for this specialization. The basket consists of total 2 core courses in the Ist and IInd semester (Department Elective – 1 and Department Elective – II), and two elective courses in the IInd semester (Department Elective – II and Skill Enhancement Course – II). The scheme of minor specialization in sports analytics is presented in Table 1. In Table 1, Cr refers to credit of the subject, L: T: P refers to contact hours required for Lecture: Tutorial: Practical. However, if a candidate opts for a minor specialization in sports analytics, then the candidate is required to pursue a seminar in the first semester consisting of 2 credits. The seminar will focus on an overview of one sport from various categories, such as water, athletics, combat, racquet, and team, applications of data analysis, AI techniques required in that sport, and focus on analytical techniques. The course scheme consists of a total of 7 subjects (2 core and 5 electives). Two core courses included in Group A and Group B. Candidate must select electives from Group C and Group D. Each subject consists of 4 credits (Figs. 1 and 2).

Fig. 1. Distribution of core and elective subjects

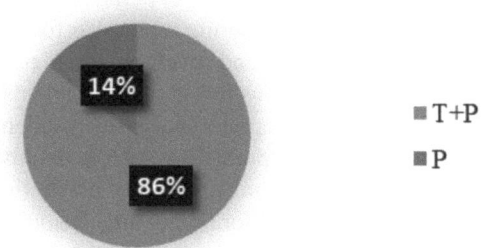

Fig. 2. Distribution of courses based on teaching hours

Table 1. Scheme of Minor Specialization in Sports Analytics

Group	S. No.	Subject Name	Type/Area	Cr	L	T	P
Group A	1	Artificial Intelligence for Sports	Department Elective – 1 as Core Subject	4	3	0	2
Group B	2	Data Analytics in Strength and Conditioning	Department Elective – 2 as Core Subject	4	2	0	4
Group C	3	Sports Performance Analytics	Department Elective – 3 as Elective Subject	4	2	0	4
	4	AI for Sports Surfaces and Equipment		4	3	0	2
Group D	5	Sports Business Analytics	Skill Enhancement Course – 2 as Department Elective	4	3	0	2
	6	Measurement and Assessment for Sports		4	2	0	4
	7	Capstone Project		4	0	0	8

4 Curriculum Courses

The brief detail of all the seven courses included in Minor Specialization basket is given below.

4.1 Artificial Intelligence for Sports

It includes foundations of AI in Sports, AI for recovery monitoring, tactical AI in sports, AI in fan engagement and ethics, AI trends and innovations.

4.2 Data Analytics in Strength and Conditioning

It includes introduction to strength and conditioning, movement patterns and anatomy, conceptual understanding of motor qualities, analytics in strength and conditioning, methods to analyze motor components and recent research.

4.3 Sports Performance Analytics

It includes data analytics in sports, performance metrics and training load, statistical models in sports analytics, predictive analytics, real-world data applications for coaches and athletes.

4.4 AI for Sports Surfaces and Equipment

It includes introduction to AI for sports surfaces, types of sports surfaces, future of sports surfaces, wearable technology – I including internal measures, wearable technology – II including devices, track players movement through sensors, impact of surfaces, assessing training techniques, environmental conditions.

4.5 Sports Business Analytics

It includes data architectures in sports analytics, data engineering and visualization, fan and social media analytics, revenue optimization and financial modeling in sports, deep learning for sports video analysis, AI models for injury prediction and recovery monitoring, building AI-driven virtual coaching systems, ethical, legal, and strategic aspects of sports analytics.

4.6 Measurement and Assessment for Sports

It includes basic understanding of tests measurement and assessment in sports, scientific authenticity, physical fitness, sports skills, various tests to monitor psychological assessment.

4.7 Capstone Project

The capstone project is a comprehensive endeavor. This course helps students to apply advanced analytical, computational, and domain-specific knowledge to real-world sports challenges in sports domain. It involves data collection, data preprocessing, developing and validating models, and implementing solutions such as dashboards or applications. The project emphasizes practical skills, ethical considerations, and professional communication, culminating in a detailed report and presentation. This course includes hands-on experience of theoretical concepts with tools like Python, Tableau, and machine learning frameworks to address pressing issues in sports performance analysis, injury prevention, fan engagement or sports event management. It helps students to apply statistics or business analytics skills and demonstrate mastery of the fundamental methods of data analysis. In the course, they will learn to apply their skills, receive feedback and learn new techniques.

5 Potential Modes of Industry Collaboration

This course will help students in solving real-world problems in the sports sector by collaborating with various industries through different modes.

5.1 Internship and Projects

Sports companies can work in partnerships with DTU to offer internship and projects to the students such as Nike, IPL franchise, Puma, ISL teams, Sporty Solutionz. Students can apply learning of this course such as machine learning, artificial intelligence, data analytics to improvise sportsperson performance, and surfaces through real performance datasets, injury datasets, and prediction dataset.

5.2 Guest Lectures

The reputed experts from sports federation and tech companies can be invited for guest lectures, workshops, short-term training programs, and laboratory sessions at DTU for sports students.

5.3 Sponsored Research

The collaboration can be proposed between DTU and industry for conducting sponsored research through funding.

5.4 Hackathons

DTU will collaborate with reputable sports companies to sponsor and mentor analytical challenges at DTU through hackathons. It will help in identifying talent, problem solving skills, and promote innovation for injury risk prediction, injury prevention techniques, and match outcome prediction.

6 Conclusion

The Minor Specialization in Sports Analytics at DTU is an initiative that is helpful for future sports, and it will also help post-graduate students to address real-world challenges in sports. This course focused on the professional to business perspective in the sports field. Further, the outcome of this course is producing more sports entrepreneurs and data engineers in the sports field. The courses include data privacy and ethical AI design, and legal considerations with respect to privacy of athlete health data. This curriculum is a strong model to develop such courses in other education institutions. It will help in academic-industry collaboration through projects, internships, sponsorship, and guest lectures, and hackathons. It will provide more job opportunities in the sports sector for coaches, athletes, executives, and professionals. It is well known that the sports industry is continuously growing, by integrating and fostering innovation, and performance excellence.

References

1. Kazan Action Plan - UNESCO Digital Library
2. Tan, X., Abbas, J., Al-Sulaiti, K., Pilar, L., Shah, S.A.R.: The role of digital management and smart technologies for sports education in a dynamic environment: employment, green growth, and tourism. J. Urban Technol. **32**(1), 133–164 (2025)
3. Steiner, E., Pittman, M., Boatwright, B.: When sports fans buy: contextualizing social media engagement behavior to predict purchase intention. Int. J. Sport Commun. **16**(2), 136–146 (2023)
4. Wang, L.F., Yim, B., Song, D. and Zhang, Y.: Promoting the path of educational activities in the international table tennis federation museum. Int. J. Sports Mark. Sponsorship (2025)
5. Pielichaty, H., Zhu, X., Sterling-Morris, R.E.: Belonging' within White male-dominated sports business management programmes. Sport Educ. Soc. **30**(3), 383–395 (2025)
6. Nanda, F.A., Hartati, H., Syamsuramel, S.: Development of web-based media for sports physical tests. In: International Seminar of Sport and Exercise Science (ISSES 2024), pp. 4–13 (2025)
7. Zamanpour, E.: Design and examination of psychometric features of sports intelligence measurement tool. J. Psychol. Sci. (2025)
8. Mallen, C., Dingle, G., McRoberts, S.: Climate impacts in sport: extreme heat as a climate hazard and adaptation options. Managing Sport Leisure **30**(2), 207–224 (2025)
9. Mase, L.Z., Gustina, D., Zahara, A., Supriani, F., Chaiyaput, S., Syahbana, A.J.: The joint method of ground response and structural dynamic analyses for building inspection under a large megathrust earthquake. Transp. Infrastruct. Geotechnol. **12**(1), 1–32 (2025)

Examining the Impact of Traffic on Shot Attempts in Ice Hockey

Miles Pitassi[✉], Evan Iaboni[✉], Fauzan Lodhi[✉], and Tim Brecht[✉]

Cheriton School of Computer Science, University of Waterloo, Waterloo, Canada
{mpitassi,e2iaboni,flodhi,brecht}@uwaterloo.ca

Abstract. In ice hockey, traffic in front of the net during a shot attempt (skaters in or near the area between the puck and the posts) may impact the shot outcome. In some cases, traffic may impede the goaltender's ability to see the puck, possibly increasing the probability of a goal. In other cases, traffic may prevent the puck from even reaching the goal. In this paper, we use puck and player tracking data from the National Hockey League to determine the number of skaters creating traffic and examine the relationship between traffic and shot outcomes.

1 Introduction

Shot quality in ice hockey is influenced by many factors including distance, angle, and preceding events [18,20,22]. One often-discussed but less precisely quantified factor is *traffic*, which we define as the presence of skaters in or near the *shooting lane*: a triangular area between the puck and the posts ("near" is defined precisely in Sect. 4). Traffic is widely believed to be a key factor in scoring chances but its effects are nuanced and to this point have not been studied in detail. In this paper, using puck and player tracking (PPT) data from the National Hockey League (NHL), we examine the impact of traffic on shot outcomes.

In our initial exploration of traffic and its relationship to shot outcomes for this paper, we observed that traffic levels were strongly correlated with shot location. Specifically, when grouping shot attempts by traffic level (N) and computing average distance and angle within each group, we found a strong positive correlation between N and shot distance ($r = 0.94$) and a moderate negative correlation between N and shot angle ($r = -0.64$). This indicates that shot attempts with more traffic tend to be taken from farther away and towards the center of the ice. Because traffic is correlated with shot location, any attempt to measure the impact of traffic must first control for the confounding effects of shot location. To do this, we group similar shot attempts together by dividing the offensive zone into regions based on distance and angle, as shown in Fig. 1 (see Sect. 5 for details on how these regions are defined). We then examine how traffic impacts shot outcomes within each region.

Each region is labelled with their distance range in feet and their orientation: "c" represents center-angle shot attempts, "o" represents off-center shot attempts (which are only included in the larger distance regions), and "w" represents wide-angle shot attempts. Figure 2 shows the relationship between traffic and goals for each region, illustrating the number of goals (y-axis) by traffic level within each of ten distance-angle regions (x-axis), ordered by distance and then angle. This figure highlights how analyzing traffic can reveal new insights into shot attempts. For example, in all close-range regions (within 23 ft), the number of goals scored with no traffic exceeds the combined total of goals scored with traffic. While this figure doesn't show how many shot attempts are in each traffic-region group (we explore that in Sect. 6), it provides a preview of the types of patterns we examine.

One reason for the high number of goals from close range may be the limits of human reaction time. Prior research suggests that the fastest male humans have an average reaction time of 0.22 s [4], which means a 70 mph shot attempt would need to be at least 23 ft away for the goaltender to react in time. Across all regions, the highest number of goals occurs in the 9–23c region with zero traffic. For long-range shot attempts (23+ ft), the relationship becomes more nuanced. Section 6 explores this further, including the impact of traffic based on whether it is dominated by attacking or defending skaters.

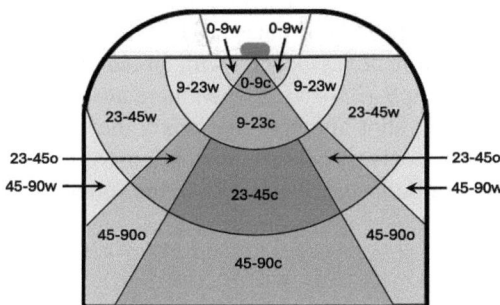

Fig. 1. Shot location regions used to control for distance and angle in traffic analysis.

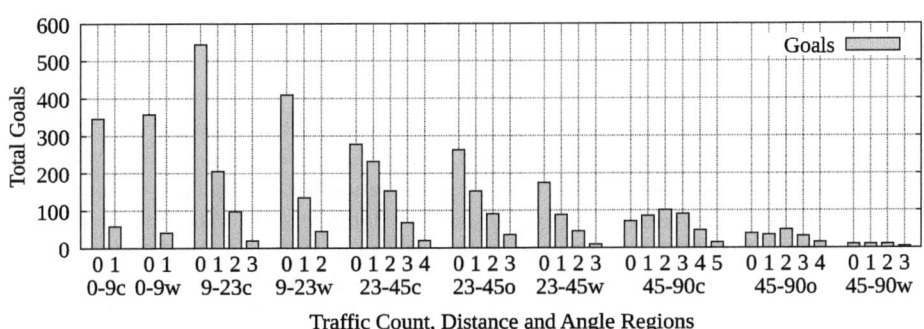

Fig. 2. Goals scored across traffic levels and location regions.

Before we can analyze the impact of traffic, we must first reliably detect and augment each shot attempt, including determining its start and end time and using a single location per player to identify who is creating traffic. Based on this work, we offer two methodological contributions for analysts working with PPT data:

- We develop a process to combine shot attempts with precise timestamps by matching official NHL shot data to PPT data. This process uses an inference algorithm to recover shot attempts not originally identified in the PPT data, and augmenting for tips, deflections and timing inaccuracies.
- We create an algorithm that divides the offensive zone into regions that minimize within-region variation in distance and angle, allowing us to study the impact of traffic independently from distance and angle.

For NHL front office staff, coaches, players, and fans, our analysis offers several empirical findings that provide insight into how traffic affects shot outcomes:

- We find that for all regions, increased levels of traffic significantly increases the percentage of shot attempts that are blocked and reduces the chance of a shot attempt resulting in a shot on goal or a goal.
- For long-range shot attempts (45–90 ft), 38% of shot attempts are blocked, compared to 29% for all shot attempts in our dataset. However, among long-range shots on goal, higher levels of traffic are associated with an increased likelihood of a goal.
- For mid-range shot attempts (23–45 ft), when there are more defenders than attackers in the shooting lane, shot attempts that reach the goaltender are significantly more likely to result in goals, suggesting that defensive traffic may unintentionally screen the goaltender or deflect shot attempts in the shooter's favor.

2 Related Work

Traffic Research in Football (Soccer). Traffic-related research in football provides relevant context for this paper. For example, a 2015 study using tracking data from a professional football league (via Prozone, now Stats Perform [21]), found that defender presence in the shooting lane significantly reduced goal likelihood [10]. However, its contribution to the predictive model used in the paper was limited, suggesting that its effect may be confounded by factors like shot location. A 2016 study computed *dangerousity*, the probability of a goal during possession, for 64 Bundesliga games (2014/15). As an input to their model, they proposed a more detailed traffic metric, *shot density*, which incorporates not only the number of players in the lane but also their proximity to the shooter and whether they are attackers or defenders [9]. While the authors did not isolate the effect of shot density in their model of dangerousity, the study provides a valuable framework for analyzing player presence and placement in traffic. However, we currently do not focus on specific placement within the shooting lane as we recognize that different positions in the shooting lane in ice hockey have varying impacts such as being close to the shooter potentially increasing the chance of blocking the shot attempt, while being near the goaltender may enhance the likelihood of obstructing the goaltender's view. These studies highlight how to account for the number and type of players in traffic. However, ice hockey poses unique challenges not present in football, such as faster puck and player movement, the difficulty of tracking and blocking a small, fast-moving puck, a higher degree of physical contact, and a smaller, enclosed playing surface. These factors underscore the necessity for tailored approaches to fully understand how traffic influences shot outcomes in ice hockey.

Related work in football has also examined how defender positioning can unintentionally impair goalkeeping performance. Using virtual reality to simulate free-kick scenarios, López-Valenciano et al. [11] found that defensive walls, though intended to reduce scoring, can actually hinder a goalkeeper's reaction by occluding their view of the ball. This raises an interesting parallel for ice hockey, where skaters in the shooting lane may likewise reduce goaltender effectiveness by unintentionally obstructing the goaltender's view, a possibility we explore later in our analysis.

Traffic-Related Research in Ice Hockey. A common proxy for traffic in ice hockey is the use of shot types typically associated with traffic such as *tips* and *deflections* [1]. Tips occur when a puck traveling towards the net is redirected via the stick with the goal of changing the puck's direction while not adding momentum to the puck. Deflections occur when a puck traveling towards the net is redirected via the body or skates. While these shot types can signal the presence of traffic, they offer only an indirect measure of its broader impact. One study analyzing NHL power plays during the 2015–16 season took a more expansive approach by manually identifying *screens* through video review, defining them as shot attempts where the goaltender's view was obstructed [14]. The study found that screened shot attempts made up 24.9% of total shot attempts and 21.3% of goals. Although this manual approach enables detailed insights, it lacks scalability and is difficult to evaluate as no comparable data points exist in league-wide datasets. More recently, the NHL introduced an "Opportunity Analysis" model designed to evaluate the quality and context of each shot attempt [2]. This model incorporates variables such as player positioning, shot type, and game situation, as well as counts of attacking and defending skaters within fixed distances of the shot cone (the area between the puck and the net), as well as a flag for "Possible Goalie Vision Block". However, the model's outputs are not available on a per-shot basis and the impact of these traffic-related variables are not reported, making it difficult to assess their influence on shot outcomes.

3 Determining Shot Timing and Duration

In this section, we describe our process for identifying and augmenting shot attempts, and then determining a shot start and end time in which to determine traffic.

3.1 Identifying Shot Attempts

Each official shot attempt is available from the NHL via their Application Programming Interface (API) [12], but this data is only recorded in whole second granularity using scoreboard time. In contrast, PPT data provides locations every hundredth of a second. Since the puck can travel about 103 ft per second (for a 70 mph shot), the coarse resolution of scoreboard time is insufficient for aligning tracking data with shot attempts. As a result, we construct our shot dataset using a combination of *detected shot attempts* based on the NHL's physics-based shot event model applied to PPT data, and *inferred shot attempts* which we derive from puck touch data, which represents the NHL's effort to determine every instance of a player making contact with the puck.

Detected Shot Attempts. The NHL detects shot attempts using the PPT data and an automated, physics-based event classification algorithm that identifies when a player directs the puck toward the net. Detection is extremely difficult and consequently, errors may occur such as passes being misclassified as shot attempts. To improve accuracy, the NHL incorporates a manual shot reviewing process where human reviewers verify and correct PPT shot event data. Unfortunately, even after manual review, discrepancies remain where an official shot attempt is not recognized by the shot detection system or where attributes such as the shooter or shot location differ between sources. To address this, we compare the PPT data with the official NHL API data and infer the timing of undetected shot attempts using puck touch data, as described in the following section.

Inferred Shot Attempts. We identify the timestamps (or release times) of undetected shot attempts in the PPT data by comparing NHL API data and puck touch data. From the NHL API data, we can obtain the scoreboard time, shooter, and shot outcome. To match it with a puck touch, we implemented a matching algorithm, adapted from a method originally developed by the NHL's Research and Development team [13]. This algorithm attempts to match official shot attempts and puck touches by considering the touch's timing, location, and the puck's incoming and outgoing direction and velocity.

3.2 Augmenting Shot Attempts

Potential Start Time Inaccuracies. After these steps, we obtain a dataset of shot attempts with estimated release timestamps. Some timestamps originate from the NHL's physics-based shot detection model, while others are derived from the end of a puck touch. In an attempt to verify and possibly improve shot release times, we use techniques from previous work on pass and shot timing corrections [16]. Specifically, we verify that the puck is within four feet of the shooter's location at the moment of release. If not, we adjust the timestamp, which is required for 20.8% of all shot attempts.

Handling Compound Shot Attempts. Tips and deflections occur when a player redirects an incoming puck with their stick, skates, or body. We refer to both the initial play action and the tip or deflection as a *compound shot attempt*. The player who tipped or deflected the puck is credited as the shooter, so the initial play action is not officially recorded as a shot attempt. However, we are interested in the initial play action as the tip or deflection is usually the result of traffic, and the initial play action reflects the traffic leading to the tip or deflection. Thus, we replace tipped and deflected shot attempts with their corresponding initial play actions. First, we identify shot attempts with type "tip" or "deflection" in the NHL API data. For each of these shot attempts, we trace the event back to the initial play action by finding the last recorded puck touch by a teammate of the player who tipped or deflected the shot attempt. We then calculate traffic based on the initial play action while preserving the final shot outcome (e.g., goal, save, block, or miss).

3.3 Results of the Shot Identification and Augmentation Process

We analyze data from 891 NHL games played during the 2024–25 season, up to February 9, 2024 (the 4 Nations Face-Off break). We define a shot attempt as *matched* if it is either detected by the NHL's shot classification algorithm or inferred from puck touch data. After excluding games where fewer than 50% of official shot attempts were matched, 870 games remain in our dataset. Across these games, there were 103,948 official shot attempts, with a total of 80,473 detected shot attempts, 12,887 inferred shot attempts, and 10,588 shot attempts for which we were unable to determine a timestamp. The remaining 870 games in our dataset had at least 75% of the official shot attempts matched, and more than half of the games have 90% or more matched. We examined the unmatched shot attempts, but missing data (e.g., location) limited analysis. Table 1 summarizes the shot-matching results.

Table 1. Summary of shot identification and matching process for 870 games in dataset.

Matching Stage	Shot Count	Percentage
Total official shot attempts in dataset, after removal of some games	103,948	100.0%
Matched via NHL shot event data ("detected shot attempts")	80,473	77.4%
Matched via puck touch data ("inferred shot attempts")	12,887	12.4%
Total matched shot attempts (with PPT data)	**93,360**	**89.8%**
Unmatched shot attempts (no PPT timestamp available)	10,588	10.2%

3.4 Shot Duration Considerations

We now turn to the question of whether or not a window of time should be used when determining traffic and if so, how to determine an appropriate window. Capturing traffic only at the exact moment of the shot release may miss skaters who obstruct the shot attempt later in its trajectory or skaters seen by the shooter or goaltender just prior to the shot attempt. Consider Fig. 3 which presents three screenshots from the broadcast of a shot attempt during the Utah Hockey Club versus Chicago Blackhawks game on October 8, 2024. In the top picture, Utah player #22 (Jack McBain) is positioned in front of the goaltender. As Utah player #11 (Dylan Guenther) prepares to shoot, McBain likely obstructs the goaltender's view of the puck and Guenther's shooting lane. By the time the shot attempt is released in the middle picture, McBain has moved out of the shooting lane. There appears to be little traffic, aside from the two players near the left post. In the bottom picture, Chicago player #8 Ryan Donato, whose body was not in the shooting lane at the time of the shot release, blocks the shot attempt. Although Donato's stick may appear to be in the shooting lane in the middle picture, we are unable to capture it as each player's location is tracked using a single LED embedded in their sweater, as detailed in the next section. This example, along with many others like it, highlights the need to consider measuring traffic over a window of time rather than just at the moment of release.

Given these observations, we define two time windows around each shot attempt: a *pre-release window* which captures the shot wind-up and decision-making phase and

Fig. 3. Sequence of screenshots showing traffic before, during, and after a shot attempt. The top and middle frames are 29 broadcast frames apart (≈ 0.48 s) and the middle and bottom are 14 frames apart (≈ 0.23 s). Utah vs. Chicago, October 8, 2024.

a *post-release window* approximating the shot attempt duration. Our video review of sample shot attempts suggests that 0.5 s reasonably approximates the time required to release a puck, as well as for the duration of a shot attempt. However, even a short post-release window (e.g., 0.1 s) often included post-shot events such as rebounds which are unrelated to the initial attempt. For this reason, for the analysis in this paper, we use only a 0.5-second pre-release window and omit the post-release window. This approach also best reflects the information available to the shooter at the time of the shot attempt which is ideal for shooter analysis and shot prediction models.

Selecting a suitable approach for a shot duration window is challenging as different choices carry tradeoffs that may impact the results. We explore the impact of these considerations in Sect. 7. Although we find that different choices may yield slight variations in specific results, the overall trends remain consistent.

4 Defining and Calculating Traffic

Once we have determined the time window in which to measure traffic for each shot attempt, the next challenge is identifying whether skaters are in or near the shooting lane during the shot window using their location data. Each player wears a single light-emitting infrared diode (LED) positioned in their sweater between the top of their back and right shoulder, as illustrated in Fig. 4. Because this is the only tracked point on the player, we are unable to determine the position of their limbs or stick. To account for this, we introduce a buffer around the shooting lane to define the broader *traffic lane*, which represents skaters that may be obstructing the shot attempt even if their LED is not directly within the shooting lane.

Fig. 4. LED embedded in a player's jersey.

Figure 5 illustrates a power-play goal by Ottawa player #18 (Tim Stützle) at 5:52 of the 1st period in Ottawa's game against Florida on October 10, 2024. The left image shows the game broadcast at the time of the shot, while the right image shows a frame from an animation we produced. The shooting lane is highlighted in blue, and the traffic lane is defined as the combined area of the blue triangle and adjacent green rectangles. Skaters detected within the traffic lane are shown in red. The goaltender is not shown as we ignore them in traffic calculations.

Fig. 5. Traffic lane analysis for a goal scored during the Florida Panthers vs. Ottawa Senators game on October 10, 2024. The yellow skater is the shooter, red skaters are in the traffic lane, and all blue skaters are outside of the traffic lane. (Color figure online)

After examining a sample of shot attempts, we found that placing rectangles two feet outside the shooting lane reasonably matched our visual assessments of traffic.

Typically, if a skater's LED is within two feet of the shooting lane, they are likely obstructing the shot attempt with their body. Conversely, if a skater's LED is not within this range, they are unlikely to significantly impact the shot attempt, though exceptions do occur. Similar to the shot duration window, we recognize the potential variability in results based on the chosen buffer size (green rectangle). To address the variability, we perform a sensitivity analysis in Sect. 7 where we repeat our computations with different buffer sizes. That section illustrates that while the magnitude of impact depends on the assumptions made, the direction and significance of the impact does not.

5 Dividing the Offensive Zone into Regions

As shown in our initial analysis (Sect. 1), traffic levels are strongly correlated with shot location. To accurately evaluate the impact of traffic on shot outcomes independently from shot distance and angle, we group similar shot attempts together. To do this, we start by dividing the offensive zone into ten regions based on shot distance and angle. We chose ten regions because we wanted a large enough number of regions to have a limited range of distance and angle within each region while keeping the number of regions small enough so that they could be represented graphically in a meaningful way. We analyzed the impact of the number of regions and found that while the quantitative results may change, the qualitative results still stand.

To define the regions, we implemented an algorithm that minimizes within-region variation in shot distance and angle. The goal is for shot attempts within each region to have similar distances and angles but vary in traffic. Shot attempts taken from below the goal line or outside the offensive zone are excluded from this analysis. Distance and angle variation within each region are defined below. $90°$ is used as the denominator for angle as it represents the maximum possible shot angle deviation, while the maximum distance within each region normalizes variation relative to the range within each region. AvgDistance[0..N] and AvgAngle[0..N] represent the average distance and angle for each level of traffic from 0 to N within the region.

$$\text{Distance Variation} = \frac{\max(\text{AvgDistance}[0..N]) - \min(\text{AvgDistance}[0..N])}{\max(\text{AvgDistance}[0..N])} \quad (1)$$

$$\text{Angle Variation} = \frac{\max(\text{AvgAngle}[0..N]) - \min(\text{AvgAngle}[0..N])}{90} \quad (2)$$

To minimize variation in shooting location within each region, the algorithm evaluates all combinations formed using four distance regions and either two or three angle regions per distance region for a total of 10 regions. Specifically, the two closest distance regions are divided into two angle segments, while the two farthest are divided into three, resulting in a total of 4,088,304 combinations. For each combination, we compute the variation in distance and angle within each region using the formulas above. We then sum the distance and angle variation for each region and evaluate each combination based on the highest such sum among its regions. The ideal set of regions is the one that minimizes this maximum within-region variation. This approach ensures that no single region has disproportionately high spatial variance which could otherwise confound the effects of traffic with those of shot location.

Resulting Regions. Unless otherwise specified, all numeric ranges in the rest of the paper are expressed with exclusive lower bounds and inclusive upper bounds (e.g., 9–23 refers to (9–23]). We omit brackets and parentheses for readability. The resulting regions shown in Fig. 1 separate the shot attempts into four distance ranges: 0–9, 9–23, 23–45, and 45–90 feet. The two smaller-distance ranges are further divided into two angle ranges: center (within 0-37° of the meridian line) or wide (37-90°). The two larger-distance ranges are divided into three angle ranges: center (within 0-29° of the meridian), off-center (29-45°), and wide (45-90°). Across all regions, the maximum distance variation of any region was 12% of the maximum distance in each region, and the maximum angle variation was 6% of 90°. We note that this approach assumes that the traffic lane is symmetric for the left and right sides of the net. Factors such as goaltender and player handedness may introduce asymmetries in how traffic forms. While we do not explicitly account for these possibilities, they represent an area for future exploration.

6 Results: Traffic Versus Shot Attempt Outcomes

In Fig. 6, the x-axis represents each of the ten distance-angle regions, ordered by distance (smallest to largest) and then by angle (center to wide). Within each region, shot attempts are grouped by traffic. We refer to each of these bars as a traffic-region group. The y-axis shows the percentage of all shot attempts that result in one of four outcomes: goals (Goal%), shots on goal that are saved by the goaltender (Saved%), blocked shot attempts (Blocked%), and missed shot attempts (Missed%). Unlike the traditional save percentage (SV%) metric, which is used to track goaltenders' saves as a percentage of the shots on goal they face, Saved% here denotes saves among all shot attempts. Goal% (blue) and Saved% (green) together form the shots on goal (SOG) percentage. The rightmost group, labeled "All SA", aggregates all shot attempts (SA) across regions for comparison. To ensure meaningful statistical comparisons, we exclude any traffic-region group that has less than 100 shots on goal.

Fig. 6. Shot outcome distributions across traffic levels and location regions.

To assess whether the patterns observed in our results are statistically significant, we apply two types of tests. When analyzing ordered traffic levels (e.g., 0, 1, 2, 3 players) against a binary outcome such as Goal vs. No Goal, we apply the Cochran-Armitage trend test [3,5]. This test evaluates linear trends across the ordered groups. For comparisons between two traffic groups (e.g., attacking versus defending traffic), we use the chi-squared test [15], which compares two independent proportions. Throughout our analysis, we report only results that are statistically significant ($p < 0.05$) or moderately significant ($p < 0.1$). In tables, statistically significant results are marked with an asterisk (*) and moderately significant results with a dagger (†). Given the number of statistical tests performed, false discoveries are a concern [17]. However, the consistency of results across regions and outcomes lends support to the broader patterns we observe.

Traffic Versus the Percentage of Shot Attempts that Miss the Net (Missed%). To our surprise, the percentage of shot attempts that miss the net (Missed%) remains fairly constant across regions and traffic levels, showing that players' frequency of missing the net is not strongly affected by location or traffic. The overall Missed% is 22.4%, varying by only 1.5% across traffic levels and 2.5% across locations (values are the coefficients of variation). The only traffic-region groups where Missed% deviates noticeably are zero-traffic shot attempts in the 9–23w and 23–45w regions. We believe that this may be due to these being common one-timer locations where players make quick shot attempts from sharp angles, slightly increasing the possibility of missing the net.

Traffic Versus the Percentage of Shot Attempts that are Blocked (Blocked%). The percentage of shot attempts that are blocked (Blocked%) increases with traffic, particularly for mid-to-long range shot attempts ($p < 0.05$ for all traffic-region groups). Notably, 29% of all shot attempts in our dataset are blocked and 91% of those blocked shot attempts are by players on the defending team. 29% is a remarkably high proportion, especially given the modern trend toward fewer total shot attempts as teams prioritize puck possession and higher-quality scoring chances [6]. This finding has important implications for shot prediction or expected goal (xG) models, which are now widely used in ice hockey analytics [8,18–20,22]. Most public xG models either exclude blocked shot attempts entirely or include them in limited ways because the NHL does not release shot location data for blocked attempts. As a result, *nearly one-third of all shot attempts are often ignored in these models*, which may limit their completeness and accuracy. In future work it would be interesting to study whether including blocked shot attempts in xG models would increase their accuracy.

Interestingly, a notable number of shot attempts with zero recorded traffic are blocked. This might seem counterintuitive but can be explained by the limitations of traffic detection. Many shot attempts are blocked by sticks rather than bodies, or by players stepping into the traffic lane during the shot attempt's flight. Capturing all such instances as traffic would require larger buffer and window sizes which could result in including players who do not meaningfully affect the shot attempt. We investigate the impact of different choices for buffer size and time window duration in Sect. 7.

Traffic Versus the Percentage of Shot Attempts that Result in a Shot on Goal (SOG%) and Goal (Goal%). As shown in Fig. 6, the percentage of shot attempts that are on goal (SOG%) declines as traffic increases ($p < 0.05$ for all traffic-region groups), indicating that traffic reliably reduces the likelihood of a shot attempt reaching the goaltender. Notably, the percentage of shot attempts that result in a goal (Goal%) is highest for shot attempts from the 0–9c and 9–23c regions when there is no traffic. For NHL front office staff, coaches and players, this highlights the value of prioritizing shot attempts in these traffic-region groups. In general, while Goal% also tends to decrease as traffic increases, this trend is only statistically significant in regions 0–9c, 0–9w, and 9–23c. Because traffic makes it hard to get a shot on goal, very few goals are scored in high-traffic situations, making it difficult to assess how traffic affects scoring itself. To better understand this, we focus next on Goal% of SOG, which captures whether traffic helps or hurts once a shot attempt reaches the goalie, more directly reflecting the possible effects of tips, deflections, and screens. Note that Goal% of SOG corresponds to what the NHL has historically referred to as "shooting percentage", though we use the term Goal% of SOG as it more clearly describes the metric.

Traffic Versus the Percentage of Shots on Goal that Result in a Goal (Goal% of SOG). To better understand the quality of shot attempts that are on goal, we examine the proportion of shots on goal that result in a goal (Goal% of SOG). Figure 7 presents these results with 95% confidence intervals. In Table 2, we summarize the direction of the trend in Goal% of SOG as traffic increases for each region. To provide context for interpreting these values, Fig. 8 shows the total number of shot attempts in each traffic-region group, which helps explain the size of the confidence intervals.

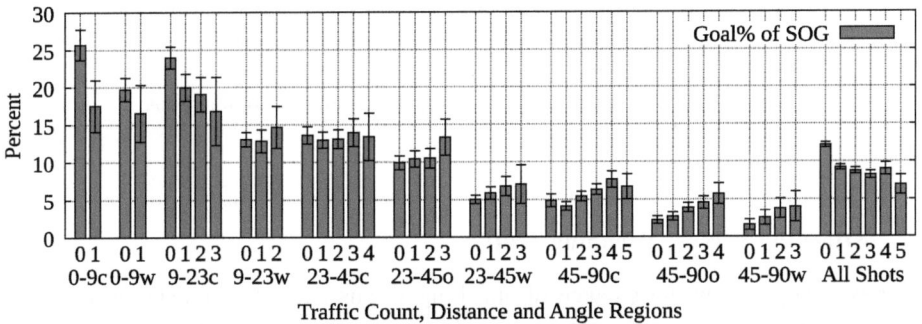

Fig. 7. Goal% of SOG across traffic levels and location regions with 95% confidence intervals.

Table 2. Trend and p-value of Goal% of SOG across traffic levels for each region. Only regions with significant ($p < 0.05$) or moderately significant ($p < 0.1$) trends are shown. * indicates statistical significance, † indicates moderate significance. ↑ indicates an increasing trend and ↓ indicates a decreasing trend.

0–9c	9–23c	45–90c	45–90o	45–90w
(↓) 0.01*	(↓) 0.04*	(↑) 0.01*	(↑) 0.01*	(↑) 0.05†

We focus on regions with significant or moderately significant trends. In the 0–9c and 9–23c regions, Goal% of SOG decreases as traffic increases, suggesting a decline in shot quality. This means that even when shot attempts from these regions reach the goaltender, they are less likely to result in goals if there is traffic. Teams may benefit from avoiding making heavily contested shot attempts in these regions. Instead they might seek ways to create or move into space to shoot with minimal traffic at a similar location. In contrast, for long-range shot attempts (45+ feet), Goal% of SOG increases with traffic. This indicates that while traffic reduces the chance of the shot attempt reaching the goaltender, it improves scoring success when it does, possibly due to tips, deflections or screens. This highlights a strategic tradeoff: for long-range shot attempts, traffic reduces total shot attempt success but increases the probability of scoring for the shot attempts that do reach the goaltender. As a result, teams may benefit from deliberately placing traffic in front of the net for long-range attempts, accepting fewer total shots on goal in exchange for more dangerous ones.

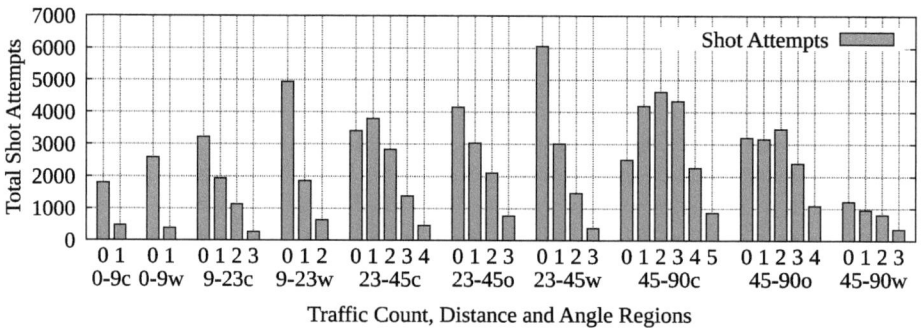

Fig. 8. Total shot attempts across traffic levels and location regions.

6.1 Attacking Versus Defending Traffic

We next analyze how the balance of attacking versus defending players affects shot outcomes. A heavy defensive presence could lead to more blocked shot attempts and fewer goals, while a strong offensive presence might create tips, deflections, screens, and more goals. However, the opposite could also be true as defensive players may screen their own goalie and offensive players might inadvertently block shot attempts. One possible approach would be to analyze each unique combination of attacking and defending players in traffic. However, this quickly results in a large number of categories (e.g., 2 attackers and 1 defender, 3 defenders and 0 attackers, etc.), many of which are rare and thus yield unreliable comparisons. Instead, in Fig. 9, we categorize shot attempts in each region into three groups: (1) more attacking than defending traffic (A), (2) more defending than attacking traffic (D), and (3) equal attacking and defending traffic (E). This figure and subsection aim to highlight differences between attacking and defending traffic rather than the overall magnitude of traffic. Consequently, shot attempts with one attacking player are treated the same as those with two or three attacking players, and shot attempts with zero traffic are excluded from this analysis.

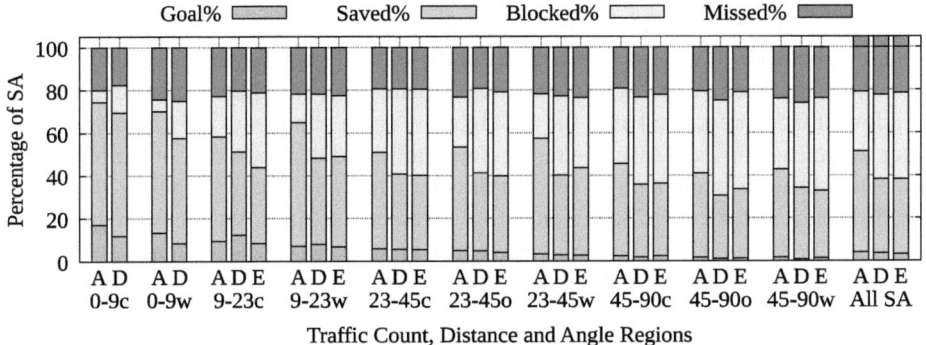

Fig. 9. Shot outcome distributions for attacking versus defending traffic.

For each region, when defending traffic exceeds attacking traffic (D), SOG% is lower and Blocked% is higher compared to when the reverse is true (A) ($p < 0.05$ for all traffic-region groups). This is expected as defenders are typically positioned to block shot attempts whereas attackers aim to tip, deflect, or screen shot attempts. However, the effect on goal scoring is less straightforward. Similar to the previous section, to investigate this, we focus on the proportion of shots on goal that result in goals (Goal% of SOG). This allows us to evaluate whether offensive or defensive traffic is more beneficial (or harmful) to scoring in each region. Results are shown in Table 3.

In all mid-range regions (9–45 ft), shot attempts with more defending traffic (D) have a higher Goal% of SOG (shooting percentage) than those with more attacking traffic (A) ($p < 0.1$ for all four traffic-region groups). This may be because defenders unintentionally screen their goalie or redirect the puck in unpredictable ways. For coaches, this highlights the risk in collapsing defenders to attempt to block mid range shot attempts as defensive traffic may sometimes impair the goalitender more than it helps. The optimal defensive approach may vary by shot location: limiting mid-range screens while emphasizing blocks on both short-range and long-range attempts.

Table 3. Effect of traffic balance on Goal% of SOG for each region. Each cell reports the p-value from a chi-squared test comparing shot attempts with more defending traffic (D) to those with more attacking traffic (A), testing whether D has a significantly higher Goal% of SOG. Only regions with statistically ($p < 0.05$) or moderately significant ($p < 0.1$) differences are shown. * denotes statistical significance, † denotes moderate significance.

9–23c	9–23w	23–45c	23–45o	23–45w
0.00*	0.01*	0.06†	0.07†	0.09†

7 Sensitivity

To examine the impact that our choice of buffer size and pre-release and post-release windows have on the results in this paper, we repeat the analysis using alternate buffer sizes and shot attempt duration windows. For buffer size, we compared our 2-foot default to both a 0-foot and 4-foot buffer. For the shot attempt duration window, we evaluate three alternative approaches: specifically using only the shot release timestamp, a window of 0.5 s before and after the shot release, and a window of 0.5 s extending only after the release. In each case, we recalculate traffic and compare shot outcomes across the same distance-angle regions used in the main analysis. Figure 10 shows the resulting distributions for the 9–23c region across parameter settings. We focus on this region because it is the source of the most goals (18% of all goals), making trends across configurations easier to observe. These alternative configurations sometimes show slightly steeper declines in shot success, but the overall trends remained consistent, helping to demonstrate the robustness of our findings with respect to the choice of buffer size and shot window.

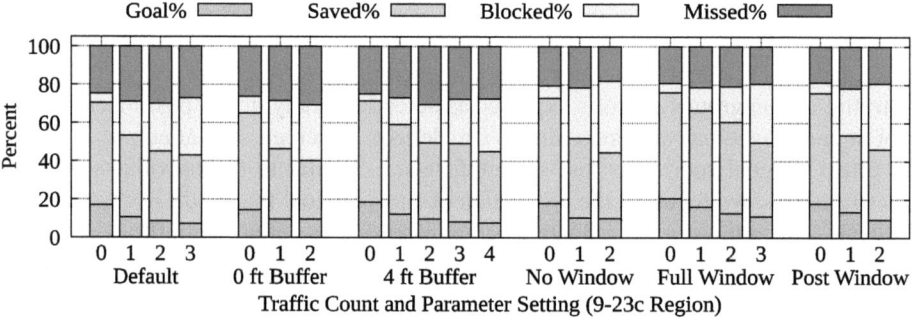

Fig. 10. Shot outcome distributions across traffic levels and parameter settings (9–23c region).

8 Discussion

This paper presents a new method for detecting and analyzing traffic in ice hockey. By leveraging PPT data including puck touch events to determine shot timing and duration, and by decoupling traffic from shot location, we introduce several techniques useful to analysts studying traffic. In analyzing the impact of traffic on shot outcomes, we also offer insights that may be valuable to NHL front office staff, coaches, players, and fans. While our techniques offer a new way to quantify traffic, they do have some limitations. First, the tracking data does not contain explicit shot start and end times for all shot attempts. The data is not official and detecting the point of release is difficult and may not be 100% accurate (despite our improvements). This inaccuracy may impact our findings. As well, to approximate shot duration, we construct a timing window based on empirical estimates. While our sensitivity analysis suggests the findings are robust to reasonable variations, having true shot start and end markers would improve precision. Second, our definition of traffic considers only the number of skaters in the traffic lane without considering player orientation, posture, or stick position. This is a necessary simplification due to the use of a single LED per player. Additionally, our analysis does not account for a skater's specific position within the shooting lane as being close to the shooter potentially increases the chance of blocking the shot attempt, but being near the goaltender may enhance the likelihood of obsructing the goaltender's view. To address these simplifications, in subsequent research, we propose two new metrics. The first is net visibility, and is defined as the fraction of the net that can be seen from the perspective of the puck. The seconds is net reachability and is defined as the fraction of the net that could be reached by the puck. These metrics are computed using a combination of PPT data and video analysis (image processing) [7]. This approach captures full player body positioning and uses a rasterization technique to account for players within the shooting lane [7].

Despite these limitations, our work offers a foundation for future research on the impact of traffic in ice hockey. To our knowledge, this is the first study in ice hockey to systematically quantify the impact of traffic using puck and player tracking data. We hope that these contributions not only advance academic understanding but also offer practical techniques for analysts and insights for NHL front office staff, coaches, players, and fans seeking to gain a competitive edge.

References

1. All About the Jersey: Tips, Deflections, and Other Not-So-Common Shots (2024). https://www.allaboutthejersey.com/2024/12/13/24320430/tips-deflections-other-not-so-common-shots-new-jersey-devils-data-scoring-shot-counting-noesen-haula
2. Amazon Web Services: NHL and AWS Unveil Opportunity Analysis: A New Hockey Statistic (2023). https://aws.amazon.com/sports/nhl-opportunity-analysis/
3. Armitage, P.: Tests for linear trends in proportions and frequencies. Biometrics **11**(3), 375–386 (1955). https://doi.org/10.2307/3001775
4. Bellis, C.J.: Reaction time and chronological age. Proc. Soc. Exp. Biol. Med. **30**, 801–803 (1933). https://doi.org/10.3181/00379727-30-6682
5. Cochran, W.: Some methods for strengthening the common χ^2 tests. Biometrics **10**(4), 417–451 (1954). https://doi.org/10.2307/3001616
6. Hockey's Arsenal: Goodbye to the Mid-Range Shot: Why NHL Teams are Shooting Differently (2023). https://hockeysarsenal.substack.com/p/goodbye-to-the-mid-range-shot. Posted on *Hockey's Arsenal*
7. Iaboni, E., et al.: New views of shots: towards measures of net visibility and reachability. In: Proceedings of the Linköping Hockey Analytics Conference, pp. 43–55 (2025)
8. Krzywicki, K.: Shot Quality Model: A Logistic Regression Approach to Assessing NHL Shots on Goal (2005). https://hockeyanalytics.com/2005/01/shot-quality/. Posted on *hockey::analytics()*
9. Link, D., Lang, S., Seidenschwarz, P.: Real-time quantification of dangerousity in football using spatiotemporal tracking data. PLOS ONE **11**(12) (2016)
10. Lucey, P., Bialkowski, A., Monfort, M., Carr, P., Matthews, I.: Quality vs quantity: improved shot prediction in soccer using strategic features from spatiotemporal data. In: MIT Sloan Sports Analytics Conference (2015)

11. López-Valenciano, A., McRobert, A., Sarmento, H., Reina, M., Page, R., Read, P.: A goalkeeper's performance in stopping free kicks reduces when the defensive wall blocks their initial view of the ball. Eur. J. Sport Sci. (2021)
12. National Hockey League: NHL API (2025). https://api-web.nhle.com/v1/
13. NHL Research and Development: Fuzzy Matching Algorithm for Shot Detection (2025)
14. Parnass, A.: How Can We Quantify Power Play Performance In Formation? (2016). https://hockey-graphs.com/2016/04/25/how-can-we-quantify-power-play-performance-in-formation, Posted on *Hockey-Graphs*
15. Pearson, K.: On the criterion that a given system of deviations from the probable in the case of a correlated system of variables is such that it can be reasonably supposed to have arisen from random sampling. Philos. Mag. Ser. 5 **50**(302), 157–175 (1900). https://doi.org/10.1080/14786440009463897
16. Radke, D., et al.: Analyzing passing metrics in ice hockey using puck and player tracking data. In: Proceedings of the Linköping Hockey Analytics Conference, pp. 25–39 (2023)
17. Ranganathan, P., Pramesh, C.S., Buyse, M.: Common pitfalls in statistical analysis: the perils of multiple testing. Perspect. Clin. Res. **7**(2), 106–107 (2016)
18. Shomer, H.: Analyzing the Game of Hockey: Evaluating my Shooter xG model (2018). https://fooledbygrittiness.blogspot.com/2018/03/evaluating-my-shooter-xg-model.html. Posted on *Fooled by Grittiness*
19. Skytte, L.: Hockey-Statistics: Building an xG Model – v. 1.0 (2022). https://hockey-statistics.com/2022/08/14/building-an-xg-model-v-1-0. Posted on *Hockey-Statistics*
20. Sprigings, D (also known as DTMAboutHeart): Expected Goals are a Better Predictor of Future Scoring than Corsi and Goals (2015). https://hockey-graphs.com/2015/10/01/expected-goals-are-a-better-predictor-of-future-scoring-than-corsi-goals. Posted on *Hockey-Graphs*
21. Stats Perform: STATS Acquires Prozone (2015). https://www.statsperform.com/press/stats-acquires-prozone/. Press release. Accessed 4 Mar 2025
22. Younggren, J., Younggren, L.: A new expected goals model for predicting goals in the NHL (2021). https://evolving-hockey.com/blog/a-new-expected-goals-model-for-predicting-goals-in-the-nhl. Posted on *Evolving-Hockey*

Predicting Penalty Kick Direction Using Multi-modal Deep Learning with Pose-Guided Attention

Pasindu Ranasinghe[1](✉) and Pamudu Ranasinghe[2,3]

[1] University of New South Wales, Sydney, Australia
pasindu.ranasinghe@unsw.edu.au
[2] Virtusa Pvt. Ltd., Colombo, Sri Lanka
[3] University of Moratuwa, Moratuwa, Sri Lanka

Abstract. Penalty kicks often decide championships, yet goalkeepers are left to anticipate the kicker's intent from subtle biomechanical cues unfolding within a narrow time window. This study presents a real-time, multi-modal deep learning framework for predicting the direction of a penalty kick—left, middle, or right—prior to ball contact. The model adopts a dual-branch architecture: MobileNetV2-based CNN extracts spatial features from RGB frames, while 2D keypoints are processed using an LSTM network with attention mechanisms. Pose-derived keypoints are further used to guide visual focus toward task-relevant regions. A distance-based thresholding method segments input sequences immediately before ball contact, providing consistent input across diverse footage. A custom dataset of 755 penalty kick events was curated from actual match videos, with frame-level annotations for object detection, penalty shooter keypoints, and the final ball placement in the goal. The model achieved 89% accuracy on a held-out test set, outperforming visual-only and pose-only baselines by 14–22%. With an inference time of 22 ms, the lightweight and interpretable design makes the model well-suited for goalkeeper training, tactical analysis, and real-time game analytics.

Keywords: Penalty kick prediction · sports analytics · multi-modal learning · computer vision · action recognition

1 Introduction

Penalty kicks are among the most critical moments in football, often shifting the momentum or deciding match outcomes in high-stakes competitions. The success of a penalty attempt traditionally depends on the kicker's skill and strategy versus the goalkeeper's reflexes and anticipation. However, the inherent variability of human performance makes the outcome of any given penalty kick difficult to predict.

Recent advances in artificial intelligence and computer vision have opened new frontiers in sports analytics, enabling data-driven decoding of complex athletic movements [1]. Deep learning-based computer vision techniques can analyse video footage frame by frame to extract subtle cues invisible to the naked eye [2]. Despite these advances, predicting the direction of a penalty kick from video remains an open challenge. This requires interpreting complex spatial dynamics—from run-up angle and body posture to foot positioning and hip rotation, all of which can influence the outcome [3].

To address these challenges, this paper proposes a multi-modal deep learning framework that integrates both visual context and the pose dynamics of the penalty taker. The architecture combines spatial features extracted from RGB frames using a MobileNetV2-based CNN, while an LSTM network models the temporal evolution of 2D body keypoints. A pose-guided spatial attention mechanism further sharpens focus on relevant player actions and scene-level cues. By fusing scene-level and biomechanical data, the model predicts the direction of a penalty kick—left, middle, or right—with an accuracy of 89%.

2 Related Work

Recent developments in football analytics increasingly rely on data-driven methods to support player evaluation, outcome prediction, and tactical decision-making [1, 3, 4]. Among these, deep learning techniques have shown significant promise in automating the visual interpretation of match footage. Convolutional Neural Networks (CNNs) are commonly employed for detecting and tracking players and the ball, while Recurrent Neural Networks (RNNs), such as Long Short-Term Memory (LSTM) networks, are used to model temporal patterns in play sequences [5–7]. For instance, Honda et al. demonstrated that combining visual features with player trajectories can significantly enhance the prediction of pass receivers in soccer [7].

In the specific context of penalty kick analysis, Chakraborty et al. [8] proposed a YOLOv4-based object detection pipeline combined with OpenCV tracking, followed by LSTM models to analyse body positioning data during penalties. Their approach achieved a mean accuracy of 79.05% one second before the kick [8]. More recently, Salazar and Alatrista-Salas introduced a dedicated penalty kick dataset and used semantic segmentation along with 3D pose estimation to train deep models for shot placement prediction [9]. Pinheiro et al. integrated body pose estimation using OpenPose to detect relevant body orientation angles that correlated with goalkeeper anticipation [3]. While prior research has shown encouraging results in penalty kick analysis, several important challenges remain unresolved. A common limitation is that many existing models have been evaluated only in offline conditions, reducing their effectiveness in real-time match scenarios. Although some studies incorporate pose estimation, this is often treated as an isolated component, rather than being deeply integrated into the prediction framework. This lack of integration limits the model's ability to capture fine-grained biomechanical cues that arise from the interaction between body posture and scene context. Evidence from sports biomechanics literature highlights that features such as trunk orientation and kicking-foot height are strong predictors of shot direction, reinforcing the need for pose-informed modelling [10]. To address these limitations, we propose a unified, real-time framework that fuses visual scene context with detailed pose dynamics. This integration supports more accurate and interpretable predictions of penalty kick direction across varied match scenarios.

3 Methodology

This study presents a multi-modal deep learning approach to predict the final ball placement in football penalty kicks. The methodology comprises several stages: (1) constructing a curated dataset from real-world match footage; (2) developing object detection and pose estimation models to extract input features; (3) training a hybrid CNN–LSTM architecture; and (4) evaluating its performance under various parameter configurations.

3.1 Data Set Creation

Penalty Kick Event Dataset
A total of 154 match highlight videos were collected from broadcasting platforms and publicly available datasets. These videos featured key match moments—including goals, fouls, and penalty kicks—from both international fixtures and top-tier club competitions. In addition, 12 full match recordings were sourced from online sports archives. From this collection, individual penalty-kick events were manually identified and extracted to isolate only the relevant action sequences involving the penalty shooter. For each event, the final position of the ball—whether it entered the left, middle, or right side of the goal, as viewed from the goalkeeper's perspective—was manually annotated. In total, 755 distinct penalty-kick scenarios were compiled.

The extracted data were further processed to align with the input requirements of the proposed neural network model. Each video clip was segmented into a fixed-length sequence, beginning from the moment the referee signalled the penalty kick and ending just before the player made contact with the ball. The endpoint of each sequence was determined using a distance-based threshold between the kicker's foot and the ball. To support this process, a customised object detection model was developed to accurately identify and track key elements within each frame.

Object Detection Model (YOLOv8) Training Dataset
A dedicated dataset was prepared to support the training of a custom object detection model designed to identify key elements involved in penalty kick scenarios. Approximately 4,000 RGB frames were extracted from the collected video clips. Each frame was manually annotated with bounding boxes corresponding to four object classes: penalty shooter, goalkeeper, net (the goal), and ball. To improve model robustness and generalisation, data augmentation techniques were applied, including rotation, blurring, scaling, shearing, and adjustments to brightness and saturation. This process expanded the dataset to 6,300 annotated frames. The final dataset was split into three subsets: 70% for training and 15% each for validation and testing.

Penalty Kick Approach Sequence Dataset (Model Input 01)
The object detection results—the positions of the goal shooter, goalkeeper, ball, and net—were used to define precise frame sequences. Each sequence began when the referee signalled the penalty kick and ended just before the shooter made contact with the ball. The endpoint was determined algorithmically by tracking the distance between the kicker's foot (estimated via pose keypoints) and the ball. A distance-based threshold was

applied to ensure the sequence captured only the preparatory motion leading up to the shot.

However, since object sizes in the frame varied depending on camera angle and zoom level, a fixed pixel-based threshold did not generalise well across all clips. For instance, players appearing smaller in the frame resulted in proportionally smaller measured distances, leading to inconsistent endpoint detection. To resolve this, the distance between the starting position of the ball and the midpoint of the net—two reliably detected objects—was used as the reference, as their physical separation remains constant during the early phase of the penalty kick. The foot-to-ball distance threshold was then expressed as a ratio relative to this fixed reference, allowing for consistent sequence segmentation across varying perspectives and resolutions (Fig. 1).

To evaluate how the endpoint threshold influences the model prediction performance, 03 separate sets of video sequences were created using different normalised threshold ratios. The trajectory prediction model was trained independently on each dataset.

Pose Keypoint Dataset of the Penalty Shooter (Model Input 02)

The penalty shooter was identified using the object detection model, and their keypoints were extracted across the refined video segment using the YOLOv8-Pose algorithm—a robust single-stage, multi-person keypoint detector. YOLOv8-Pose accurately detects 17 standard body keypoints per frame (Fig. 1), corresponding to key anatomical landmarks such as the ankles, knees, hips, shoulders, elbows, wrists, neck, and head. This structured pose information was used as input to the skeletal feature branch of the proposed model.

Fig. 1. Object detection results on a video frame with lines indicating distances from the ball to the net and to the shooter's nearest foot. A zoomed-in view displays the extracted pose keypoints of the shooter.

Model Input Sequence Finalisation

The proposed neural network model required fixed-length input sequences of 8 frames, sampled from each video segment and its corresponding keypoint sequence.

To ensure that the model captured meaningful motion patterns, frames were not selected from the beginning of the sequence, as early frames often contained idle waiting periods and lacked informative biomechanical cues. Instead, frames were sampled uniformly across the entire duration of each video segment to provide a more representative view of the player's preparatory movement. For example, in a sequence containing 100 frames, 8 frames were selected at regular intervals (1, 15, 29, 43, 57, 71, 85, 99), enabling uniform temporal coverage. If the pose estimation model failed to detect keypoints in a selected frame with an average confidence score above 0.6, that frame was replaced by the nearest valid frame within the sequence.

The final dataset consisted of 755 samples. Each sample included two inputs: a sequence of 8 RGB frames and a keypoint tensor capturing the (x, y) coordinates of 17 body keypoints across the same 8 frames. The dataset was then divided into three subsets: 70% for training, 15% each for validation and testing.

3.2 Neural Network Architecture

The model processes two synchronised input streams extracted from each penalty kick sequence: (1) a sequence of RGB video frames with shape (B, 8, 224, 224, 3), where B is the batch size and 8 represents the temporal dimension; and (2) a corresponding sequence of 2D body keypoints with shape (B, 8, 17, 2), representing the (x, y) coordinates of 17 anatomical joints per frame. These inputs are passed through a custom hybrid deep learning architecture implemented using the TensorFlow Functional API. The architecture consists of four main components: the Spatial Feature Branch, Skeletal Feature Branch, Pose-Guided Spatial Attention Module, and the Late Fusion and Classification Head (Fig. 2). The design enables parallel processing of visual and pose information, with attention mechanisms guiding the model to focus on task-relevant spatial and temporal features. The outputs from both streams are then integrated through a late fusion module, which generates the final prediction across three goal direction categories.

1. **Spatial Feature Branch – Visual Data Processing**

    The spatial stream is responsible for analysing the visual dynamics of the penalty kick from RGB video frames. Each frame is individually processed using a MobileNetV2 convolutional neural network wrapped in a time-distributed layer to preserve temporal consistency. This produces spatial feature maps for each frame. These features are then refined using a pose-guided attention mechanism that highlights areas of interest based on the kicker's body posture. The refined features undergo global average pooling to produce concise descriptors, which are then passed through a multi-head self-attention layer to capture temporal dependencies. An LSTM layer further processes this sequence to summarise the visual stream into a single vector representation.

2. **Skeletal Feature Branch – 2D Body Keypoints Processing**

    The skeletal stream complements the visual analysis by modelling the biomechanics of the penalty shooter using pose keypoints. In each frame, the 2D keypoints are flattened into vectors, forming a temporal sequence that represents the player's motion pattern. A multi-head attention layer is then applied to capture key temporal

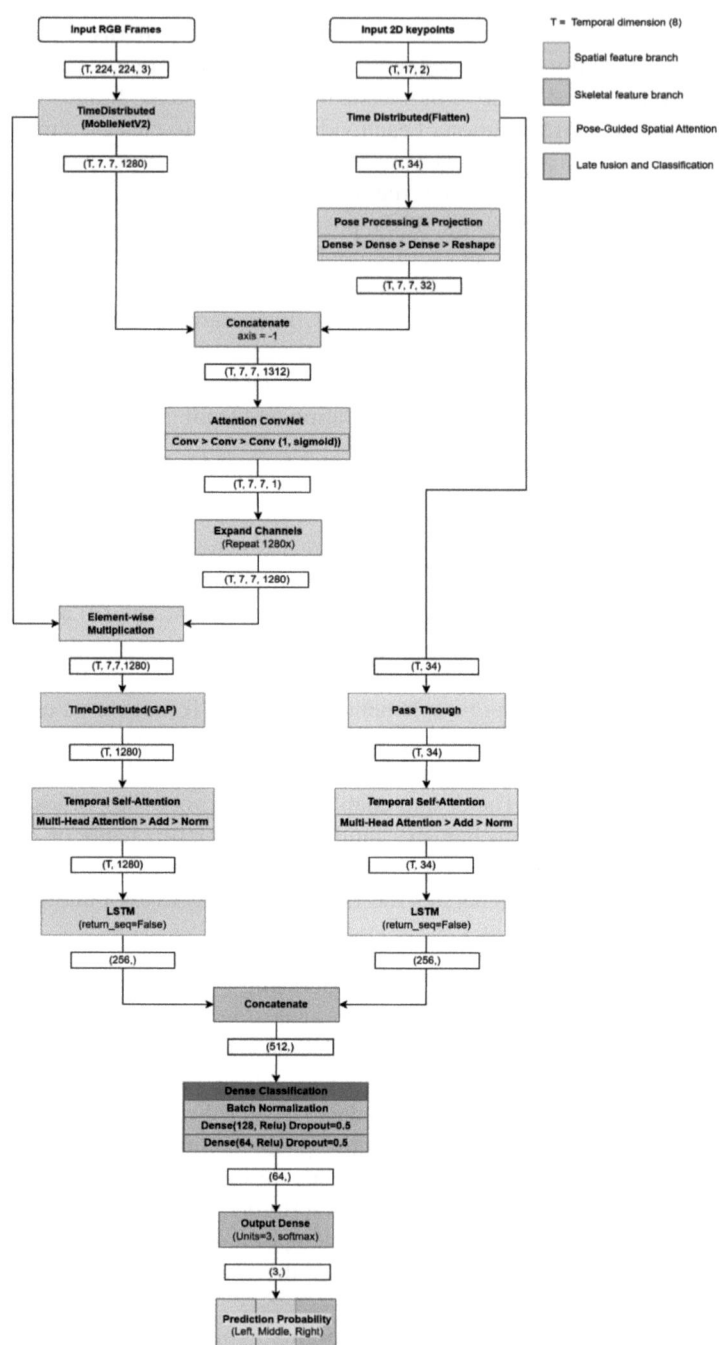

Fig. 2. Proposed multi-modal architecture: visual and pose inputs are processed in parallel with spatial and temporal attention, then fused for final kick direction prediction.

dependencies in the movement, such as foot orientation, leg swing, or hip rotation. The attention-weighted sequence is then processed by an LSTM network, which outputs a summary vector that encodes the overall pose dynamics across the sequence.

3. **Pose-Guided Spatial Attention Module**

   This module bridges the spatial and skeletal streams by allowing the pose information to influence the visual attention mechanism. It generates a dynamic attention map for each frame by transforming the pose features and combining them with the visual feature maps using convolutional layers. These attention maps act as spatial filters that instruct the visual stream where to focus more precisely, often highlighting regions such as the kicking foot, the ball, the plant foot area and body orientation. By applying this pose-informed focus, the visual features become more task-specific and informative for prediction.

4. **Late Fusion and Classification Head**

   In the final stage, the outputs from the spatial and skeletal streams are combined to produce the final prediction. The two summary vectors—one from each stream—are concatenated and passed through a fusion block comprising batch normalisation, dense layers, and dropout for regularisation. This structure enables the network to learn meaningful interactions between visual and pose-based information. The final output is generated through a softmax layer, yielding the probability distribution over the three goal zones: left, middle, and right.

The final model comprised 57 million trainable parameters while balancing model complexity with performance on the penalty kick direction prediction task.

3.3 Model Training

The model was trained end-to-end using the Adam optimiser with a learning rate of 0.001. The Categorical Crossentropy loss function was used to support the multi-class classification task of predicting shot direction (left, middle, or right). To promote generalisation and prevent overfitting, the training data was shuffled at the start of each epoch. A batch size of 32 was chosen to balance computational efficiency with gradient stability. The model was trained for up to 100 epochs, with early stopping applied based on validation loss. Training was terminated if no improvement was observed over 10 consecutive epochs, and the model checkpoint with the lowest validation loss was saved for final evaluation.

4 Results and Evaluation

4.1 Object Detection Model

The custom-trained YOLOv8 object detection model achieved strong performance on the test set, with an mAP@0.5 of 0.935, a precision of 0.984, and a recall of 0.916. These results indicate reliable detection of the penalty shooter, ball, goalkeeper, and net, providing accurate inputs for tracking and subsequent analysis.

4.2 Penalty Kick Direction Prediction Model

To assess the impact of temporal sequence length, models were trained using three different normalised foot-to-ball distance thresholds. These thresholds were defined as ratios relative to the fixed distance between the kicker and the net, ensuring consistency across varying camera perspectives. Higher threshold values captured only the early stages of preparation, while lower thresholds included the full approach phase, potentially providing richer motion cues for trajectory prediction. Table 1 summarises the number of training iterations and the corresponding prediction accuracy on the test set for each threshold configuration. All models initially demonstrated effective learning; however, signs of overfitting were observed in each scenario, and training was halted via early stopping before reaching the maximum of 100 epochs. The final model checkpoints were then evaluated on a held-out test set. Figure 3 presents the confusion matrices for each threshold, offering a more detailed view of the model's classification performance on the test set, which consisted of 113 samples.

Table 1. Model performance across varying foot-to-ball distance thresholds

Distance threshold (Normalised ratio)	Training iterations	Testing accuracy (%)
0.15	77	89.38
0.25	72	76.11
0.35	75	60.18

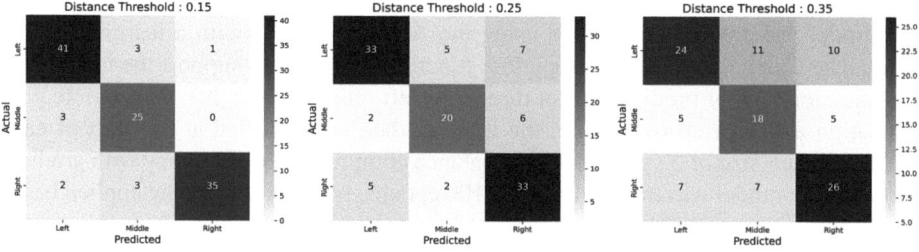

Fig. 3. Confusion matrices for different foot-to-ball distance threshold configurations

To evaluate the contribution of each model component, ablation studies were conducted using the dataset generated with a distance threshold of 0.15. Four model variants were compared:

1. Visual-only model: Used only the RGB frame sequence as input (spatial branch).
2. Pose-only model: Used only the player's keypoints as input (skeletal branch).
3. Dual-branch without pose-guided attention: Combined both input streams but excluded the pose-guided attention module.
4. Final proposed model: Full architecture with both spatial and pose branches, along with the pose-guided attention.

Table 2. Ablation study comparing test accuracy across different model configurations

Model configuration	Test accuracy (%)
Visual-only (spatial feature branch)	75.22
Pose-only (skeletal feature branch)	68.14
Dual-branch (no pose-guided attention)	82.30
Final proposed model (with attention)	89.38

The results, summarised in Table 2, show that while each individual branch contributes useful information, the visual-only configuration achieved a test accuracy of 75%, whereas the pose-only stream, based solely on body keypoints, reached 68%, indicating that visual cues alone are more informative than pose features when used in isolation.

However, when both streams were fused in the dual-branch model without attention, the accuracy increased to 82%, confirming that the integration of visual and biomechanical features leads to more robust predictions. The final proposed model, which incorporates pose-guided spatial attention, further improved accuracy to 89%. This highlights the added value of the attention mechanism in enabling the network to selectively focus on task-relevant areas of the input. The attention map generated by the pose-guided mechanism (Fig. 4) visually confirms the model's ability to concentrate on the most relevant areas of the scene during a penalty kick.

Fig. 4. Pose-guided spatial attention visualised as a heat-map overlay for a representative penalty-kick frame

4.3 Inference Pipeline

To evaluate the model's inference performance, the trained model was deployed on an NVIDIA RTX 4080 GPU and tested using real-world penalty kick footage. The inference framework, illustrated in Fig. 5, processed a continuous video stream. It dynamically trimmed the footage based on a user-defined distance threshold by monitoring the relative distance between the kicker's foot and the ball, and then passed the extracted input sequence to the model to generate the prediction.

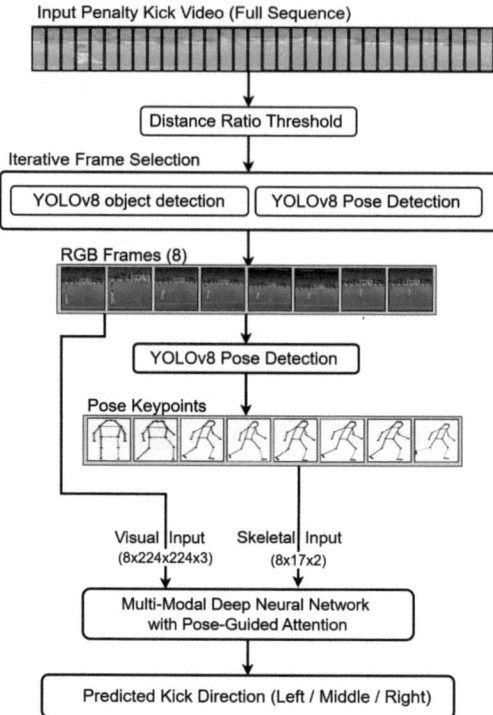

Fig. 5. End-to-end inference pipeline for penalty kick direction prediction using multi-modal inputs

To enable this process, a custom-trained YOLOv8 object detection model was used to detect key elements in each video frame, including the penalty taker, ball, goalkeeper, and goal net. Based on these detections, the distance from the ball to the net was calculated. At the same time, pose keypoints were extracted using YOLOv8-Pose to identify the kicker's foot closest to the ball. The distance from this foot to the ball was measured, and the ratio of foot-to-ball distance to ball-to-net distance was continuously monitored. When this ratio reached the user-defined threshold, the corresponding video segment was extracted. From this segment, 8 frames were selected to represent the player's final approach. For each frame, the 2D body keypoints of the penalty shooter were obtained. If a frame lacked reliable keypoints, it was replaced with the nearest valid frame to ensure a complete and consistent set of 8 RGB frames and their corresponding pose data.

These synchronised inputs—video frames and player keypoints—were then fed into the trained multi-modal neural network. The model produced a probability distribution over the three goal zones: left, middle, and right, as shown in Fig. 6. The entire inference process was completed in just 22 ms.

Fig. 6. Output from the trained model showing the probability distribution across the three goal zones

5 Discussion

The testing results highlight the effectiveness of the proposed multi-modal deep learning architecture for penalty kick direction prediction. By integrating visual context and player biomechanics through RGB frames and pose keypoints, the model captures detailed spatio-temporal patterns that are often overlooked by traditional approaches. The ablation studies confirm this advantage: models trained on only one modality—either visual or skeletal—achieved significantly lower accuracy compared to the combined model. This demonstrates that the spatial scene elements and player motion dynamics contribute complementary information essential for accurate prediction.

The inclusion of attention mechanisms further enhances performance. The pose-guided spatial attention and temporal self-attention modules allow the model to selectively focus on the most informative regions and moments during the penalty kick. As shown in Fig. 4, the pose-guided attention mechanism does not merely focus on the kicking foot but also highlights surrounding context such as the ball–foot interaction zone, the plant-foot position, upper-body orientation, and the relative positioning of the goalkeeper and goal. This context-aware attention mimics how human observers analyse action sequences and enables the model to effectively exploit the interplay between biomechanics and scene geometry.

With an inference time of 22 ms per segment, the model is well-suited for real-time applications. It maintains a modest parameter count of 57 million, which is relatively low compared to similar dual-branch architectures with attention mechanisms. This contributes to its overall lightweight design, ensuring a balance between accuracy and computational efficiency, making it practical for goalkeeper training, match analysis, and tactical decision-making.

This work relies on a distance-based thresholding method to define the endpoint of each input sequence, capturing the moment just before the player strikes the ball. Unlike fixed-duration windows, which assume uniform camera perspectives and player speeds, the proposed approach adapts dynamically by monitoring the relative distance between the foot and the ball, normalised by the kicker–goal distance. This ensures consistent capture of the most informative preparatory phase—when body posture, foot angle, and momentum signal shot direction, regardless of run-up style or video perspective.

To further evaluate the impact of input sequence segmentation on performance, the model was tested with three different distance thresholds—0.15, 0.25, and 0.35—each representing how close the kicker's foot must be to the ball. The results show a clear trend: the smaller the threshold (the closer to the kick), the higher the prediction accuracy, with the 0.15 ratio achieving the best test accuracy of 89%. This setting captures the most critical biomechanical indicators, such as trunk rotation, kicking foot angle, and upper-body posture, all of which are strong cues for shot direction. At larger thresholds (0.25 and 0.35), the input sequence includes more of the approach phase, which often contains generic motion patterns or idle movement. As a result, the model receives less directly relevant information for the prediction task, leading to reduced test accuracy—76% and 60%, respectively.

However, it is noteworthy that even with these lower accuracies, the model still performs significantly above chance, indicating that useful predictive cues exist earlier in the run-up as well. Movements such as the approach angle, pace, posture, and weight distribution begin forming early and provide initial indicators of the kicker's intent, even if they are less explicit than the cues near the final strike. This insight proves valuable in real-world applications, where an early but approximate prediction can still influence strategic decision-making. For instance, even a moderately confident early prediction generated by our model, based solely on the initial phase of the run-up, can help a goalkeeper begin shifting their weight or adjusting their stance, thereby gaining critical milliseconds in reaction time.

Since the model analyses both pose dynamics and visual context over time, it can learn player-specific behavioural cues such as approach angle, trunk orientation, and foot positioning that tend to correlate with shot direction. These cues, even when captured early in the sequence, offer useful signals. Goalkeepers can be trained using these early predictions to understand not only where a player is likely to shoot, but also when to commit to a movement. For instance, if the model consistently predicts "right" based on certain shoulder alignments or stride patterns two steps before the kick, training can focus on timing dives as soon as these patterns emerge. This enables the development of player-specific anticipation strategies, helping goalkeepers learn to respond differently depending on the unique habits of each player.

On the offensive side, coaches can use the model's attention visualisations to highlight which early-phase movements are most predictive of direction. This can guide players in refining or disguising their approach—for example, adjusting foot placement or body posture to make their intended direction less detectable. In both cases, the model's ability to fuse spatial, temporal, and biomechanical data provides a rich foundation for strategic training grounded in real-world match behaviour.

6 Limitations

While the proposed multi-modal deep learning architecture demonstrates strong performance in predicting penalty kick direction, several limitations remain. The model's effectiveness depends heavily on clear visibility of the goalpost, ball, goalkeeper, and especially the shooter. Variability in camera angles, occlusions, or zoom levels can hinder detection and pose estimation accuracy, affecting performance in real-world broadcast conditions. Additionally, the model currently classifies shots into only three broad categories—left, middle, or right—limiting its tactical utility. Future work could explore finer-grained predictions, such as shot height, ball spin, or impact location within the goal, possibly supported by ball tracking or segmentation models.

Another limitation concerns domain generalisability. The model was trained on curated video segments, and its performance across varied leagues, player styles, or live broadcast conditions remains uncertain. This study establishes an initial step by creating a dedicated penalty kick dataset; however, expanding it with more diverse examples and real-time match scenarios will be necessary to improve generalisability and support broader applicability.

7 Conclusion

This study introduced a multi-modal deep learning framework for predicting penalty kick direction by combining visual context with pose-based biomechanical information. The dual-branch architecture integrates a MobileNetV2-based CNN for visual processing and an LSTM with attention mechanisms for temporal modelling. A pose-guided spatial attention module enhances the model's ability to focus on task-relevant regions, while a distance-based thresholding strategy ensures consistent input segmentation prior to ball contact. The model achieved 89% accuracy with a 22 ms inference time, offering a lightweight and efficient solution suitable for real-time deployment, with promising applications in goalkeeper training, match strategy development, and broader sports analytics.

Disclosure of Interests. The authors have no competing interests to declare that are relevant to the content of this article.

References

1. Zheng, F., Al-Hamid, D.Z., Chong, P.H.J., Yang, C., Li, X.J.: A review of computer vision technology for football videos. Information **16**, 355 (2025)
2. Sharma, V., Gupta, M., Pandey, A.K., Mishra, D., Kumar, A.: A review of deep learning-based human activity recognition on benchmark video datasets. Appl. Artif. Intell. **36** (2022)
3. Pinheiro, G.D.S., Jin, X., Costa, V.T.D., Lames, M.: Body pose estimation integrated with notational analysis: a new approach to analyze penalty kicks strategy in elite football. Front. Sports Act. Living **4** (2022)
4. Host, K., Ivašić-Kos, M.: An overview of human action recognition in sports based on computer vision. Heliyon **8**, e09633 (2022)
5. Cioppa, A., et al.: A context-aware loss function for action spotting in soccer videos (2020)
6. Qiu, Z., Yao, T., Mei, T.: Learning spatio-temporal representation with pseudo-3D residual networks (2017)
7. Honda, Y., Kawakami, R., Yoshihashi, R., Kato, K., Naemura, T.: Pass receiver prediction in soccer using video and players' trajectories. In: 2022 IEEE/CVF Conference on Computer Vision and Pattern Recognition Workshops (CVPRW), pp. 3502–3511 (2022)
8. Chakraborty, D., Kaushik, M.M., Akash, S.K., Zishan, M.S.R., Mahmud, M.S.: Deep learning-based prediction of football players' performance during penalty shootout. In: 2023 26th International Conference on Computer and Information Technology (ICCIT), pp. 1–6 (2023)

9. Mauricio Salazar, J.A., Alatrista-Salas, H.: Football penalty kick prediction model based on kicker's pose estimation. In: ACM International Conference Proceeding Series, pp. 196–203
10. Secco Faquin, B., Teixeira, L.A., Coelho Candido, C.R., Boari Coelho, D., Bayeux Dascal, J., Alves Okazaki, V.H.: Prediction of ball direction in soccer penalty through kinematic analysis of the kicker. J. Sports Sci. **41**, 668–676 (2023)

A Bayesian Dual-Skill Framework for Roster-Based Cycling Race Outcome Prediction

Denis Rize[1](✉), Paulo Saldanha[2], and Robert Moskovitch[1]

[1] Software and Information Systems Engineering, Ben Gurion University of the Negev, Beer Sheva, Israel
rize@post.bgu.ac.il, robertmo@bgu.ac.il
[2] Department of Kinesiology, McGill University, Montreal, Canada
paulo.saldanha@powerwatts.com

Abstract. Professional road cycling is a team sport where cyclists serve in different tactical roles, yet most predictive models focus solely on individual performance. This paper introduces VeloRost, a Bayesian dual-skill framework that separately models cyclists' capabilities as leaders and supporting helpers. Using the TrueSkill rating system, we develop three methods for quantifying helper contributions and aggregate them into a roster strength score combined with each cyclist's leader skill to predict race outcomes. We evaluated our framework through direct ranking using the skill estimation and statistically enhanced learning across seven seasons of cycling data. Results demonstrate that modeling helper skills significantly outperforms state-of-the-art method, achieving NDCG@10=0.443, highlighting the important role of helpers in race outcomes.

Keywords: Sports Analytics · Machine Learning · Recommendation System

1 Introduction

Professional road cycling has evolved into a global sport with events like the Tour de France, Giro d'Italia, and Vuelta a España drawing millions of viewers and generating significant economic impact [15]. A professional cycling team typically consists of approximately thirty cyclists, but in each race, teams assign a roster of typically 6 to 8 cyclists. Teams aim to maximize Union Cycliste Internationale (UCI) point accumulation during the race season to achieve higher rankings annually through both one-day races and multi-day stage races. Unlike many team sports with formalized positions, cycling features two primary tactical roles

---

Supplementary Information The online version contains supplementary material available at https://doi.org/10.1007/978-3-032-06167-6_15.

of leader and helper (often called "domestique") [16]. Leaders are cyclists designated to compete for victory, whose performances are directly measured through race outcomes, such as finishing positions and UCI points gained. Helpers sacrifice personal ambitions to support the team's leader(s) through various contributions: shielding leaders from wind, pacing them up climbs, fetching supplies, or positioning them strategically for crucial race moments. While helpers' efforts are often invisible in final race statistics, a leader's success frequently depends on their teammates' effectiveness.

The complexity of cyclist performance extends beyond roles to race terrain diversity. Professional competitions unfold across diverse terrains, from flat sprint stages to high-alpine mountain passes, each requiring specialized physiological attributes. This terrain diversity means cyclists' effectiveness varies across race profiles, and their tactical roles can change throughout a season based on course characteristics and team objectives. Current predictive models face significant limitations by focusing solely on individual performance while neglecting team composition and tactical roles. Standard performance metrics like finishing positions or UCI points fail to account for competition strength or teammate support. When two cyclists of similar individual ability compete, the one supported by stronger teammates often holds a significant advantage. To address these challenges, this study introduces VeloRost, a terrain-specific Bayesian framework that separately models cyclists' capabilities as leaders and supporting teammates. Our approach recognizes that race outcomes depend not only on individual capability but also on teammate support quality, enabling roster-aware predictions without requiring explicit role assignments. The paper's contributions are the following:

1. We introduce VeloRost, a Bayesian framework that models cyclists' dual capabilities as leaders and helpers, maintaining distinct skill distributions for each role across different race profiles.
2. Comprehensive evaluation on seven seasons of professional cycling races (2017–2023) demonstrating that modeling helper contributions significantly improves predictive performance compared to state-of-the-art method.

2 Background

2.1 Cycling Analytics

Professional road cycling has evolved significantly towards data-driven approaches, driven by advances in wearable technology and analytics. While cycling remains fundamentally a team sport that relies on various tactics [14], it uniquely emphasizes individual physiological capabilities more than other team sports. Many studies have leveraged machine learning to extract valuable insights from these physiological metrics. Zignoli et al. [24] developed non-invasive oxygen uptake (VO_2) estimation methods, while environmental factors have also been integrated into performance modeling, with Kataoka and Gray [10] predicting power outputs in race conditions and Millour et al. [17] extending this approach by incorporating weather variables.

In a more practical application, recent studies [20,21] developed a visualization platform for professional teams that support data-driven coaching decisions. Beyond physiological aspects, researchers have begun addressing cycling's tactical dimension, for example, a recent study by Sagi et al. [22] developed an analytical system that captures team managers' decision-making patterns to optimize race lineup selections. Predicting cyclist performance in upcoming races presents different challenges, as it requires data from multiple teams whose performance data are usually kept private within the team for competitive reasons. Most sensor-based approaches rely on information collected within a single team, making data from rival teams inaccessible. To overcome this limitation, researchers leverage publicly available sources like Pro Cycling Stats (PCS)[1]. Previous research in cycling analytics has pursued various predictive strategies. Recent works [9,23] developed talent identification models to forecast a cyclist's professional potential by analyzing youth category rankings, while Kholkine et al. [11] focus on predicting individual cyclists' relative race times. However, these methods treat cyclists as isolated individuals, neglecting competitive context and relative performance. To address this shortcoming, Kholkine et al. [12] proposed a learn-to-rank methodology rooted in information retrieval [13], deploying a boosted tree model that uses a pairwise loss function to minimize ranking errors between every pair of participating cyclists.

Furthermore, most existing methods suffer from two key limitations. First, they rely on absolute outcome metrics (positions, points, time gaps), which implicitly assume these measures are consistent and directly comparable across all races and seasons. Such metrics ignore the quality of the opposition, leading to skill estimates that can over- or under-value cyclists depending on the strength of the start list. Second, they disregard team composition and tactical roles, overlooking how helpers influence leaders' performance. Janssens and Bogaert, [8] partially addressed the first issue by using TrueSkill ratings to normalize for competition level across race clusters. Their team-level variant, however, either unfairly penalizes teammates' performance by ranking the team based on their average positions or neglects multiple team members' point-scoring scenarios when using only the best finisher. More fundamentally, the model still treats all cyclists as interchangeable leaders, assigning equal weight regardless of their tactical role or contribution, and does not incorporate roster strength into final predictions.

Our VeloRost framework addresses these limitations by: (1) maintaining separate leader and helper skill distributions, (2) weighting helper contributions based on their race-specific impact, and (3) integrating both components into a roster-aware performance score that directly informs downstream predictions.

2.2 Statistical Rating Systems

In sports, statistical rating systems are a valuable tool for fair matchmaking and tournament qualification by assigning each player a latent skill inferred from

---

[1] https://www.procyclingstats.com.

past results while accounting for opposition strength the classical Elo model for chess [3] represents players with a single rating value updated after each game outcome. However, Elo struggles with multi-entrant events and cannot distinguish individual contributions to team results. Herbrich et al. [6] address these limitations with TrueSkill, a fully Bayesian rating system for multiplayer and team-based games. TrueSkill models each player's skill as a Gaussian distribution $p(s_i) = \mathcal{N}(\mu_i, \sigma_i^2)$, where μ_i represents the system's current estimate of the player's skill, and σ_i^2 quantifies uncertainty in this estimate.

New players receive initial values $\mu_0 = 25$ and $\sigma_0 = \frac{25}{3}$ to allow significant rating changes from early outcomes. The system explicitly tracks and updates the uncertainty in skill ratings, meaning that as more evidence (game outcomes) is gathered, the uncertainty typically decreases, and the skill estimate becomes more precise. Crucially, TrueSkill is designed to infer individual skills even from team-based results and can accommodate games with any number of competing entities or teams (e.g., cycling races). The system begins with a prior belief about a player's skill that is represented by their initial skill distribution. For new players entering the system, TrueSkill assigns initial default values to their latent skill. In the original paper [6], the authors set the initial mean (μ_0) to be 25, and the initial standard deviation (σ_0) is set to $\frac{25}{3}$. These default values ensure that new players have a broad skill distribution, allowing their ratings to change significantly based on their initial game outcomes.

The system distinguishes between underlying skill s_i and actual performance p_i in a specific game. Performance is modeled as a noisy realization of skill $p(p_i) = \mathcal{N}(p_i; s_i, \beta^2)$, where β^2 reflects the inherent randomness or variability associated with game outcomes that is not directly attributable to differences in skill (e.g., luck, weather conditions). For team-based games, TrueSkill aggregates individual performances into team performance through a weighted sum $t_k = \sum_{i \in A_k} w_{ik} \cdot p_i$, where A_k represents players on team k and w_{ik} is player i's contribution weight. The assumption behind TrueSkill's weighted sum for team performance stems from the recognition that players don't contribute equally to team outcomes in real-world scenarios. Since individual performances are Gaussian and their weighted sum is also Gaussian, team performance follows a Gaussian distribution. Given a game outcome $r = (r_1, \ldots, r_K) \in \{1, \ldots, K\}$ where r_K is the resulting rank of team K in the played game, and team assignments A, player skills are updated using Bayes' theorem $p(\mathbf{s}|r, A) \propto P(r|\mathbf{s}, A) \cdot p(\mathbf{s})$ where $p(\mathbf{s}) = \prod_{i=1}^n p(s_i)$ is the prior probability distribution of all players' skills and $P(r|\mathbf{s}, A)$ is the likelihood of observing the outcome r given skills $\mathbf{s}$ and team assignment A. The normalization component $P(r|A)$ ensures the posterior is a valid probability distribution. Computing this posterior distribution exactly is analytically intractable due to the high-dimensional nature of the integration problem [1]. Since exact computation is intractable, TrueSkill uses Expectation Propagation (EP) [18,19] on a factor graph to approximate the posterior through iterative message passing.

For ranking players based on their skill rating, Herbrich et al. [6] suggest a conservative skill estimate $\mu_i - \kappa\sigma_i$ where the penalty factor κ down-weights high-uncertainty ratings. The original work used $K = 3$, which represents a skill level that the system is approximately 99.7% confident the player's true skill exceeds. TrueSkill has gained popularity across sports analytics with extensions for home advantage and score differences [7], and weighted in-game events to provide more nuanced player rankings beyond final game outcome [2].

3 Methodology

3.1 Dual Skill Estimation

In cycling races, every cyclist appears twice in the tactical hierarchy: (i) as a potential **leader** who tries to win the race, (ii) as a **helper** whose work may improve the leader's result. To estimate cyclist skill at each role using historical race results, we adapt the TrueSkill rating system (see Sect. 2.2) by maintaining two latent Gaussian skills for each cyclist, one as a leader $s_i^{L} \sim \mathcal{N}(\mu_i^{L}, \sigma_i^{2L})$ and one as a helper $s_i^{H} \sim \mathcal{N}(\mu_i^{H}, \sigma_i^{2H})$. Both initialized with uninformative priors $\mathcal{N}(25, (25/3)^2)$ as suggested by Herbrich et al. [6], to avoid bias introduced by the authors. Our framework performs separate Bayesian updates for each skill type using the same TrueSkill mechanism but with different team performance aggregation strategies. For any cyclist i in role $r \in \{L, H\}$, race performance is drawn from $p_i^r \sim \mathcal{N}(s_i^r, \beta^2)$, where β^2 captures performance variability due to daily form and environmental factors.

For leader skill modeling, we treat every participant as an individual competitor, setting team performance equal to personal performance $t_i^L = p_i^L$. This reflects our definition of leadership capability as the ability to achieve strong individual results, measured through finishing position. Bayesian updates compare individual performances across all participants, with cyclists who finish higher than expected (relative to their current skill estimate) receiving positive updates to their leader skill distribution, while those who underperform experience negative adjustments. For the helper skill update, we shift focus from individual results to the team's roster aggregated results, recognizing that helpers' contributions manifest through collective success rather than personal achievements. Since cycling races rank individuals rather than teams, we define a team's result based on total UCI points earned by all roster members during the race. To estimate the team's roster performance in a race for helper skill updates, we aggregate each roster cyclist's helper skills through a weighted sum:

$$t_k^H = \sum_{i \in A_k} w_{ik} \cdot p_i^H \quad (1)$$

where A_k denotes team k's roster, w_{ik} represents cyclist i's contribution weight to the roster overall performance, and $\sum_{i \in A_k} w_{ik} = 1$. This weighted formulation differentiates cyclists' contributions, where cyclists with higher weights receive proportionally larger helper skill updates. Furthermore, recognizing that cyclists typically specialize in specific race profiles, we maintain these two skill distributions for each cyclist separately for each race profile. We use five official

PCS terrain-based race profiles—Flat, Hills with a flat finish, Hills with an uphill finish, Mountains with a flat finish, and Mountains with an uphill finish, and add one additional specialized profiles of Time Trial (TT). We distinguish between terrain-based profiles that reflect physical course characteristics and race format category of TT that represent competitive structure, capturing additional skill attributes such as aerodynamic positioning.

The practical benefit of maintaining separate skill distributions becomes apparent when predicting outcomes for upcoming races. Rather than simply ranking cyclists by individual capability, we construct roster-level performance estimates that combine each potential leader's individual skill with their roster teammates' helper skill. For a cyclist i serving as leader in team k's roster, we compute the expected roster performance as:

$$t_k = p_i^L + \sum_{j \in A_k \setminus \{i\}} p_j^H \qquad (2)$$

where the first term represents the chosen leader's individual performance and the summation captures the collective support provided by teammates in their helper roles. The probability of roster k achieving victory is then determined by comparing t_k against other teams' roster performances. This formulation enables the simulation of different roster compositions and tactical assignments. By systematically varying which cyclist serves as leader and observing the resulting performance distributions, teams can identify the best leader assignments for specific race profiles and evaluate the marginal value of different helper combinations.

Probabilistic comparisons become computationally challenging with many participating teams, a more direct approach to rank a team's roster can be achieved by adapting TrueSkill's individual conservative skill estimate $mu - \kappa\sigma$ into a roster-level estimation. The conservative estimate provides a skill level that the system is approximately confident the player's true skill exceeds, with confidence determined by the parameter κ. To adapt this conservative skill estimate into roster-level estimation, we aim to combine the individual leadership skill of cyclist i with roster A_k helper skills into a single deterministic rating representing roster strength. However, simply summing the leader's conservative estimate with all helpers' conservative estimates would over-emphasize helper contributions, as there are typically multiple helpers but only one leader per roster. In reality, while helpers provide crucial support, the leader remains the primary determinant of the team's race success.

To address this imbalance, we use the mean of helpers' conservative estimates rather than their sum, which yields our flexible Roster-Adjusted Score (RAS) for cyclist i with roster A_k:

$$\text{RAS}_i = (\mu_i^L - \kappa\sigma_i^L) + \lambda \frac{1}{|A_k| - 1} \sum_{j \in A_k \setminus \{i\}} (\mu_j^H - \kappa\sigma_j^H) \qquad (3)$$

We introduce a tuneable parameter $\lambda \in [0, 1]$ in the formula, to weight the helper roster component relative to individual leadership capability. This parameterization allows the model to adapt to different racing contexts where some races require more intensive team support than others.

3.2 Modeling Helper Contribution

The team's roster performance aggregation in Eq. 2 relies on contribution weights w_{ik} that determine how each cyclist's helper performance influences team results and control the magnitude of Bayesian skill updates. When team k achieves strong results, cyclists with higher weights w_{ik} receive proportionally larger positive updates to their helper skill distributions. The challenge lies in determining these weights from observable race outcomes, since helpers' tactical support often remains invisible in final statistics.

We use race outcome metrics as proxies for tactical contribution through the concept of "depth of contention" - how long a helper remains in proximity to decisive race action. Helpers positioned near the front or losing minimal time at crucial moments have effectively supported their leader longer than those dropping significant time early. We develop three approaches to estimate raw contribution weights, each representing different assumptions about the relationship between observable performance and tactical value. For each race observation, we assign each cyclist i in team k's roster A_k a raw weight $\tilde{w}_i$ derived from one of the following methods:

(1) **Uniform-All Weighting:** This approach makes no assumptions about the relative value of each helper's work, acknowledges cycling as a team sport where contributions take diverse forms. Each cyclist receives equal raw weight $\tilde{w}_i = 1$, ensuring all cyclists receive skill updates regardless of personal results.

(2) **Position-Based Weighting:** The finishing position of each cyclist provides a straightforward, ordinal measure of their contention depth, reflecting relative performance differentials between cyclists. This approach assumes that helpers finishing closer to the front provide more tactical assistance. For cyclist i finishing position r_i in a race with N participants $\tilde{w}_i = 1 - \frac{r_i - 1}{N - 1}$. This gives winners full credit ($\tilde{w}_i = 1$) and last finishers zero credit ($\tilde{w}_i = 0$), with each position step representing equal incremental value.

(3) **Time-Gap Based Weighting** Time gaps offer a physiologically grounded measure of contention depth, reflecting absolute performance differentials between cyclists. We develop a direct probabilistic transformation that converts standardized time gaps into contribution weights. First, raw time gaps typically follow a right-skewed distribution where most cyclists finish together in the peloton with outliers trailing behind. We apply a logarithmic transformation $\Delta t_i = \log(\text{gap}_i + 1)$ where adding one prevents numerical issues with zero gaps for tied finishers, converting the skewed distribution into approximately Gaussian form. Second, we standardize these log-transformed gaps using z-scores $z_i = \frac{\Delta t_i - \mu_{\Delta t}}{\sigma_{\Delta t}}$ where $\mu_{\Delta t}$ and $\sigma_{\Delta t}$ are race-specific parameters, ensuring comparability across diverse race contexts. Finally, we transform z-scores directly into probabilistic weights using the complement of the standard normal cumulative distribution function $\tilde{w}_i = 1 - \Phi(z_i)$. To handle races

with very tight finishing clusters where time gap variance is minimal, we add small rank-based increments to distinguish between closely positioned cyclists. This formulation ensures that cyclists performing better than average ($z_i < 0$) receive weights $\tilde{w}_i > 0.5$, while those performing worse receive weights $\tilde{w}_i < 0.5$, with the transformation preserving relative performance hierarchy through probabilistically interpretable contributions.

After computing raw weights using any of the three methods described above, we apply a final normalization step to ensure proper roster-level aggregation. For each race observation, we assign each cyclist i in team k's roster A_k a raw contribution weight $\tilde{w}_i$ derived from their respective weighting method, then normalize these weights within the team's roster $w_{ik} = \frac{\tilde{w}_i}{\sum_{j \in A_k} \tilde{w}_j}$, ensuring $\sum_{i \in A_k} w_{ik} = 1$ so roster performance reflects relative rather than absolute contributions.

3.3 Statistically Enhanced Learning

While the RAS (Eq. 3) provides direct ranking, incorporating our dual-skill latent variables into a machine learning framework offers significant advantages by leveraging rich contextual features alongside skill estimates. The Statistically Enhanced Learning approach [4] can capture more nuanced predictive relationships, as the RAS formula represents a simplified aggregation that doesn't account for race-specific conditions, historical performance patterns, and seasonal form variations. However, directly incorporating roster helper skills into machine learning models presents challenges since roster sizes vary across races, potentially creating feature vectors with many null values.

We address this through a hybrid approach: for each cyclist i participating in a given race, we directly include their leader skill parameters μ_i^L and σ_i^L as features, preserving full distributional information. For their roster teammates' helper skills, we apply conservative estimates $\mu_j^H - \kappa \sigma_j^H$ to each teammate j, then compute four aggregate statistics (minimum, maximum, mean, variance) of these conservative helper estimates. This enables the model to understand not just average helper quality but also roster consistency, providing richer context for race outcome prediction. Since we make predictions across the entire racing calendar, we employ two distinct κ values, one tuned for the complete racing calendar and another specifically for WorldTour events, creating two representations of confidence in the roster helper strength. In mixed-competition races spanning multiple tiers, more conservative estimates (higher κ) help prevent the model from over-weighting less-observed cyclists who may have inflated skill estimates. Conversely, in WorldTour races where all participants are elite cyclists with usually substantial racing histories, less conservative estimates may preserve finer distinctions between high-level helpers that could be decisive for race outcomes.

Beyond the dual-skill latent variables, our feature set captures multiple dimensions of racing context and cyclist performance history that could improve race outcome prediction. Following Janssens and Bogaert [8], we include race profile-specific capabilities while enriching features with race context across UCI tiers. Table 1 summarizes the complete feature set, which we organize into several key categories reflecting different aspects of predictive information.

This includes demographic and race context information, historic peak performance with recency indicators, points accumulation trends, and recent form dynamics through finishing positions in similar race contexts.

Table 1. Feature set implemented in the learn-to-rank model

Category	Feature Description	# Features
Cyclist information	Age and career length	2
Race context	One day vs stage race, race profile, race tier, distance, elevation	6
Historic best results	Highest past ranks and time since those results	6
Points Trend (3-yr)	UCI points for the previous three seasons and their linear trend line slope	4
Form	Mean UCI points from prior and current season (all races, by race profile)	6
Recent form	Finishing positions in last 1–10 races within same race profile	10
Leader-skill parameters	μ_i^L and σ_i^L	2
Helper aggregates (all races)	Min, max, mean, var of $\mu_j^H - \kappa_{\text{all}}\sigma_j^H$	4
Helper aggregates (WorldTour)	Min, max, mean, var of $\mu_j^H - \kappa_{\text{wt}}\sigma_j^H$	4

For the final prediction model, VeloRost employs a learn-to-rank framework with pairwise loss rather than conventional point-wise regression. While standard regressors could predict individual performance metrics such as finishing position or points, our task fundamentally involves ranking cyclists relative to each other within specific races. The learn-to-rank paradigm naturally aligns with this objective by treating each race as a query and participating cyclists as documents with relevance scores corresponding to their actual finishing positions. This framework's key advantage over point-wise approaches lies in its direct optimization of ranking quality measures, while learning to distinguish between better and worse performers within the same competitive context.

4 Evaluation

The goal of the empirical study is two fold: (1) to examine whether explicitly modeling helper abilities improves predictive power compared to approaches relying solely on leader performance, and (2) to identify the configuration that best captures teammates' contributions in modern road cycling, optimizing our VeloRost framework for the most accurate roster-aware race predictions.

Research questions

1. For direct ranking by Roster-Adjusted Score, what are the best values for the uncertainty penalty κ and the roster weight λ on a full seasonal racing calendar prediction?
2. Does adding the team's roster strength to the cyclists' leader score improve race outcome prediction?
3. Which weighting method (Uniform-All, Positional, or Time-Lag) is best for a full seasonal racing calendar prediction?
4. Which prediction approach (direct roster-adjusted ranking or statistically enhanced learning) produces superior results for race outcome prediction?

4.1 Evaluation Metrics

We used two predictive performance metrics to measure prediction quality. Firstly, we measure ranking quality using Normalized Discounted Cumulative Gain at rank 10 (NDCG@10), which emphasizes the correct prediction order of the highest-placed cyclists. We define a linear gain for finishing position i as $g_i = max(11 - \text{pos}(i), 0)$, so that $g_1 = 10$, $g_2 = 9$, up until $g_{10} = 1$. The Discounted Cumulative Gain at 10 is then:

$$\text{DCG@10} = \sum_{i=1}^{10} \frac{g_i}{\log_2(i+1)} \quad (4)$$

where g_i is the cyclist predicted in the ith position. We then normalize this raw score by dividing it by the ideal DCG@10 (IDCG@10), which applies the same formula to the true top-10 ordering, yielding NDCG@10 in the range $[0,1]$. A higher NDCG@10 indicates better alignment of the predicted ranking with the actual finish, especially at the top positions. We also report top-10 accuracy, defined as the proportion of cyclists in the predicted top 10 who indeed finish within the race's actual top 10.

4.2 UCI Races Data

Professional cycling teams pursue UCI points across all competition tiers. WorldTeams regularly participate in lower-classified races (ProSeries, Class1, Class2) beyond their WorldTour obligations to accumulate additional points. We must include all race classes to accurately model performance across the entire racing season, rather than focusing exclusively on WorldTour events. We collected data from PCS covering all UCI point-awarding events from 2017–2023 across the complete men's elite hierarchy (WorldTour, ProSeries, Class1, Class2). For each race, we recorded metadata (date, category, format), course characteristics, and UCI point allocations. Cyclists' data includes finishing position, time gaps, points earned, team affiliation, and cyclists' attributes. Our dataset comprises approximately 5,000 race days (1,100 WorldTour, 901 ProSeries, 647 Class1, 2,400 Class2) involving 13,220 distinct cyclists and over 535,000 cyclist-race observations.

4.3 Experimental Plan

We train and evaluate our models on seven seasons of ProCyclingStats data: 2017–2021 for training (3.34k races), 2022 for validation (877 races), and 2023 for testing (839 races, including 159 WorldTour). Cyclist skills (both leader and helper) are updated sequentially in chronological order, so that each new race outcome is fed into the VeloRost skill estimation components, starting from an uninformative prior for every skill component. Before any given race, a cyclist's skill reflects all prior races up to that date under the race category. To address research questions 2, we establish state-of-the-art baseline for each prediction approach, building upon Janssens and Bogaert [8] TrueSkill adaptation for cycling. For each prediction method, we implement the baseline that will be referenced as Leader Only, which ranks cyclists exclusively by their individual leader conservative skill estimate ($\mu_i^L - \kappa\sigma_i^L$), disregarding any roster support component. This corresponds to setting $\lambda = 0$ in the RAS Eq. 3 or omitting all helper aggregate features in the statistically enhanced learning. Evaluation metrics are averaged across the entire test set to enable comprehensive comparison, with additional segmentation specifically for WorldTour events to measure model effectiveness in cycling's most prestigious competitions. To assess the significance of our findings, we conduct pairwise comparisons for each baseline against VeloRost helper-weighting methods (Uniform-All, Positional, Time-Lag) using paired-sample t-tests with the Holm procedure [5] for multiple comparisons.

4.4 Experiment 1 Direct Roster-Adjusted Ranking

For the first experiment, we rank cyclists directly using their leader and helper skills through the RAS formula (Eq. 3), without incorporating auxiliary data or training a machine learning model. To answer the first research question about the best parameter values of this formula, we tune the uncertainty penalty $\kappa \in \{1, 2, \ldots, 30\}$ and $\lambda \in (0, 1]$ using increments of 0.05 on the 2022 validation set. We evaluate direct ranking performance on 2023 data, conducting statistical testing for each comparison across all penalty values ($N = 30$) to identify consistently significant improvements regardless of the penalty level chosen. The best κ and λ values for race outcome prediction are derived separately for the complete calendar and WorldTour subset.

4.5 Experiment 2 Statistically Enhanced Learning

We integrate the dual-skill latent variables into a machine learning framework using the approach described in Sect. 3.3. We transform roster helper skills into fixed-length features using aggregate statistics of conservative helper estimates, employing two κ values to create features representing different confidence levels for the complete racing calendar and WorldTour events. We implement this using XGBoost LambdaMART with NDCG@10 optimized loss function. Each helper-weighting method generates distinct roster-strength features, allowing direct comparison of their effectiveness. We conduct hyperparameter tuning for the

XGBoost model using grid search on the validation set, employing the values detailed in Appendix Table 3. We evaluate these tuned models against baselines using the statistical testing procedure provided earlier across the entire racing calendar and WorldTour subset.

5 Results

5.1 Direct Roster-Adjusted Ranking

The first experiment evaluates the predictive performance of dual-skill modeling using the direct roster-adjusted ranking approach. Figure 1 shows that across all uncertainty penalty values $\kappa \in \{1, \ldots, 30\}$, each helper-weighting method (Uniform-All, Time Lag, Positional) consistently outperforms the baseline in NDCG@10, whether over all 2023 races or WorldTour events. However, no clear superiority emerges among the three helper-weighting methods themselves. All methods exhibit performance degradation at both low κ values (under-penalizing uncertainty) and high κ values (over-conservative estimates). For the entire racing calendar, Uniform-All achieved the best result using $\kappa = 7$ and $\lambda = 0.15$ (NDCG@10=0.388), followed by Time-Lag with $\kappa = 8$ and $\lambda = 0.12$ (NDCG@10=0.385), and Positional using $\kappa = 7$ and $\lambda = 0.10$ (NDCG@10=0.378).

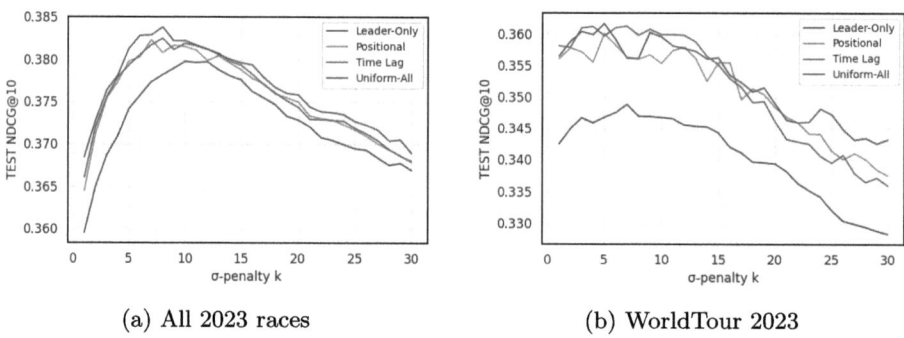

(a) All 2023 races (b) WorldTour 2023

Fig. 1. Mean NDCG@10 for the three helper-weighting methods and the baseline across uncertainty penalties κ.

Statistical analysis detailed in Appendix Table 4, confirms significance for all helper-weighting methods over the Leader-Only baseline ($p < 5 \times 10^{-4}$). Notably, within WorldTour events, Positional weighting delivers the most significant improvements, suggesting ordinal position captures important teammate support aspects in elite competition.

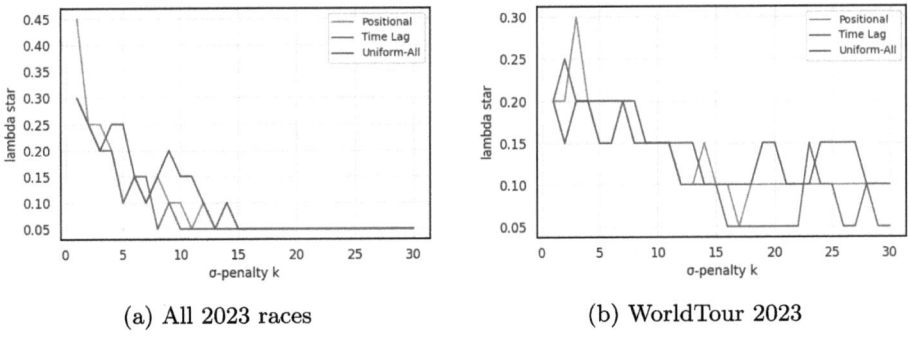

Fig. 2. Lambda values for the three helper-weighting methods across uncertainty penalties κ.

Figure 2 examines the best roster-weight parameter λ across penalty values, revealing patterns that align with the peak performance regions observed in Fig. 1. The analysis compares the best λ values across all three helper methods. For all races (panel a), Uniform-All demonstrates higher λ values across most κ ranges, with all helper influence starting high ($\lambda \in [0.30, 0.45]$ at $\kappa = 1$) and gradually declining to the minimum at $\kappa = 15$. Notably, within the best performance region ($\kappa \in [5, 8]$) where NDCG@10 peaks, Uniform-All holds the higher influence ($\lambda \in [0.10, 0.25]$). In WorldTour events (panel b), all helper skills maintain similar λ values throughout the best performance region ($\kappa = 4$ to 8), with some fluctuation including peaks around $\kappa = 3$ and $\kappa = 8$ ($\lambda \approx 0.2$).

Both race contexts demonstrate consistent patterns of helpers skill-based roster component maintains high influence within the high-performance regions where NDCG@10 peaks. These experimental findings suggest that modeling helper skills and combining them into the roster strength component significantly enhances the predictive performance of the direct ranking approach using our roster-adjusted score across all race contexts.

5.2 Statistically Enhanced Learning

Experiment 2 evaluates the statistically enhanced learning pipeline by comparing how the three helper-weighting methods perform and the state-of-the-art baseline when integrated as roster-strength features within the learning-to-rank framework. Using κ values determined from validation ($\kappa = 7$ for complete calendar, $\kappa = 3$ for WorldTour), Table 2 presents that Uniform-All yields the best overall results (0.443 NDCG@10, 0.405 Top-10 accuracy), while Time-Lag excels in WorldTour races (0.451 NDCG@10). Statistical analysis detailed in Appendix Table 5 shows that all helper methods significantly outperformed the Leader-Only baseline ($p < 2 \times 10^{-3}$).

Table 2. Mean NDCG@10 and Top-10 Accuracy (T@10) of each helper weighting method and baseline on the entire 2023 test season and WorldTour races subset.

Method	All Races		WorldTour Races	
	NDCG@10	T@10	NDCG@10	T@10
Leader-Only	0.424	0.392	0.414	0.379
Uniform-All	**0.443**	**0.405**	0.445	**0.394**
Positional	0.434	0.399	0.436	0.393
Time Lag	0.436	0.402	**0.451**	**0.394**

Analysis of our statistically enhanced learning results indicates that the Uniform-All method consistently delivers superior performance across the entire professional calendar, while Time-Lag excels specifically within WorldTour events. This is similar to the direct roster-adjusted ranking findings, where across the entire race calendar, equal-share credit yielded the strongest roster-aware estimates, but at the highest level, a more discriminating approach (Positional) further enhances predictive performance, suggesting a fundamental relationship between helper's weighting strategy effectiveness and race context. The performance gap between roster-aware methods and Leader-Only baseline increases in higher race classes, demonstrating that roster support information becomes more valuable with well-observed cyclists. Mean NDCG@10 improves from 0.421 (Class 2) to 0.463 (WorldTour), with the largest jump at ProSeries level (+6.6%). Another key consideration for the helper-weighting methods' efficiency is the predicted race profile. Appendix Table 6 shows the averaged results for each helper-weighting method across distinct race profiles.

Each weighting method emphasizes different aspects of rosters' helpers' contributions, leading to varying estimation accuracy depending on the race profile characteristics. For instance, in Flat races, the Time Lag method achieves the highest NDCG@10 (0.595), which might be because it better differentiates teammates riding in distinct groups by penalizing significant gaps while assigning similar weights to cyclists within a small time lag. Conversely, Uniform-All weighting outperforms in challenging race profiles such as Mountains with an uphill finish (NDCG@10 = 0.495), where equal credit distribution could better reflect the shared nature of helpers' efforts in demanding terrains, which is not observed in the final race outcome.

6 Conclusion and Discussion

This study introduced VeloRost, a Bayesian framework designed to model helpers' contributions in professional road cycling by separately representing each cyclist's capabilities as both leader and helper. We developed three distinct weighting methods (Uniform-All, Position-Based, and Time-Lag) to quantify helper contributions and incorporated them into a roster-aware performance score. Applied in two ranking approaches, we found that the statistically enhanced machine learning model delivered higher accuracy than the direct roster-aware scoring, regardless of which weighting method was used.

Additionally, our extensive evaluation across seven seasons of professional cycling data demonstrated that modeling helpers' skills and incorporating them into a roster-aware representation significantly outperforms state-of-the-art method. Both our direct roster-aware score and statistically-enhanced learning approaches showed substantial improvements over the baseline, with all three helper-weighting methods. The Uniform-All method provided the most consistent performance across the full professional calendar (NDCG@10=0.443), while context-specific advantages emerged for other weighting methods. For example, Time-Lag excelled in WorldTour races (NDCG@10=0.451) and flat terrain, while Positional showed strengths in hill stages.

These findings point toward promising directions for future research. A hybrid approach that dynamically selects the most suitable helper-weighting method based on cyclist participation history, race level, and race profile could further enhance predictive performance. Another promising avenue involves integrating our roster-aware performance estimations into broader analytical applications, such as team roster optimization and strategic planning tools. Finally, employing datasets that segment races into smaller tactical intervals could enable a more precise modeling of helpers' contributions within individual races, potentially leading to even finer-grained insights into cycling team dynamics and strategy.

Acknowledgments. This study was funded by the SYLVAN ADAMS FAMILY FOUNDATION ISRAEL.

References

1. Cooper, G.F.: The computational complexity of probabilistic inference using Bayesian belief networks. Artif. Intell. **42**(2–3), 393–405 (1990)
2. D'Astous, J., Browne, R.: Expanding the trueskill algorithm using in-game events (2022)
3. Elo, A.E., Sloan, S.: The rating of chessplayers: past and present (1978)
4. Felice, F., Ley, C., Groll, A., Bordas, S.: Statistically enhanced learning: a feature engineering framework to boost (any) learning algorithms. arXiv preprint arXiv:2306.17006 (2023)
5. García, S., Fernández, A., Luengo, J., Herrera, F.: Advanced nonparametric tests for multiple comparisons in the design of experiments in computational intelligence and data mining: experimental analysis of power. Inf. Sci. **180**(10), 2044–2064 (2010)
6. Herbrich, R., Minka, T., Graepel, T.: Trueskill^{TM}: a Bayesian skill rating system. In: Advances in Neural Information Processing Systems, vol. 19 (2006)
7. Ibstedt, J., Rådahl, E., Turesson, E., vande Voorde, M.: Application and further development of trueskill^{TM} ranking in sports (2019)
8. Janssens, B., Bogaert, M.: Perforank: cluster-based performance ranking for improved performance evaluation and estimation in professional cycling. Mach. Learn. **114**(1), 1–30 (2025)

9. Janssens, B., Bogaert, M., Maton, M.: Predicting the next pogačar: a data analytical approach to detect young professional cycling talents. Ann. Oper. Res. **325**(1), 557–588 (2023)
10. Kataoka, Y., Gray, P.: Real-time power performance prediction in Tour de France. In: Brefeld, U., Davis, J., Van Haaren, J., Zimmermann, A. (eds.) MLSA 2018. LNCS (LNAI), vol. 11330, pp. 121–130. Springer, Cham (2019). https://doi.org/10.1007/978-3-030-17274-9_10
11. Kholkine, L., Schepper, T., Verdonck, T., Latré, S.: A machine learning approach for road cycling race performance prediction. In: Brefeld, U., Davis, J., Van Haaren, J., Zimmermann, A. (eds.) MLSA 2020. CCIS, vol. 1324, pp. 103–112. Springer, Cham (2020). https://doi.org/10.1007/978-3-030-64912-8_9
12. Kholkine, L., et al.: A learn-to-rank approach for predicting road cycling race outcomes. Front. Sports Active Living **3**, 714107 (2021)
13. Liu, T.Y., et al.: Learning to rank for information retrieval. Found. Trends® Inf. Retrieval **3**(3), 225–331 (2009)
14. Mignot, J.-F.: Strategic behavior in road cycling competitions. In: Van Reeth, D., Larson, D.J. (eds.) The Economics of Professional Road Cycling. SEMP, vol. 11, pp. 207–231. Springer, Cham (2016). https://doi.org/10.1007/978-3-319-22312-4_10
15. Mignot, J.-F.: The history of professional road cycling. In: Van Reeth, D., Larson, D.J. (eds.) The Economics of Professional Road Cycling. SEMP, vol. 11, pp. 7–31. Springer, Cham (2016). https://doi.org/10.1007/978-3-319-22312-4_2
16. Mignot, J.F.: Strategic behavior in road cycling competitions. In: The Economics of Professional Road Cycling, pp. 227–251. Springer, Cham (2022)
17. Millour, G., Plourde-Couture, F., Domingue, F.: Cycling modelling under uncontrolled outdoor conditions using a wearable sensor and different meteorological measurement methods. Int. J. Sports Sci. Coaching **18**(4), 1102–1112 (2023)
18. Minka, T.P.: Expectation propagation for approximate Bayesian inference. arXiv preprint arXiv:1301.2294 (2013)
19. Minka, T.P.: A family of algorithms for approximate Bayesian inference. Ph.D. thesis, Massachusetts Institute of Technology (2001)
20. Moskovitch, R., Sinai, P., Rize, D., Holohan, L., Saldanha, P.: The velodromeprocyclists data analytics. In: International Sports Analytics Conference and Exhibition, pp. 167–172. Springer, Cham (2024)
21. Rize, D., Sinai, P., Holohan, L., Saldanha, P., Moskovitch, R.: Visualization of professional cyclists analytics. SN Comput. Sci. **6**(3), 289 (2025)
22. Sagi, M., Saldanha, P., Shani, G., Moskovitch, R.: Pro-cycling team cyclist assignment for an upcoming race. PloS One **19**(3) (2024)
23. Van Bulck, D., Vande Weghe, A., Goossens, D.: Result-based talent identification in road cycling: discovering the next eddy merckx. Ann. Oper. Res. 1–18 (2023)
24. Zignoli, A., et al.: Estimating an individual's oxygen uptake during cycling exercise with a recurrent neural network trained from easy-to-obtain inputs: a pilot study. PLoS ONE **15**(3), e0229466 (2020)

Horse ReIDing: Addressing Re-Identification in Horse Racing Scenarios

Luca Francesco Rossi[1,2](✉), Andrea Sanna[1], Federico Manuri[1], and Mattia Donna Bianco[2]

[1] Department of Control and Computer Engineering (DAUIN), Politecnico di Torino, Corso Duca degli Abruzzi 24, 10129 Turin, TO, Italy
[2] netventure R&D S.r.l., Via della Consolata 1/bis, 10122 Turin, TO, Italy
lucafrancesco.rossi@polito.it

Abstract. Re-identification (ReID) tasks, traditionally employed in person tracking across diverse camera views, face unique challenges in the domain of horse racing due to frequent occlusions, dynamic motion, and varying environmental conditions. This study addresses these complexities by developing a custom pipeline and dataset for jockey ReID, specifically collected from horse racing footage. A ResNeXt-based architecture is employed to process input data, with additional experiments exploring the inclusion of segmentation mask information for improved performance. Empirical evaluations demonstrate the model's efficacy in both closed-set and open-set scenarios, showcasing significant gains in mean Average Precision (mAP) and top-k Cumulative Matching Characteristic (CMC) rank metrics when segmentation masks are incorporated. Comparative analysis across different ResNeXt configurations underscores the robustness and scalability of the proposed approach, contributing as a pioneering framework for ReID in high-motion sports contexts and advancing the application of computer vision technologies in horse racing scenarios.

Keywords: Re-identification · Computer vision · Artificial intelligence · Sport · Broadcast footage · Horse racing

1 Introduction

Person Re-Identification (ReID) is the task of recognizing whether a person seen in one camera view or image has appeared in another, effectively matching people across non-overlapping camera views [1]. This problem is important in many real-world settings such as video surveillance, security, and smart-city applications, where one might track a person of interest through a network of cameras [2]. In a sense, it can be thought of as a specialized image-retrieval problem: given a "query" image of a person, the system searches a database of other images to find the same individual [3]. A central difficulty is that the same entity can have very dissimilar representations in different camera views or at distinct times, with changes in lighting, viewpoint, pose or clothing making matching way more complicate [4]. Early approaches to person ReID relied on hand-crafted appearance features, i.e., color histograms, texture descriptors or edge-based features, along with traditional matching metrics [5]. However, as large labeled datasets became available starting from last decade, deep learning methods – especially convolutional neural networks, or CNNs – became the standard, automatically learning powerful appearance features end-to-end from the data. Since then, deep architectures have dominated the literature, greatly outperforming earlier hand-engineered methods. In the past few years, even more flexible Transformer-based [6] architectures have emerged: these models (originally developed for language processing) can learn to focus on different parts of an image and have quickly set new state-of-the-art records in ReID tasks. In sports, person ReID can therefore be used to track players across multiple broadcast cameras, enabling richer analytics and improved viewing experiences [7]: in such a way, coaches can monitor each player's movements and performance over time, and broadcasters can overlay real-time statistics and highlights onto live video. However, despite the growing research attention devoted to widely followed sports such as soccer [8], basketball [9], or hockey [10], relatively limited investigation has been directed toward the domain of horse racing. This study thus aims to bridge such a gap by addressing the re-identification task within the context of horse racing, a highly-dynamic environment characterized by its fast-paced nature and frequent occlusions.

The paper is organized as follows: Sect. 2 provides a comprehensive review of related work relevant to computer vision applications specifically tailored for horse racing; Sect. 3 details the proposed methodology, including the architectural design and implementation specifics for incorporating segmentation masks into ReID models; Sect. 4 presents an empirical evaluation of the proposed approach, analyzing its performance gains relative to baseline models; and Sect. 5 concludes the study by summarizing key contributions and outlining potential directions for future research.

2 Related Work

Binning et al. [11] addressed the challenges of horse racing video analysis, specifically focusing on the detection and re-identification (ReID) of jockeys through helmet identification. Horse racing presents unique challenges such as occlusion, motion blur, and varying illumination, which complicate the detection and tracking of both horses and jockeys. The authors propose a helmet detection framework [12] that aims to address these issues and serve as a preliminary step for ReID. The evaluation dataset features 12 jockeys, but the study focuses on five distinct helmet classes for semi-automated ground truth labeling. The model was trained on data from one camera and tested on the same camera as well as unseen data from others, showing good performance with higher accuracy for simple helmet designs compared to more intricate ones. Hedayati et al. [13] presented a novel approach for tracking jockeys in a crowded horse racing environment, focusing on the challenges of occlusion and group dynamics. The primary goal is to track jockeys at the turning point of the race, where their motion changes direction [14]. Tracking in this scenario is particularly difficult due to the high levels of occlusion, as jockeys race closely together, often obscuring one another. The proposed solution leverages the behavior of jockeys as they move in a relatively uniform manner, even when partially obscured, to improve tracking accuracy. By integrating group dynamics into the tracking system, the approach helps estimate the positions of jockeys who are not fully visible. The system consists of three main components: sampling, localization, and data association. Sampling involves capturing motion and color cues from the jockey's cap at multiple levels—point, object, and group—to track the jockeys even under occlusion. Localization uses two strategies—object-based localization when the jockey's cap is clearly visible, and group-based localization when occlusion occurs. The latter estimates the position of a jockey by considering the movement of the entire group of jockeys. Data association is employed to link new detections with existing tracks [15], ensuring consistent tracking even when multiple objects overlap. Empirical results on video footage of horse races show that, by incorporating group dynamics, the method effectively handles occlusion, achieving a high correct tracking ratio. Wing et al. [16] similarly examined the challenges associated with multi-object tracking (MOT) in the context of horse racing, given the well-known difficulties of occlusion, trajectory overlap, frequent camera switching, and motion blur caused by horses sprinting. The authors introduce an augmentation-based multi-object tracking method (GMOT) designed to address these challenges. GMOT leverages an auxiliary classifier generative adversarial network (ACGAN) to augment horse racing data [17], thereby enhancing target detection and re-identification by generating synthetic images that simulate difficult scenarios, such as blurred sprinting and occlusion. Building upon the FairMOT framework [18], which employs an anchor-free approach with separate branches for position detection, GMOT is particularly well-suited for handling issues such as camera switching and occlusion, which typically hinder traditional MOT methods. The model is trained on both day- and night-time race datasets, with additional synthetic training data generated through ACGAN to further strengthen the model's ability to cope with real-world complexities in horse racing. Experimental results indicate that GMOT outperforms FairMOT in a variety of challenging horse racing conditions, demonstrating superior accuracy in tracking. The interested reader is finally referred to [19] for a broader analysis on how AI is impacting the world of horse racing, aside from computer vision applications only.

3 Method

Due to the lack of availability of horse racing data for computer vision, a custom ReID dataset is privately collected from multiple videos provided by The Hong Kong Jockey Club. Specifically, six full-race replays from the Sha Tin Racecourse

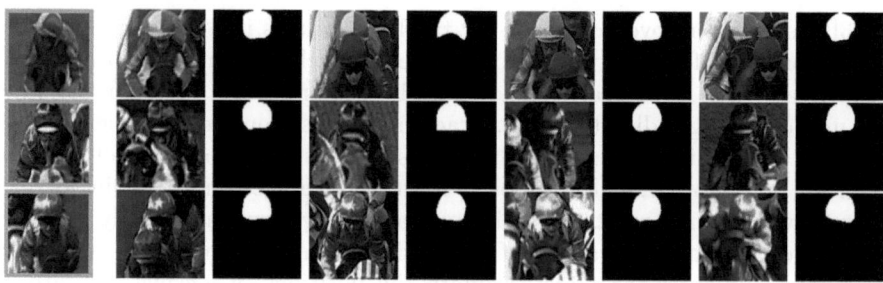

Fig. 1. Qualitative visualization of dataset samples with paired helmet masks.

Table 1. Summary statistics concerning the collected horse racing ReID dataset.

Race	Racecourse	Time	Condition	Jockeys	Frames	Images	Split
#1	Sha Tin	Day	Grass	11	89	954	Train
#2	Sha Tin	Day	Grass	14	74	1014	Train
#3	Sha Tin	Night	Grass	14	86	1171	Train
#4	Sha Tin	Night	Grass	14	78	1059	Train
#5	Sha Tin	Night	Dirt	12	76	875	Train
#6	Sha Tin	Night	Dirt	12	107	1235	Train
#7	Happy Valley	Day	Grass	12	104	1216	Test

under different environmental conditions are used for training, while a full-race from the Happy Valley Racecourse is separately used for testing. To construct such a dataset, a semi-automatic pipeline is instantiated in the following way:

1. a frame is extracted every single second from each full-race video;
2. bounding boxes for jockeys' helmets are retrieved by deploying a GroundingDINOv1.5 model [12];
3. helmets' segmentation mask are determined via a Segment Anything Model 2 (SAM2) [21] by prompting the previously-detected boxes;
4. hybrid manual-automatic refinement [22] is performed to edge cases, followed by track identity assignment;

Since silks (i.e., the brightly colored shirts worn by jockeys) convey as much information concerning identification as helmets do, the final image for each jockey is cropped by zooming out from helmet detection only. Roughly following the heuristics of the canon of proportions in ancient Greek sculpture, if one defines the helmet's bounding box in its *tlwh* format, that is, by the (x, y)-coordinates of its top-left corner, width and height, optimal empirical results for the final image are retrieved by cropping the original one at $< x_{helmet} - w_{helmet}, y_{helmet}, 3 \times w_{helmet}, 3 \times h_{helmet} >$, eventually clipped to image borders, which captures the whole jockey's upper body. Figure 1 qualitatively depicts some output examples from such pipeline, while Table 1 provide a more quantitative summary concerning the constructed dataset. Due to the real-time requirements often imposed by computer vision applications for sports, a ResNeXt architecture [23] is instantiated.

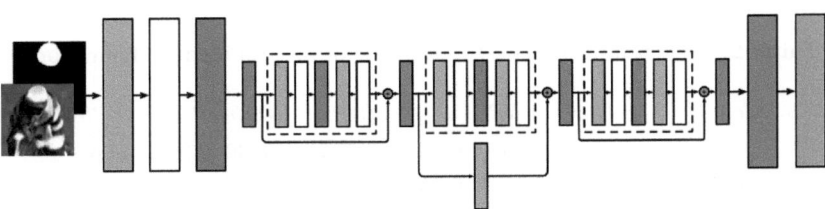

Fig. 2. Illustrative visualization of the residual architecture, with the final layer embedding the high-dimensional representation of the input.

ResNeXt is a lightweight convolutional neural network architecture that builds on the principles of ResNet [24] by introducing a new structural dimension called "cardinality", which refers to the number of parallel transformations within each block. While maintaining the simplicity and modularity of ResNet through repeating blocks with identical topology, ResNeXt enhances representational power by aggregating multiple homogeneous transformations (each operating on a low-dimensional embedding of the input) through summation. For this study, input images have been reshaped as 128×128 before being fed to the ResNeXt model, with the final classification head computing a 256-dimensional embedding representation. Following literature [25], the network is optimized for a total of 10 epochs jointly using cross-entropy and hard-mining triplet loss [26]. Similarly to other works in occluded ReID [27], the inclusion of segmentation mask information is investigated to assess whether it could guide the model towards focusing on more relevant parts of the image [28–31]. To do so, the architecture is extended in order to accept a dilated representation of the input image, obtained by concatenating the jockey's helmet mask to the RGB channels.

4 Evaluation and Discussion

In a practical setting, it is expected that such a model will be deployed over two different scenarios: known jockeys and unknown jockeys. The former consists in a deployment scenario where all athletes have already been clustered during training (closed-set), while the latter deals with generalization, that is, how well the model behaves when encountering jockeys never seen before (open-set). For such a reason, Race #4 is employed for validation during the training phase, as it features the same jockeys as Race #3, albeit recorded by a different patrol camera. On the contrary, the Happy Valley race features completely different, never-seen-before jockeys, and it is used to test its generalization capabilities concerning open-set ReID. Following comon practices for ReID tasks, such test race is further subject to a random 90-10 split in order to provide the standard gallery-query setting. Three different ResNeXt configurations are considered: depth-50, depth-101 and depth-152. Pretrained on the ImageNet-1k dataset [33], these configurations are fine-tuned on the collected dataset both with and without injecting helmet's segmentation mask inside the input tensor.

Table 2. Empirical results on the test set for all proposed configurations.

Method	Depth	mAP	R@1	R@5	R@10	R@20	FPS
Vanilla	50	0.7690	0.9580	0.9830	0.9830	0.9920	906.8
	101	0.8380	0.9580	0.9830	1.0000	1.0000	894.4
	152	0.8560	0.9500	1.0000	1.0000	1.0000	882.4
Segmentation	50	0.7550	0.9330	0.9750	0.9920	0.9920	685.5
	101	**0.8960**	0.9670	1.0000	1.0000	1.0000	676.2
	152	0.8890	**0.9750**	**1.0000**	**1.0000**	**1.0000**	670.4

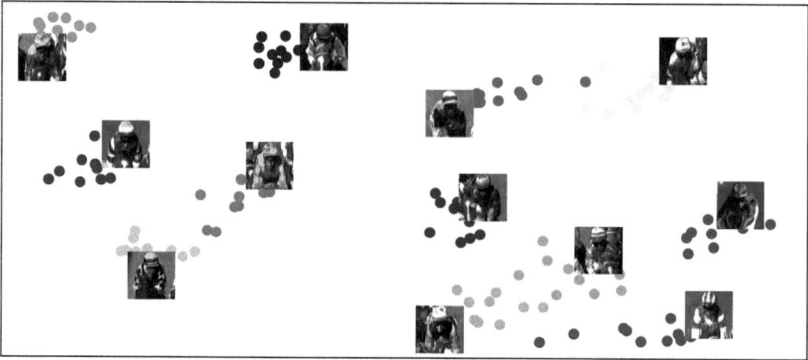

Fig. 3. Qualitative t-SNE [32] visualization of query jockeys clustering.

Table 2 lists mean Average Precision (mAP) and top-k Cumulative Matching Characteristic (CMC) rank metrics (R@k) on the test set. Aside from the most shallow configuration, empirical results validate the advantage of incorporating segmentation knowledge for ReID purposes, with boosts up to 6.92% in mAP and 2.63% in R@1. FPS (averaged over 10 test iterations) on customer-level hardware are also reported in Table 2, confirming the suitability of the proposed architecture for real-time purposes.

5 Conclusion and Future Work

This study demonstrates the effectiveness of integrating ReID dynamics in the context of horse racing for jockey ReID in broadcast footage, with the incorporation of helmet segmentation masks significantly improving re-identification performance. Future work could explore expanding the dataset to include a more diverse range of racing environments and conditions to enhance model generalization. Additionally, the integration of temporal information and the exploration of transformer-based architectures may further improve the robustness of the system, paving the way for broader applications in sports analytics and beyond.

Disclosure of Interests. The authors have no competing interests to declare that are relevant to the content of this article.

References

1. Asperti, A., Fiorilla, S., Nardi, S., Orsini, L.: A review of recent techniques for person re-identification. Mach. Vis. Appl. **36**(1), 25–53 (2025). https://doi.org/10.1007/s00138-024-01622-3
2. Kansal, K., Wong, Y., Kankanhalli, M.: Privacy-Enhancing Person Re-identification Framework – A Dual-Stage Approach. In: 2024 IEEE/CVF Winter Conference on Applications of Computer Vision (WACV), pp. 8528–8537. Waikoloa, HI, USA (2024). https://doi.org/10.1109/WACV57701.2024.00835
3. Saad, R.S.M., Moussa, M.M., Abdel-Kader, N.S., Farouk, H., Mashaly, S.: Deep video-based person re-identification (deep vid-reid): comprehensive survey. EURASIP J. Adv. Signal Process. **2024**(1), 63–111 (2024). https://doi.org/10.1186/s13634-024-01139-x
4. Ming, Z., et al.: Deep learning-based person re-identification methods: A survey and outlook of recent works. Image and Vision Computing **119**, 104394 (2022). https://doi.org/10.1016/j.imavis.2022.104394
5. Chahla, C., Snoussi, H., Abdallah, F., Dornaika, F.: Learned versus handcrafted features for person re-identification. Int. J. Pattern Recognit Artif Intell. **34**(4), 2055009 (2020). https://doi.org/10.1142/S0218001420550095
6. Vaswani, A., et al.: Attention is all you need. In: Proceedings of the 31st International Conference on Neural Information Processing Systems (NIPS'17), pp. 6000–6010. Curran Associates Inc., Red Hook, NY, USA (2017). https://papers.nips.cc/paper/7181-attention-is-all-you-need
7. Fujii, K.: Computer vision for sports analytics. machine learning in sports, SpringerBriefs in Computer Science, 21–57 (2025). https://doi.org/10.1007/978-981-96-1445-5_2
8. Deliège, A., et al.: SoccerNet-v2: A dataset and benchmarks for holistic understanding of broadcast soccer videos. In: 2021 IEEE/CVF Conference on Computer Vision and Pattern Recognition Workshops (CVPRW), pp. 4503–4514. Nashville, TN, USA (2021). https://doi.org/10.1109/CVPRW53098.2021.00508
9. Van Zandycke, G., Somers, V., Istasse, M., Del Don, C., Zambrano, D.: DeepSportradar-v1: computer vision dataset for sports understanding with high quality annotations. In: Proceedings of the 5th International ACM Workshop on Multimedia Content Analysis in Sports (MMSports '22), pp. 1–8. Association for Computing Machinery, New York, NY, USA (2022). https://doi.org/10.1145/3552437.3555699
10. Koshkina, M., Elder, J. H.: Towards long-term player tracking with graph hierarchies and domain-specific features. In: 2025 IEEE/CVF Winter Conference on

Applications of Computer Vision Workshops (WACVW), pp. 1175–1185. Tucson, AZ, USA (2025). https://doi.org/10.1109/WACVW65960.2025.00140
11. Binning, W., Rahmaniboldaji, S., Dong, X., Gilbert, A.: Detection and re-identification in the case of horse racing. In: Information Sciences (2024). https://openresearch.surrey.ac.uk/esploro/outputs/conferenceProceeding/Detection-and-Re-Identification-in-the-case/99925861902346
12. Ren, T., et al.: Grounded SAM - assembling open-world models for diverse visual tasks. In: arXiv cs.CV (2024). https://arxiv.org/abs/2401.14159
13. Hedayati, M., Cree, M. J., Scott, J. B.: Tracking jockeys in a cluttered environment with group dynamics. In: Proceedings Proceedings of the 2nd International Workshop on Multimedia Content Analysis in Sports (2019). https://doi.org/10.1145/3347318.3355518
14. Hedayati, M., Cree, M.J., Scott, J.B.: Scene structure analysis for sprint sports. In: 2016 International Conference on Image and Vision Computing New Zealand (2016). https://doi.org/10.1109/IVCNZ.2016.7804429
15. Murty, K.G.: An algorithm for ranking all the assignments in order of increasing cost. In: Operations Research 16 (1968). http://www.jstor.org/stable/168595
16. Ng, W.W.Y., Liu, X., Yan, X., Tian, X., Zhong, C., Kwong, S.: Multi-object tracking for horse racing, In: Information Sciences 638 (2023). https://doi.org/10.1016/j.ins.2023.118967
17. Odena, A., Olah, C. Shlens, J.: Conditional image synthesis with auxiliary classifier GANs. In: Proceedings of the 34th International Conference on Machine Learning (2017). https://proceedings.mlr.press/v70/odena17a.html
18. Zhang, Y., Wang, C., Wang, X., Zeng, W., Liu, W.: FairMOT - on the fairness of detection and re-identification in multiple object tracking. In: Int. J. Comput. Vis. 129, 3069–3087 (2021). https://doi.org/10.1007/s11263-021-01513-4
19. Colle, P.: What AI can do for horse-racing? In: arXiv cs.LG (2022). https://arxiv.org/abs/2207.04981
20. Ren, T., et al.: Grounding DINO 1.5: Advance the "Edge" of Open-Set Object Detection. arXiv 2405.10300, cs.CV (2024). https://arxiv.org/abs/2405.10300
21. Ravi, N., et al.: SAM 2: Segment Anything in Images and Videos. arXiv 2408.00714, cv.CV (2024). https://arxiv.org/abs/2408.00714
22. Kirillov, A., et al.: Segment Anything. arXiv 2304.02643, cv.CV (2023). https://arxiv.org/abs/2304.02643
23. Xie, S., Girshick, R., Dollár, P., Tu, Z., He, K.: Aggregated residual transformations for deep neural networks. In: 2017 IEEE Conference on Computer Vision and Pattern Recognition (CVPR), pp. 5987–5995. Honolulu, HI, USA (2017). https://doi.org/10.1109/CVPR.2017.634
24. He, K., Zhang, X., Ren, S., Sun, J.: Deep residual learning for image recognition. In: 2016 IEEE Conference on Computer Vision and Pattern Recognition (CVPR), pp. 770–778. Las Vegas, NV, USA (2016). https://doi.org/10.1109/CVPR.2016.90
25. Yuan, Y., Chen, W., Yang, Y., Wang, Z.: In Defense of the triplet loss again: learning robust person re-identification with fast approximated triplet loss and label distillation. In: 2020 IEEE/CVF Conference on Computer Vision and Pattern Recognition Workshops (CVPRW), pp. 1454–1463. Seattle, WA, USA (2020). https://doi.org/10.1109/CVPRW50498.2020.00185
26. Xuan, H., Stylianou, A., Liu, X., Pless, R.: Hard negative examples are hard, but useful. In: Vedaldi, A., Bischof, H., Brox, T., Frahm, J.-M. (eds.) ECCV 2020. LNCS, vol. 12359, pp. 126–142. Springer, Cham (2020). https://doi.org/10.1007/978-3-030-58568-6_8

27. Sun, Z., et al.: Multiple pedestrian tracking under occlusion: a survey and outlook. IEEE Trans. Circuits Syst. Video Technol. **35**(2), 1009–1027 (2025). htttps://doi.org/10.1109/TCSVT.2024.3481425
28. Meng, H., Zhao, Q.: A lightweight model based on co-segmentation attention for occluded person re-identification. In: Proceedings of 2021 Chinese Intelligent Automation Conference. Lecture Notes in Electrical Engineering, vol 801., pp. 692–701. Springer, Singapore (2022). htttps://doi.org/10.1007/978-981-16-6372-7_74
29. Zhou, Y., Qin, Y., Wang, S., Huang, Z., Zhang, D.: Person re-identification based on instance segmentation and pose estimation. In: IEEE 6th International Conference on Pattern Recognition and Artificial Intelligence (PRAI), pp. 584–589. Haikou, China (2023). htttps://doi.org/10.1109/PRAI59366.2023.10332122
30. Qin, P., Gao-hua, C.: Occluded person re-identification based on foreground segmentation and multi-scale feature fusion. In: Proc. SPIE 13552, International Conference on Physics, Photonics, and Optical Engineering (ICPPOE 2024), 135522D (2025). https://doi.org/10.1117/12.3060412
31. Liu, Y., Wang, Z., Zhang, W., Li, Z.: DGSN: learning how to segment pedestrians from other datasets for occluded person re-identification, Image and Vision Computing **140**, 104844 (2023). https://doi.org/10.1016/j.imavis.2023.104844
32. van der Maaten, L., Hinton, G.: Visualizing Data using t-SNE. J. Mach. Learn. Res. **9**(86), pp. 2579–2605 (2008). http://jmlr.org/papers/v9/vandermaaten08a.html
33. Deng, J., Dong, W., Socher, R., Li, L., Li, K., Fei-Fei, L.: ImageNet: a large-scale hierarchical image database. In: 2009 IEEE Conference on Computer Vision and Pattern Recognition, pp. 248–255. Miami, FL, USA (2009). https://doi.org/10.1109/CVPR.2009.5206848

Barbell Trajectory Tracking for Performance Analysis During Snatch Movement in Weightlifting

Dhairya Shah[1], Christopher Taber[2], Tolga Kaya[2], Eva Maddox[3], and Mehul S. Raval[1(✉)]

[1] Ahmedabad University, Ahmedabad, India
{dhairya.s4,mehul.raval}@ahduni.edu.in
[2] Sacred Heart University, Fairfield, CT, USA
{taberc,kayat}@sacredheart.edu
[3] Old Dominion University, Norfolk, VA, USA
emaddox@odu.edu

Abstract. Olympic-style weightlifting involves complex and technical movements where accurate tracking of barbell motion is crucial for performance analysis. In this paper, we present a computer vision based framework that first corrects for perspective distortion caused by varying camera height and distance, then employs a rule-based algorithm to classify snatch trajectories into four distinct types. Preliminary investigation on 6000 frames suggests 70% classification accuracy. Building on these labels, eight key barbell kinematic variables were calculated and utilized three—vertical peak height (Y_{max}), initial horizontal setup (X_1), and bar drop efficiency (Y_{catch}) to generate a consolidated 0–4 performance score, mapped to five qualitative categories from "Very Bad" to "Excellent". This two fold approach, comprising trajectory classification and score calculation, was validated by a sports scientist, ensuring its reliability in helping athletes optimize lifting techniques by providing insights into barbell trajectory patterns.

Keywords: Computer Vision · Deep Learning · Injury Risk Prediction · Olympic Weightlifting · Sports · Trajectory Analysis

1 Introduction

1.1 Background

Weightlifting is a competitive strength sport featured in the Olympic Games, where athletes attempt to lift maximum weights in two highly technical lifts: the snatch and the clean and jerk. These lifts are not only demonstrations of pure strength but also demand exceptional speed, flexibility, coordination, and balance. In the snatch, the athlete continuously lifts the barbell from the ground to an overhead position, requiring a wide grip and precise coordination to stabilize the weight overhead [1]. The clean and jerk is a two-phase lift; the 'clean' involves raising the barbell from the floor to the shoulders, followed by the 'jerk', where the athlete propels the barbell overhead, typically using a split stance to catch and stabilize the weight [2]. Figure 1 shows both types of lifts: snatch, clean-and-jerk.

(i) (ii)

Fig. 1. Lifts: i. A snatch being performed in competition [3], ii. A clean and jerk being performed in competition [4].

Each athlete is allowed three attempts in both the snatch and the clean and jerk. The best successful lift from each category is combined to form the athlete's total score. The athlete with the highest total in their weight class is ranked highest. In case of a tie, the lighter athlete is ranked higher [5]. Weight classes are divisions based on body mass, created to ensure fair competition among lifters of similar size. For example, as of the 2024 International Weightlifting Federation (IWF) classifications, men compete in 55 kg to 109+ kg, and women from 45 kg to 87+ kg [6]. These classes encourage inclusivity, structure competitions, and allow athletes of various body types to compete on a level playing field. While Olympic Weightlifting and Powerlifting are strength sports, Olympic weightlifting differs significantly from powerlifting. Olympic lifts emphasize explosive power, speed, flexibility, and technical precision, whereas powerlifting focuses more on maximum static strength in three lifts: squat, bench press, and deadlift [7].

A barbell is a long metal bar used in weight training and Olympic weightlifting, onto which varying weights (plates) are loaded. Standard Olympic barbells differ slightly between men's and women's categories. A men's barbell typically weighs 20 kg, measures 2.2 m long, and has a shaft diameter of 28 mm [8]. In contrast, a women's barbell weighs 15 kg, is 2.01 m long, and has a shaft diameter of 25 mm [8]. In the Olympics, the outer diameter of standard Olympic bumper plates is 45 cm, as defined by the IWF for all plates weighing 10 kg and above [9]. These plates are also colour-coded for easy visual identification: red for 25 kg, blue for 20 kg, yellow for 15 kg, and green for 10 kg [10]. Accurate measurement of barbell trajectory in Olympic weightlifting is essential for optimizing technique, understanding movement patterns, and preventing injuries. Deviations from the ideal path can indicate mistimed pulls or poor positioning, leading to inefficient lifts or increased injury risk [11,12].

Studying kinematics for a weightlifter's performance and injury risk analysis is crucial. Kinematics in weightlifting refers to the study of motion without accounting for the forces behind it. It involves tracking the barbell and the lifter's body throughout the lift to assess performance. Key elements include the barbell trajectory, which reflects movement efficiency and balance; joint angles, which show how well the athlete transitions through phases; body positioning, especially of the torso and hips for power generation; and the timing of movement phases such as the first pull, transition, second pull, and catch [13,14].

Figure 2 shows the four types of barbell trajectories in the case of a snatch lift. Barbell trajectories are classified based on horizontal displacement relative to a vertical reference line [15]. Type 1 trajectory exhibits a "toward-away-toward" pattern, where the barbell initially moves toward the lifter, then away, and back toward the lifter, crossing the vertical reference line during the "away" phase. Type 2 trajectory also follows a "toward-away-toward" pattern but does not cross the vertical reference line at any point during the lift. Type 3 trajectory follows an "away-toward-away-toward" pattern, involving multiple crossings of the vertical reference line. Type 4 trajectory may begin with a "toward" phase, as in Type 1 or 2 trajectories, or an "away-toward" phase, as in the Type 3 trajectory. The defining feature of the Type 4 trajectory is an intervening "away-toward" phase between the first "toward" phase and the final "away-toward" phase.

1.2 Literature Review

Various methods have analyzed barbell trajectories, offering unique cost, accuracy, and practicality trade-offs. Video-based analysis is among the most accessible techniques, relying on frame-by-frame tracking from standard or high-speed cameras. Still, it is time-consuming and susceptible to human error [17]. Motion capture systems using infrared cameras and reflective markers provide high-precision three-dimensional data but are expensive, require calibration, and are limited to laboratory environments [18]. Linear position and velocity transducers offer real-time data with high sampling rates, such as Tendo units [19] or GymAware [20]. However, they typically measure only vertical displacement and cannot capture horizontal or rotational movement [21]. Smartphone applications leverage onboard sensors or video algorithms to estimate barbell velocity and trajectory; they are highly accessible but generally lack the precision of dedicated systems [22]. Mathematical and computational modeling allows for the simulation of barbell motion using kinematic and dynamic equations, providing insights into force and torque. However, the models simplify assumptions that may not reflect actual lift conditions [23]. Wearable sensors and accelerometers provide portable, real-time feedback on barbell motion, yet they may suffer from signal noise, misalignment, and require careful calibration [24].

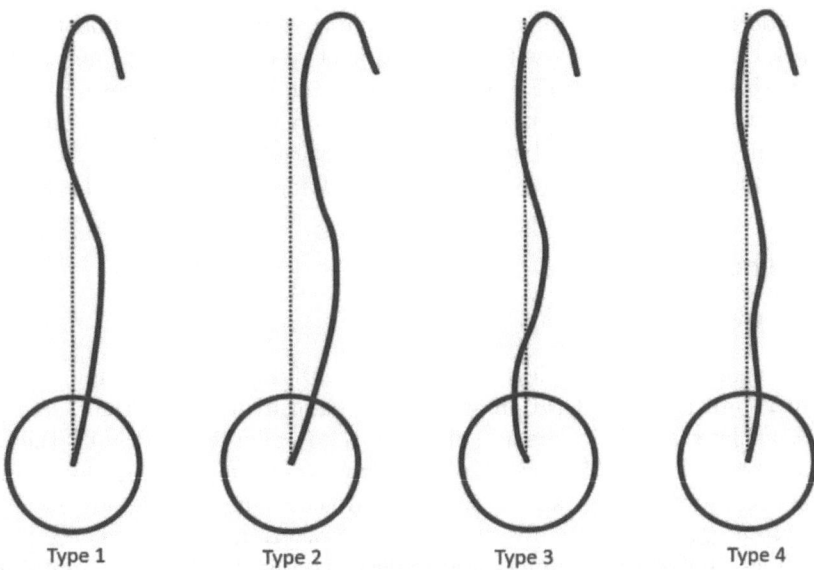

Fig. 2. Barbell trajectory types determined by the horizontal displacement pattern and vertical reference line crossing [16].

Traditional methods for assessing barbell trajectory typically involve manual video analysis, which can be subjective, labor-intensive, and prone to human error [25]. As a result, sports scientists have increasingly turned to automated approaches for motion analysis. Automation improves objectivity and reproducibility, offering high temporal resolution and reducing workload [26]. Computer vision techniques have emerged as particularly effective tools in this domain, enabling accurate motion tracking without physical markers, which is ideal for minimally invasive performance assessment [27]. Markerless motion capture systems, powered by deep learning and pose estimation models, allow real-time analysis with reduced setup time and cost compared to traditional marker-based systems [28]. Despite these advances, most existing systems are sensitive to environmental variables and camera placement, limiting their generalization. In elite weightlifting, where minor technical differences can significantly impact performance, accurate analysis of barbell motion is essential. Studies at the 2015 World Weightlifting Championship and 2017 Pan-American Weightlifting Championship used standardized GoPro HERO4 Black camera setups to reduce measurement errors. Still, they did not account for variations in camera viewpoint, such as height and distance from the platform [16].

Considering the above limitations, the paper develops a method to classify and track barbell trajectory without the impact of camera height and distance. The paper's contributions are: i. Creating a snatch lift video dataset and developing a barbell trajectory classification algorithm based on barbell movement. ii. A module to calculate the barbell kinematic variables from video analysis. iii. Suggest a metric based on kinematic variables for performance analysis.

2 Methods

2.1 Participants

The participants were senior-level athletes, as in Olympic weightlifting, the International Weightlifting Federation and USA Weightlifting define the "Senior" age group as athletes aged 15 years and older, with no upper age limit until the "Masters" category begins at 35 years. The dataset was collected during a local competition in the United States, and it was an open-access meet. These are entry-level events, often organized by clubs or regional associations. They are accessible to many lifters, including beginners and experienced athletes. Local competitions are ideal for gaining experience and qualifying for higher-level events. A total of 44 athletes participated in the study, consisting of 28 males (mean bodyweight = 89.88 kg) and 16 females (mean bodyweight = 73.27 kg). In this paper, a random sample of the data from the total collected dataset was used for analysis. This study was approved by the Sacred Heart University Institutional Review Board, approval number IRB-FY2025-241, in April 2025.

2.2 Data Acquisition and Annotation

The data was collected during the East Coast Gold Spring Fling USA Weightlifting-sanctioned regional meet on April 12th, 2025, at Virginia Beach, Virginia, USA. A GoPro HERO10 Black action camera was employed for video acquisition due to its compact design, image stabilization, and high-resolution recording capabilities. The HERO10 features a 23MP sensor and supports video capture up to 5K at 30 frames per second (fps) or 4K at 60 fps, enabling detailed analysis of fast barbell movements [29]. The camera was placed at 3.35 m from the barbell, and the videos were recorded at 2160p resolution. A controlled environment was established to minimize external factors affecting data quality. Subsequently, the weightlifter recorded a series of snatch lift attempts. The data collected was recorded, ensuring a diverse sample that included different weight categories and skill levels. The data included over 200 trials, with over 100000 video frames from various angles, distances, and heights. Some sample frames are shown in Fig. 3. Our study requires camera placement in a lateral plane in alignment with the barbell axis, and 10 videos from the dataset have been chosen accordingly. An expert labeled each sample video as Types 1 to 4 by observing the trajectory.

Fig. 3. Sample data frames from the video dataset: i. Athlete starting the lift, ii. Athlete in the middle of the lift, iii. The athlete finished the lift.

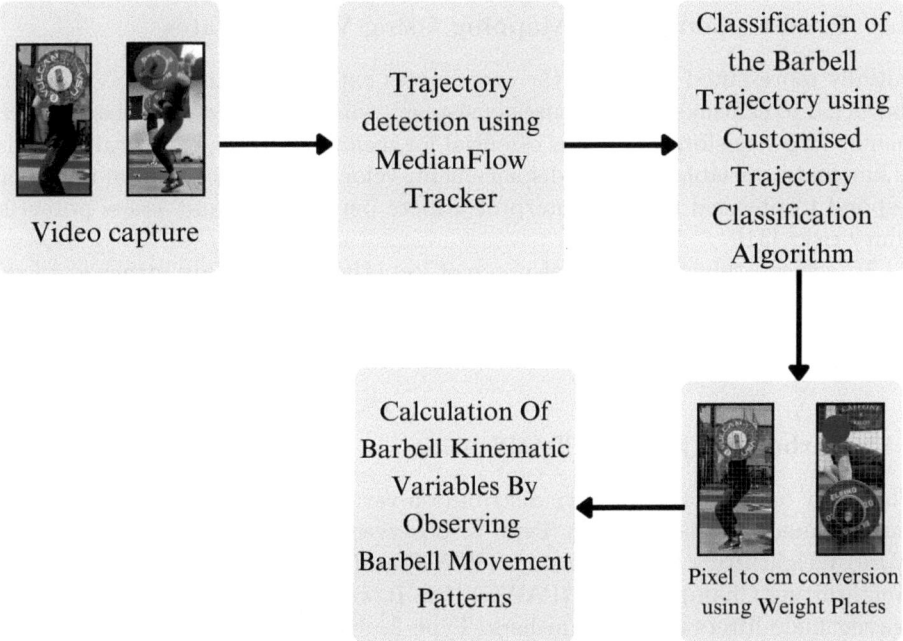

Fig. 4. The flow for video processing and calculation of barbell kinematic variables.

2.3 Video Processing and Frame Extraction

Figure 4 illustrates the overall workflow for video processing and barbell trajectory detection. After collecting the data, the videos were resized to (1920 × 1080) to standardize the dataset's width and height. Furthermore, as it was challenging to decide dynamically when the lift starts and ends, we trimmed the videos from the start and the end in a way that only contained the motion of the lift to avoid redundant frames. After that, the MedianFlow tracking algorithm [30] is deployed to track the movement of the snatch lift performed by the athlete. MedianFlow operates by monitoring a set of points using forward-backward error estimation. It first estimates the trajectory of each point from frame to frame, then validates the tracking consistency by comparing the forward path with the backward path, filtering out unreliable points. This makes the algorithm robust to partial occlusions and abrupt motions. The extracted coordinates obtained by tracking the snatch lift are stored in a CSV file. Subsequently, we utilize these coordinates to compute kinematic variables of the snatch lift.

2.4 Pixel to Centimeter Mapping Using Weight Plates

Initially represented in pixels, the coordinates extracted from the CSV file were converted to centimeters to ensure consistency with real-world distance measurements. This transformation was essential for meaningful biomechanical analysis, as kinematic variables such as displacement, velocity, and acceleration must correspond to physical units to interpret athlete performance and assess potential injury risks [31].

To achieve this, a known reference object, the weight plate attached to the barbell, was used for calibration. A pixel-to-centimeter conversion factor was calculated by identifying the number of pixels corresponding to the visible plate diameter in the video frame.

2.5 Barbell Trajectory Classification

The study on barbell trajectory distribution varies across weight categories. The most common trajectory was Type 3, observed in 53% of lifters at the 2015 World Weightlifting Championships (WWC) and 59% at the 2017 Pan-American Weightlifting Championships (PAWC) [16]. It was particularly prevalent among heavier male lifters and top finishers. Type 2, which does not cross the vertical reference line, was the second most frequent, representing about 30% of male and female lifters. In contrast, Type 1 was less common, appearing in 13% of lifts at WWC and 8% at PAWC. The rarest trajectory was Type 4, accounting for just 6% and 3% at WWC and PAWC, respectively.

Initially, we considered classifying the barbell trajectory by training a Machine Learning (ML) algorithm. However, we decided against this approach for several reasons. First, we lacked a sufficiently large dataset to train an ML model reliably. The prevalence of Types 2 and 3 trajectory causes a data imbalance and leads to bias in Machine learning (ML) algorithms. Developing and

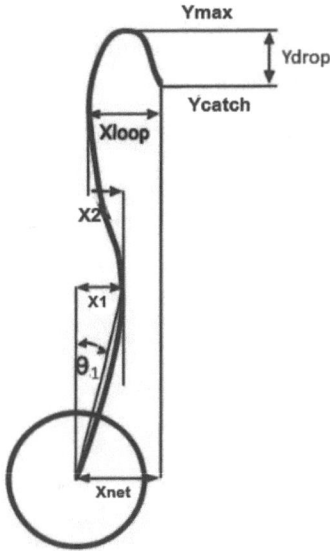

Fig. 5. Barbell kinematic variables of displacement [16].

teaching an accurate model would have been computationally intensive and time-consuming. This was unnecessary for our objective, as the classification could be effectively achieved by observing barbell movement patterns. As a result, we adopted a rule-based approach by analyzing the horizontal displacement patterns to classify barbell trajectories and determine whether the barbell crosses the vertical reference line at critical points. The classification is based on key motion parameters extracted from tracking data and described by Algorithm 1. If no matching type is found, the algorithm adds them to "unknown" for the review.

2.6 Calculation of Barbell Kinematic Variables

Eight key barbell kinematic variables are associated with the snatch lift as shown in Fig. 5. These variables help us determine whether the lift was performed accurately or whether there were chances of potential injuries. We consider the center of the barbell rod to be the origin of our analysis. The eight kinematic parameters are discussed as follows:

1) Y_{max} is the highest point achieved during the lift [16]. We calculate it as the highest y-coordinate during the lift.

2) Y_{catch} is the height of the catch [16]. We calculate it by finding the lowest y-coordinate post Y_{max}, which happens due to the load of the weight of the barbell before the y-coordinate starts increasing again.

Algorithm 1. Barbell Trajectory Classification

1: **Input:** Standardized barbell trajectory coordinates
   $C = [(x_1, y_1), (x_2, y_2), \ldots, (x_n, y_n)]$
2: **Output:** Classified trajectory type
3: Initialize $x\_negative \leftarrow$ False {Tracks if x becomes negative}
4: Initialize $x\_cycles \leftarrow 0$ {Counts oscillations in x direction}
5: Initialize $increasing\_y \leftarrow$ True {Tracks if y is increasing}
6: Initialize $phase \leftarrow$ "increasing" {Tracks whether x is increasing or decreasing}
7: **for** $i = 2$ **to** n **do**
8:     $x \leftarrow C[i, 0]$
9:     $y \leftarrow C[i, 1]$
10:    **if** $x < 0$ **then**
11:       **if** not any$(C[1:i, 0] > 0)$ **then**
12:          **Return** "Type 3"
13:       **end if**
14:       $x\_negative \leftarrow$ True
15:    **end if**
16:    **if** $i > 2$ **and** $y < C[i-1, 1]$ **then**
17:       $increasing\_y \leftarrow$ False {Y has started decreasing}
18:    **end if**
19:    **if** $increasing\_y$ **then**
20:       $prev\_x \leftarrow C[i-1, 0]$
21:       **if** $phase =$ "increasing" **and** $x < prev\_x$ **then**
22:          $phase \leftarrow$ "decreasing" {X starts decreasing}
23:       **else if** $phase =$ "decreasing" **and** $x > prev\_x$ **then**
24:          $x\_cycles \leftarrow x\_cycles + 1$ {X starts increasing again, completing a cycle}
25:          $phase \leftarrow$ "increasing"
26:       **end if**
27:    **end if**
28: **end for**
29: **if** NOT $x\_negative$ **then**
30:    **Return** "Trajectory Type 2"
31: **else if** $x\_cycles \geq 2$ **then**
32:    **Return** "Trajectory Type 4"
33: **else if** $x\_cycles = 1$ **then**
34:    **Return** "Trajectory Type 1"
35: **else**
36:    **Return** "No matching trajectory type detected."
37: **end if**

---

3) Y_{drop} is the difference between $Y_{\max}$ and Y_{catch} [16]. We calculate it by the difference of y-coordinates of $Y_{\max}$ and Y_{catch}. The less the Y_{drop}, the better the athlete has performed the lift and the lower the chances of injury.

4) X_{net} is the net horizontal displacement from the start position to Y_{catch} [16]. Since the origin is at (0,0), X_{net} is simply the x-coordinate of Y_{catch}:

$$X_{\text{net}} = x_{\text{catch}} \tag{1}$$

5) X_1 is the net horizontal displacement from the start to the most rearward position during the first displacement phase toward the lifter [16]. To calculate X_1, we ignore all the negative x-coordinates until the x-coordinate starts increasing. X_1 is the maximum value before the x-coordinates start decreasing.

6) θ_1 is the angle relative to the vertical reference line from the start position to the position at X_1 [16]. It is calculated as:

$$\theta_1 = \tan^{-1}\left(\frac{X_1}{y_{X1}}\right) \qquad (2)$$

where y_{X1} is the y-coordinate corresponding to X_1.

7) X_2 is the horizontal distance from X_1 to the most anterior position between X_1 and Y_{max} [16]. Since the x-coordinates start decreasing after X_1, we define a temporary coordinate x_{temp} as the minimum value before x-coordinates start increasing again. Then, X_2 is calculated as:

$$X_2 = X_1 - x_{temp} \qquad (3)$$

8) X_{loop} is the horizontal distance from X_2 to Y_{catch} [16]. It is given by:

$$X_{loop} = x_{catch} - x_{temp} \qquad (4)$$

2.7 Athlete Performance Scoring

To properly validate our method, we needed a way to measure performance after correct trajectory classification. Therefore, we measure three barbell kinematic parameters: Y_{max}, X_1, and Y_{drop} to generate a score for the lifting performance. Here Y_{max} shows how much vertical height the athlete generated [15], X_1 indicates the initial barbell balance and setup [32], and bar drop relates to how efficiently the athlete transitions into the catch phase [33].

After successful trajectory classification, we check whether the barbell movement stayed within expected ranges for three parameters to generate a score. For an athlete, we generate a score out of 4, where 1 is assigned for successful classification and subsequently 1 each for three metrics if they are within the typical range observed in Olympic weight lifting; the parameter is assigned 0 if it falls outside the range. For an incorrect classification, 0 is assigned and no further processing is done. Therefore, the performance is quantified on a scale of 0 (Incorrect Trajectory) to 4 (correct trajectory and all kinematics parameters are within range). We further categorize each score value into one of the five categories - Very bad to excellent, as shown in Table 1.

Table 1. Assigning categories to the score value.

Score Value	Interpretation	Score Category
0	Incorrect Trajectory	Very Bad
1	Correct Trajectory but all parameters outside range	Bad
2	Correct Trajectory but two parameters outside range	Fair
3	Correct Trajectory but one parameter outside range	Good
4	Correct Trajectory and all parameters within range	Excellent

3 Results and Discussion

Our preliminary investigation analyzed the barbell trajectories during athletes' weightlifting sessions across 6000 frames from 10 videos.

3.1 Tracker Solution For Barbell Trajectory Detection

We experimented to detect the barbell trajectory with the help of different trackers such as Boosting, MIL (Multiple Instance Learning), KCF (Kernelised Correlation Filters), TLD (Tracking-Learning-Detection), MedianFlow, GoTurn, Dlib-Tracker, CamShift, and Template Matching. In our study, we recorded videos directly in the athlete's lateral plane, introducing challenges such as background clutter and potential occlusions due to static objects in the environment.

Among the trackers evaluated, the MedianFlow tracker consistently provided the most accurate and stable performance, particularly in handling occlusions and maintaining robustness across frames, consistent with previous findings [34]. In contrast, other trackers showed significant limitations: Boosting and MIL failed to detect tracking losses, KCF struggled with rapid movements [35], and TLD, although accurate, was computationally intensive [36]. GoTurn required domain-specific training [37], while DlibTracker and CamShift were sensitive to speed variations and lighting changes, respectively [38,39]. Template Matching also proved unreliable due to sensitivity to appearance changes [40].

3.2 Trajectory Classification

Our classification algorithm achieved an accuracy of 70% detecting 7 out of 10 trajectories correctly. The misclassifications in the lifts can be primarily attributed to deviations from expected barbell trajectories. In one of the lifts, the barbell did not cross the vertical reference line—a key feature for the Type 1 trajectory type, whereas in the other, it did, leading to confusion between types. Additionally, for one of the cases, the detected trajectory did not align with any predefined types. These discrepancies likely stem from non-standard execution by the athletes, technical errors during the lift, or observational error by the annotator, ultimately leading to incorrect classification outcomes.

3.3 Kinematic Parameter Extraction

The average height measured by our measurement is 173.33 cm, which is in close accordance with the average height of 175.26 cm [16] for male weight lifters in the USA. The average values of kinematics across 6000 frames are shown in Table 2. Here we compare the average values of the kinematic parameters with the typical range [16,41,42] observed in elite weight lifters to check measurement efficacy. All these ranges are proportional to the athlete's height. For valid comparison, measurements are normalized with respect to athlete height. Y_{max} and Y_{catch} are primary vertical measures, with Y_{drop} indicating efficiency. X_{Net}, X_1, X_2, and X_{loop} describe horizontal movement and should be minimum for best techniques. Angle θ describes the bar path and the lifter's mechanics. These ranges provide a robust framework for analyzing and comparing kinematic parameters in Olympic snatch and lift trajectories. For example, efficient lifts (Type I trajectory) typically show minimal X_{Net}, small Y_{drop}, and smooth transitions between phases. We can observe from Table 2 that all average values of parameters are within the range, ensuring fidelity of CV-based measurement.

Table 2. Average values of kinematic parameters and their typical range in Olympic snatch and lift [16,41,42] normalized to average athlete height estimated from 10 videos. The average estimated height is 173.33 cm.

Parameter	Average Value	Typical range %	Typical range (cm)
Y_{max} (cm)	133.00	70–85	120–145
Y_{catch} (cm)	125.22	60–75	100–130
Y_{drop} (cm)	07.78	04–15	6.5–25
X_{net} (cm)	23.61	00–10	00–17
X_1 (cm)	05.11	02–06	03–10
θ_1 (degrees)	04.44	05–15	–
X_2 (cm)	08.17	02–08	03–14
X_{loop} (cm)	02.76	02–07	03–12

3.4 Trajectory Detection and Performance Score

The results of 10 videos are summarized in Table 3. For each video, the table compares the trajectory type identified by subject matter experts with the predicted type obtained from our algorithm. It also shows the estimated height of the athlete using computer vision and values for three kinematics parameters. The score is computed by checking whether Y_{max}, X_1, and Y_{drop} falls in the designated range as indicated in Table 2.

From Table 3, we observe that for athletes 1, 2, and 6, the trajectory could not be correctly classified and therefore, they were assigned a 0 score or a very bad category. We can see that most athletes showed good vertical barbell displacement, with Y_{max} values that matched well with their respective heights and falls in the range of 120–145 cm. This suggested that they could generate sufficient vertical force during the lift. X_1, representing the initial horizontal position, was within the ideal range (3–10 cm) for most lifts, except for athletes 5 and 7, indicating a proper starting setup for them. For athletes 5 and 7, the score is 3, or category Good is generated due to X_1 falling outside the ideal range. Similarly, the bar drop (Y_{drop}) was in the expected range of 6.5–25 cm for efficient catch mechanics in most cases. Overall, athletes 3, 4, 8, 9, and 10 had a score of 4 or excellent category lifts, as they consistently had Y_{max}, controlled X_1 displacement, and optimal bar drop values.

Table 3. Classification, barbell kinematic parameters validation, and Performance score. The boldface indicates incorrect classification or a parameter not in the range. Cat. indicates the category of the score.

Athlete	Type Given	Type Predicted	Height of Athlete cm	Y_{max} cm	X_1 cm	Y_{drop} cm	Score & Cat.
1	1	**2**	165	129.0	5.0	8.0	0 & Very Bad
2	2	**1**	170	133.5	1.5	9.5	0 & Very Bad
3	3	3	174	137.0	3.5	14	4 & Excellent
4	3	3	179	139.5	8.0	4.5	4 & Excellent
5	3	3	185	139.5	**11**	6.0	3 & Good
6	4	–	–	–	–	–	0 & Very Bad
7	2	2	170	133.5	**3.0**	6.5	3 & Good
8	1	1	177	139.0	5.0	6.0	4 & Excellent
9	1	1	178	136.0	3.5	7.0	4 & Excellent
10	1	1	162	126.0	5.5	8.5	4 & Excellent

3.5 Limitations of the Proposed Approach

The reliability of the proposed approach is based on certain assumptions. One major limitation is the dependency on the manual drawing of the bounding box annotations around the barbell. Some vision-related challenges are shown in Fig. 6. Occlusions from external objects in the background led to tracking failures or inaccuracies. Additionally, the method assumes that the camera is positioned laterally in line with the athlete performing the lift. When the videos are captured from a high angle (20° or more), this assumption breaks down. This would need preprocessing of videos to correct the affine transformation. Furthermore, tracking continuity was lost if the barbell moved out of the camera frame due to improper field of view.

Another major issue arose when another athlete performed a different movement behind the main subject, causing the tracker to lock onto the background motion instead of the barbell incorrectly. Lastly, an excessively high frame rate introduced motion blur and unnecessary frame redundancy, making it challenging to extract precise kinematic variables of the barbell. Since the approach relies on pixel-level measurements to classify trajectory patterns and calculate derived metrics, the resulting perspective distortions lead to improper trajectory classification and reduce the interpretability of the results. The performance score could be made more discriminative with the addition of features like velocity and total power in the lifts. These challenges highlight the need for careful camera placement, appropriate frame rate selection, and improved tracking methods to enhance accuracy in motion analysis.

Fig. 6. Tracking Challenges: i. Occlusion due to the background person, ii. Video taken from an angle, iii. The barbell is going out of the camera plane.

4 Conclusion and Future Work

Incorrect posture while lifting weights significantly affects athletes' performance and careers. Our proposed approach demonstrates strong predictive capabilities for identifying risk factors and achieving high precision. Preliminary tests on additional video samples suggest that the framework can be generalised to achieve similar results; however, broader validation is required.

Future work will focus on expanding the dataset and incorporating advanced detection techniques, such as automating the pixel-to-centimetre conversion and mitigating affine transformation effects through camera calibration. Manual bounding box initialization will be replaced by real-time object detection to streamline the process.

To enhance adaptability to subtle variations in lifting styles, the current rule-based classifier will be replaced with a machine learning-based model trained on a larger and more diverse dataset, improving trajectory classification accuracy. The analysis will be extended beyond snatch movements to include other Olympic lifts, such as the clean and jerk.

Deployment on mobile applications is planned to facilitate real-time analysis, increasing accessibility for athletic training and injury prevention. Finally, the performance metric will be made more holistic by incorporating velocity and power measurements.

Acknowledgment. The authors thank the weight lifters and coaching staff for cooperating and participating in this study. We are also grateful to Sacred Heart University, USA, Old Dominion University, USA, and Ahmedabad University, India, for their support in this research work.

References

1. USA Weightlifting. The lifts. Accessed 5 Apr 2025
2. International Weightlifting Federation. What are the 2 olympic lifts?. Accessed 5 Apr 2025
3. Wikipedia contributors. Snatch (weightlifting) (2024). https://en.wikipedia.org/wiki/Snatch_(weightlifting). Accessed 19 Apr 2025
4. Wikipedia contributors. Clean and jerk (2024). https://en.wikipedia.org/wiki/Clean_and_jerk. Accessed 19 Apr 2025
5. International Weightlifting Federation. Iwf technical and competition rules & regulations (2024). https://iwf.sport/technical-and-competition-rules/. Accessed 19 Apr 2025
6. International Weightlifting Federation. Bodyweight categories – IWF (2024). https://iwf.sport/bodyweight-categories/. Accessed 19 Apr 2025
7. Garhammer, J.: A review of power output studies of olympic and powerlifting: methodology, performance prediction, and comparative analysis. J. Strength Cond. Res. **7**(2), 76–89 (1993)
8. International Weightlifting Federation. Specifications for competition equipment (2023). https://iwf.sport/weightlifting_equipment. Accessed 19 Apr 2025
9. International Weightlifting Federation. Technical and competition rules & regulations (2020). https://iwf.sport/wp-content/uploads/downloads/2020/03/IWF_TCRR_2020.pdf. Accessed 4 Apr 2025
10. Vulcan Strength. Absolute training kg bumper plates & barbell set (n.d.). https://www.vulcanstrength.com/Absolute-Training-Kg-Bumper-Plates-Barbell-Set-p/v-atkgbarbump160.htm. Accessed 4 Apr 2025
11. Gourgoulis, V., Aggeloussis, N., Kalivas, V., Antoniou, P., Mavromatis, G.: Biomechanical differences of snatch technique between elite male and female weightlifters. J. Sports Med. Phys. Fitness **49**(1), 1–7 (2009)
12. Huebner, M., Perotto, A., Suchomel, T.J.: Biomechanics of olympic weightlifting: a review. Strength Cond. J. **43**(3), 56–69 (2021)
13. Baumann, W., Gross, V., Quade, K., Galbierz, P., Schwirtz, A.: The snatch technique of world class weightlifters at the 1985 world championships. Int. J. Sport Biomech. **4**(1), 68–89 (1988)
14. Garhammer, J.: A review of power output studies of olympic and powerlifting: methodology, performance prediction, and evaluation tests. J. Strength Cond. Res. **5**(2), 89–102 (1991)

15. Gourgoulis, V., Aggeloussis, N., Garas, A., Mavromatis, G.: Snatch lift kinematics and bar energetics in male adolescent and adult weightlifters. J. Sports Med. Phys. Fitness **40**(2), 126–131 (2000)
16. Cunanan, A.J., et al.: Survey of barbell trajectory and kinematics of the snatch lift from the: world and 2017 pan-American weightlifting championships. Sports **8**(9), 2020 (2015)
17. Garhammer, J.: A review of power output studies of olympic and powerlifting: methodology, performance prediction, and evaluation tests. J. Strength Cond. Res. **7**(2), 76–89 (1993)
18. Enoka, R.M.: Neuromechanics of human movement. Human Kinetics (2008)
19. Tendo unit overview (2025). https://www.tendosport.com/products/tendo-unit/overview/. Accessed 30 Apr 2025
20. Gymaware rs (LPT) - gold standard VBT device (2025). https://gymaware.com/gymaware-rs/. Accessed 30 Apr 2025
21. Weakley, J.J.S., Mann, B., Banyard, H.G., McLaren, S.J., Scott, T., Garcia-Ramos, A.: Comparison of velocity-based and percentage-based training methods for improving strength and power in collegiate athletes. Int. J. Sports Physiol. Perform. **16**(4), 516–523 (2021)
22. Balsalobre-Fernández, C., Kuzdub, M., Poveda-Ortiz, P., Campo-Vecino, J.D.: Validity and reliability of the push wearable device to measure movement velocity during the back squat exercise. J. Strength Cond. Res. **31**(9), 2610–2617 (2017)
23. Chiu, L.Z.F., Schilling, B.K.: A kinematic analysis of a weightlifting exercise using mathematical modeling. Strength Cond. J. **27**(1), 42–51 (2005)
24. Picerno, P.: Wearable inertial sensors for human movement analysis: a five-year update. J. Biomed. Sci. Eng. **10**(8), 147–156 (2017)
25. Dæhlin, T.E., Krosshaug, T., Chiu, L.Z.F.: Enhancing digital video analysis of bar kinematics in weightlifting: a case study. J. Strength Cond. Res. **31**(6), 1592–1600 (2017)
26. Colyer, S.L., Evans, M., Cosker, D.P., Salo, A.I.T.: A review of the evolution of vision-based motion analysis and the integration of advanced computer vision methods towards developing a markerless system. Sports Med. - Open **4**(1), 24 (2018)
27. Balsalobre-Fernández, C., Cuspinera, E., Campo-Vecino, J.D., Romero-Moraleda, B.: Automatic identification of vertical jumps using smartphone video analysis: a reliability and validity study. J. Strength Cond. Res. **34**(10), 2953–2959 (2020)
28. Mathis, A., Warren, R.A., Uchida, T., Suleiman, A., Mathis, M.W.: Deep learning tools for the measurement of animal behavior in neuroscience. Curr. Opin. Neurobiol. **60**, 1–11 (2020)
29. GoPro. Gopro hero9 black—more everything (2020). https://gopro.com/en/us/news/hero9-black-launch. Accessed 19 Apr 2025
30. OpenCV Team. cv::trackermedianflow class reference (2021). https://docs.opencv.org/3.4/d7/d86/classcv_1_1TrackerMedianFlow.html. Accessed 02 Apr 2025
31. Whittlesey, S.: Biomechanics of Sport and Exercise, 3rd edn. Human Kinetics (2017)
32. Isaka, T., Okada, J., Funato, K.: Kinematics analysis of the barbell during the snatch movement of elite Asian weightlifters. J. Appl. Biomech. **12**(4), 508–516 (1996)
33. Baehrle, R.E., Earle, R.W.: Fitness and wellness. McGraw-Hill Education (2000)
34. Varfolomieiev, A., Lysenko, O.: An improved algorithm of median flow for visual object tracking and its implementation on arm platform. J. Real-Time Image Proc. **11**(3), 527–534 (2013)

35. Henriques, J.F., Caseiro, R., Martins, P., Batista, J.: High-speed tracking with kernelized correlation filters. IEEE Trans. Pattern Anal. Mach. Intell. **37**(3), 583–596 (2015)
36. Kalal, Z., Mikolajczyk, K., Matas, J.: Tracking-learning-detection. IEEE Trans. Pattern Anal. Mach. Intell. **34**(7), 1409–1422 (2012)
37. Held, D., Thrun, S., Savarese, S.: Learning to track at 100 FPS with deep regression networks. In: Leibe, B., Matas, J., Sebe, N., Welling, M. (eds.) ECCV 2016. LNCS, vol. 9905, pp. 749–765. Springer, Cham (2016). https://doi.org/10.1007/978-3-319-46448-0_45
38. King, D.E.: DLIB-ML: a machine learning toolkit. J. Mach. Learn. Res. **10**, 1755–1758 (2009)
39. Bradski, G.R.: Computer vision face tracking for use in a perceptual user interface. Intel Technol. J. **2**(2), 1–15 (1998)
40. Brunelli, R.: Template Matching Techniques in Computer Vision: Theory and Practice. Wiley (2009)
41. Harbili, E., Alptekin, A.: Comparative kinematic analysis of the snatch lifts in elite male adolescent weightlifters. J. Sports Sci. Med. **13**(2), 417–422 (2014)
42. Nagatani, T., Haff, G.G., Guppy, S.N., Poon, W., Kendall, K.L.: Effect of different set configurations on barbell trajectories during the power snatch. Int. J. Sports Sci. Coach. **18**(5), 1594–1604 (2022)

A Data-Driven Imputation Scheme for Cohort Studies: A Collegiate Basketball Casestudy

Srishti Sharma[1(✉)], Vishal Barot[2], Srikrishnan Divakaran[3], Tolga Kaya[4], Christopher B. Taber[5], and Mehul S. Raval[1]

[1] School of Engineering and Applied Science, Ahmedabad University, Ahmedabad, Gujarat, India
{srishti.s1,mehul.raval}@ahduni.edu.in
[2] LDRP Institute of Technology and Research, Gandhinagar, Gujarat, India
vishal_ce@ldrp.ac.in
[3] School of Interwoven Arts and Sciences, Krea University, Sri City, Andhra Pradesh, India
srikrishnan.divakaran@kreauni.edu.in
[4] School of Computer Science and Engineering, Sacred Heart University, Fairfield, CT, USA
kayat@sacredheart.edu
[5] Department of Physical Therapy and Human Movement Science, Sacred Heart University, Fairfield, CT, USA
taberc@sacredheart.edu

Abstract. Missing data remains a critical challenge in cohort studies. This study introduces a novel missing value imputation technique that integrates feature sensitivity and factor analysis with clustering and predictive modelling to enhance accuracy, reliability, and interpretability.

The dataset comprises 42 features collected from 16 collegiate female basketball athletes over 26 weeks, including sleep and cardiac rhythms, training loads, cognitive states, travel, and countermovement jump performance. The objective is to model the impact of these contextual stressors on athletic readiness, quantified via the Reactive Strength Index modified (RSImod).

When compared to state-of-the-art the proposed methodology reduces computation time by up to 35.71% (KNN), 29.41% (EM), 21.43% (MICE), 14.29% (CART), and 7.14% (XGBoost). It reduces RMSE by up to 12.20% and MAE by up to 10.77%. Moreover, RSImod predictions on the imputed dataset showed substantial improvements, up to an 80.85% reduction in MSE and a 79.99% increase in R^2 scores. Interpretability was enhanced using SHAP (SHapley Additive exPlanations), providing actionable insights for coaches and practitioners.

Keywords: Athletic readiness · clustering · data imputation · factor analysis · feature sensitivity · prediction · SHAP explanations

1 Introduction

Cohort studies follow a group of individuals, typically sharing a common characteristic or exposure, and observe them longitudinally to assess changes and outcomes [1]. In sports science, they help understand athletes' health and readiness, optimise training protocols, prevent injuries, and enhance overall performance [2].

However, sports datasets encounter missing values due to athlete non-compliance, data collection errors, and equipment malfunctions. There are three types of missing data mechanisms: Missing Completely At Random (MCAR), Missing At Random (MAR), and Missing Not At Random (MNAR) [3]. MCAR implies that missing data is independent of any other data, while MAR indicates that missing data correlates with the observed data but not directly with the missing values [4]. In contrast, MNAR suggests that missing data depends on both observed and missing data, making unbiased estimation challenging. Data collected in cohort studies establish associations among variables based on participant characteristics, generating both MAR and MNAR data types [4]. Imputing MAR allows for a more reliable estimation of missing values without introducing additional uncertainty. Therefore, for this study, all data in the database are categorized as MAR missing data [5].

In current practice, cohort studies increasingly leverage machine learning (ML), hence, effective strategies for handling missing data are essential. These include removing instances or features with missing values (resulting in data loss), allowing the ML algorithm to handle missing values during its execution, or replacing them with estimated ones through missing value imputation (MVI), closely approximating the real values [6].

Recent research on data imputation in sports science focused on professional soccer players' training load, revealing that the Daily Team Mean method effectively mitigated inaccuracies due to missing data [7]. Simple imputation replaces missing values with means or modes from available data but may yield poor results in complex datasets [5]. Regression imputation predicts missing values using regression equations fitted on complete data [5]. Expectation-maximization (EM) iteratively estimates missing values based on existing data distributions [8]. Multiple imputations (MICE) generate multiple datasets to reflect uncertainty in missing values, aggregating results for final imputations [6,9]. K-nearest neighbour (KNN) imputation estimates missing values based on similarities with neighbouring samples [10]. Clustering imputation groups data into clusters to impute missing values within similar groups [8]. Decision tree (CART) and random forest (RF) methods build predictive models from complete data and use these to impute missing values [11]. These traditional imputation techniques have limitations, such as assuming linear relationships (Regression), sensitivity to initialization (EM), and reliance on similarity metrics (KNN). MICE requires accurate distribution assumptions. Clustering imputation may introduce bias with poorly defined clusters. CART is prone to overfitting, while RF can bias results if feature selection is inadequate [1].

Recent advancements in deep learning have significantly improved data imputation in medical and health datasets. Overcomplete Denoising Autoencoders (ODAE) and similar deep learning models, which treat missing data as noise and use denoising regularization, have shown superior performance by effectively learning hidden data representations [12]. These models achieve lower mean squared error and higher prediction accuracy, proving more reliable for imputation tasks than conventional methods [13]. However, deep learning methods are less useful in cohort studies due to their need for huge training data, inherent complexity and lack of interpretability, critical for understanding subtle nuances in longitudinal data and deriving actionable insights for personalized interventions [1,13].

This research focuses on a collegiate women's basketball team, aiming to provide actionable advice from the model outcomes. Therefore, quantifying and explaining features and predicted outcomes is crucial. We propose a data-driven MVI technique that adaptively learns from the dataset. The technique first identifies the most influential features using a hybrid feature importance analysis, followed by factor analysis – to generate latent factors compactly representing the original dataset. These steps reduce the data dimensionality (help achieve quadratic speed-up) and improve the interpretability of outcomes. It then utilizes the strength of clustering and a hybrid decision tree (DT) based ensemble-boosting model to improve imputed values' accuracy and reliability. The rest of the paper is organized as follows: Sect. 2 describes the dataset. Section 3 details the methodology adopted to implement the proposed MVI technique. Section 4 summarizes the results and discusses the interpretations drawn and Sect. 5 is the conclusion.

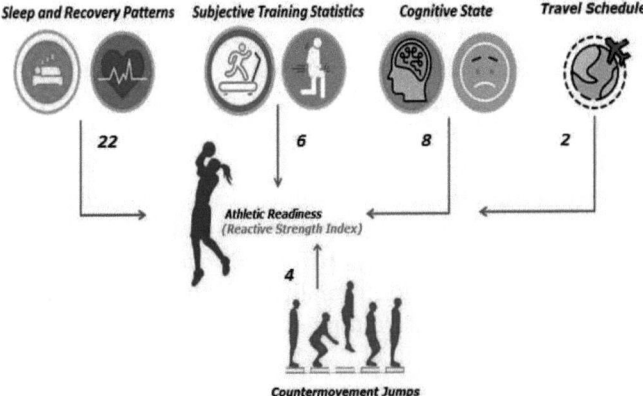

Fig. 1. Multi-modal dataset for athletic readiness modelling and prediction.

2 Dataset

This real-world dataset includes sleep-recovery patterns, subjective training load, cognitive state information, vertical jumps and travel data of collegiate women's basketball athletes. This data was collected to assess the fatigue caused by various internal and external – physical, physiological and cognitive stressors and its impact on athletic readiness. Figure 1 shows the multiple modalities included in this dataset.

2.1 Subjects

From October 2021 to March 2022, sixteen female basketball players (mean age: 21 years; average height: 174.21 cm; mean body mass: 73.98 kg) underwent comprehensive testing and monitoring. Before participation, all participants received detailed explanations of the study procedures and provided informed consent ((Institutional review board approval number XXXXXXX on DD/MM/YYYY).

2.2 Sleep and Recovery Data

All athletes were given Whoop straps to wear continuously throughout the data collection period, except during games and practice. The straps monitored daily activity and sleep, recording data using Whoop's specialized software. The study analyzed 22 metrics per athlete daily, including resting heart rate, heart rate variability, sleep parameters, and recovery metrics. Third-party testing has validated Whoop's reliability and accuracy compared to polysomnography for sleep and heart rate assessments [14].

2.3 Training Data

The training load was quantified weekly by aggregating the workload from sports practice, metabolic conditioning, strength training, and gameplay. After each session, athletes reported their perceived exertion using a 1–10 Likert scale, which was then multiplied by the session duration to compute the session rating of perceived exertion (sRPE) [14]. Total Weekly Load (TWLoad) and its standard deviation were derived from these sRPE values across the week. Additionally, the weekly resistance training load was computed by summing the total weight lifted during resistance sessions ($sets \times repetitions \times load$). Training monotony was calculated as the average daily load normalized by the weekly standard deviation of the training load. In contrast, training strain was determined by multiplying TWLoad by the monotony score [14].

2.4 Short Recovery Short Stress Questionnaire

Twice weekly, athletes utilised an online dashboard to complete a brief questionnaire assessing their emotional and mental states. The survey included eight questions; four focused on recovery and the remainder on stress, each rated on a 0–6 Likert scale [15]. Table 1 lists the internal and external load quantifying features.

Table 1. Internal and external load quantifying features.

Source	Features
Sleep and Recovery (22)	hours of sleep, hours in bed, awake hours, wake periods, sleep disturbances, resting heart rate (RHR), heart rate variability (HRV), respiratory rate, recovery, deep sleep hours, light sleep hours, restorative sleep hours, rapid eye movement (REM) sleep hours, total cycle sleep hours, sleep consistency, sleep efficiency, sleep score, sleep debt hours, total cycle nap hours, latency, sleep cycles, sleep need
Training (6)	strain, monotony, resistance training (RT) volume load, total workload (TWLoad), daily average, weekly standard deviation (SD)
Short recovery short stress (SRSS) questionnaire (8)	overall recovery (OR), overall stress (OS), negative emotional state (NES), lack of activation (LA), muscular strength (MS), physical performance capabilities (PPC), mental performance capabilities (MPC), emotional balance (EB)
CMJs (4)	peak power, jump height (JH), body weight, reactive strength index modified (RSImod)
Travel Schedule (2)	Travel (yes/no), Hours of travel

2.5 Vertical Jump Data

Subjects performed weekly countermovement jumps (CMJs) on the first practice day each week, typically on Monday or Tuesday. After a standardized general warm-up, they executed two submaximal jumps at 50% and 75% of perceived maximum effort, with 30 s of passive rest between repetitions. Dual force plate sampling at 1000 Hz recorded all jumps, with data collected and analyzed using proprietary Force Decks software. The RSImod, jump height via flight time, and peak power were the key performance indicators (KPIs) monitored during routine athlete testing and monitoring [14]. RSImod, derived from CMJ data, measures an athlete's ability to generate maximal vertical impulse quickly, integrating jump height and contact time.

2.6 Travel Data

During the season, athletes travel to different states for games. We gathered travel dates from their game schedule journal, marking them as binary 0 or 1, and recorded the hours of travel involved.

3 Methodology

We designate the feature requiring imputation as the target feature and determine the importance scores of other features using Pearson's correlation, Ran-

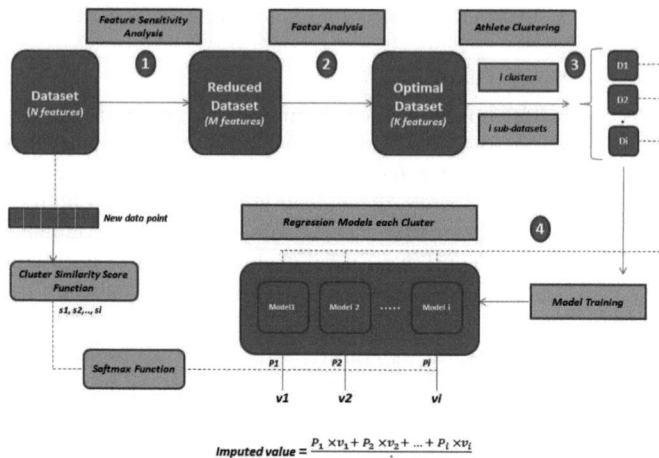

Fig. 2. Proposed methodology depicts how a new data point is imputed.

dom Forest-based feature importance, and XGBoost-based feature importance for the target feature (A). Following this, factor analysis addresses multi-collinearities and reduces dimensionality, improving data interpretation and processing efficiency (B). Athlete clustering via k-means further refines the approach by grouping similar athletes (C). Finally, our model employs XGBoost regressors trained on these clusters, using similarity scores and softmax probabilities to accurately predict missing values (D). Figure 2 shows the proposed methodology – detailing the techniques step-wise.

3.1 Feature Importance Analysis

The feature requiring imputation is designated as the target feature. We determine the importance scores of other features in the dataset relative to the target feature using three distinct techniques: Pearson's correlation [16], Random Forest-based feature importance, and XGBoost-based feature importance. Pearson's correlation captures linear relationships, while ensemble methods like Random Forest and XGBoost handle complex interactions and non-linearities [6].

We proposed a hybrid approach for computing aggregate feature importance (FI) score to offer a comprehensive and nuanced feature selection process, particularly benefiting complex datasets where individual techniques may have limitations [1].

Figure 3 presents the feature importance analysis. For each FI technique, we identify the top M ($M << N$) features based on their significance value (p-values), ensuring only relevant features are selected for model training. Subsequently, we train three models using the selected feature subsets to predict RSImod scores. To evaluate the effectiveness of each method in generating an optimal feature subset for RSImod score prediction, we assess all three models

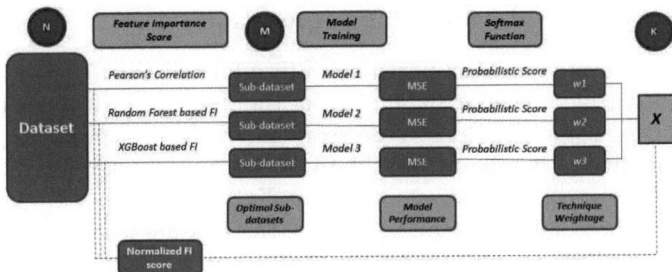

Fig. 3. Hybrid feature importance analysis technique.

based on Mean Squared Error (MSE), defined as the average sum of squared differences between the observed and predicted RSImod scores [17]. The MSE values obtained from these models are then processed using a Softmax function to generate probabilistic scores for each technique, reflecting their respective weights in the overall analysis [18]. By normalising the FI scores for all three techniques within a range of 0 to 1, we computed a weighted average feature importance score for each feature, where the probabilistic score from each technique served as the weight. We identify the top K features using this aggregated score, with K being significantly smaller than N.

3.2 Factor Analysis

We observed several multi-collinearities and linear dependencies in the dataset. To preserve the significance of these features without discarding any, we combined them into latent features through factor analysis [17]. This approach helped reduce dimensionality and simplify the interpretation of the data. We conducted factor analysis on the optimal subset of features to identify underlying latent factors that represent the dataset compactly. Each factor (F_i) is expressed as a linear combination of features (f_x) sharing common variance, with the resulting factor loadings (l_x) serving as weights [17].

$$F_i = \sum_{x=1}^{n} l_x \times f_x \quad (1)$$

3.3 Athlete Clustering

Athlete clustering is a statistical method to identify patterns and similarities by grouping athletes based on shared traits or behaviours within a dataset [19]. It allows us to leverage information from similar athletes to estimate the missing values more accurately [19]. We performed athlete clustering using the k-means clustering technique on the identified latent factors. The Silhouette score was used determine the optimal number of clusters [20].

3.4 Proposed Hybrid Model for Imputation

The dataset was systematically partitioned into four distinct sub-datasets, each representing an athlete cluster by allocating the data records of each athlete to their respective cluster. XGBoost regressor, known for its speed, regularisation capabilities, ability to handle missing data and feature interpretability, was utilized in this process [21]. Subsequently, four XGBoost regressors were trained, each on one of these sub-datasets, to predict the missing feature values.

Algorithm: Missing Value Imputation

Step 1: Cluster Assignment and Similarity Score Calculation. For a new record x, the similarity score s_i is calculated as the negative Euclidean distance between the record and the centroid of the cluster μ_i:

$$s_i = -\|x - \mu_i\| \qquad (2)$$

where $\|x - \mu_i\|$ is the Euclidean distance between the record x and the centroid μ_i of cluster i.

Step 2: Softmax Probabilities. The similarity scores are converted into probabilistic weights using the softmax function:

$$P_i = \frac{e^{s_i}}{\sum_{j=1}^{k} e^{s_j}} \qquad (3)$$

where e is the base of the natural logarithm, s_i is the similarity score for cluster i, and k is the number of clusters.

Step 3: Missing Value Imputation. The final feature value $\hat{y}$ is computed as a weighted average of the predictions from the XGBoost regressors, with the softmax probabilities as weights:

$$\hat{y} = \sum_{i=1}^{k} P_i \times \hat{y}_i \qquad (4)$$

where $\hat{y}_i$ is the feature value predicted by the XGBoost regressor trained on cluster i, and P_i is the softmax probability for cluster i.

For the imputation of missing feature value, a new record was first classified using an unsupervised machine learning model developed through the k-means clustering technique to determine the athlete's cluster affiliation. The similarity score for this record with each cluster was calculated based on its distance from the centroid of each cluster. These similarity scores were passed through a softmax function, generating probabilistic weights for the individual clusters. The final feature value was imputed as a weighted average of the predictions from the four XGBoost regressors, with the softmax probabilities serving as weights corresponding to their respective clusters.

4 Experimental Evaluation

4.1 Results

The dataset has 42 features in all and the overall rate of missing data is 32%. As this dataset's objective is modelling and predicting athletic readiness, RSImod serves as the KPI, which is the target feature. We first computed Pearson's correlation coefficient value for all other features of RSImod to identify which features caused the most variability to the RSImod score. Apart from the RSImod score, the features identified as the most significant ones -hours of sleep (0.72), TWLoad (0.68), HRV (0.60) and sleep need (0.54) were selected for imputation. The methodology and results of the preprocessing steps have been explained in terms of RSImod score imputation.

To impute the RSImod score for the weeks athletes missed their resistance and strength training sessions, we set the RSImod score as the target feature. The dataset included daily recordings for each of the 41 ($N = 41$) independent features, while the weekly CMJs provided a single RSImod score per week. To facilitate imputation, we first adjusted the dataset by averaging the daily readings for each feature into a single weekly reading. Over 26 weeks, 416 RSImod scores were recorded ($n = 416$, $mean = 0.37 \pm 0.08$) with 98 missing entries (used as a test set).

Using the three distinct techniques, we first calculated the FI score of the 41 independent features. We employed forward selection and backward elimination methods to identify the features that most significantly impacted RSImod prediction, with the threshold for the number of features set by the p-significance value. On average, the top 15 features, with a $p < 0.05$, improved model performance and were included in the reduced data subsets. We calculated the MSE of the models, and the FI scores for all techniques were normalized to a range of 0 to 1. The MSE values from these models were then processed through a Softmax function, which generated a probabilistic score for each technique, reflecting its weight in the overall analysis. We computed each feature's average importance score using these weights. Using this aggregated score and considering the p-values derived from model fits, we identified the top 20 features (please refer to Fig. 4) as the optimal subset for further analysis.

Our dataset now consisted of 21 features (20 input features and RSImod). To assess the suitability of the dataset for factor analysis, we conducted the Kaiser-Meyer-Olkin (KMO) test, which measures sampling adequacy, and Bartlett's test of sphericity, which examines the redundancy among features that could be summarized into factors. The KMO test yielded a value of 0.69 (> 0.50), and Bartlett's test resulted in a significance value of 0.000 (< 0.005). These results indicated that the dataset was suitable for factor analysis. Consequently, we performed factor analysis, which computed four latent factors. The features were then categorised into these four discrete clusters (factors) based on their correlations with the identified factors, resulting in a 5 feature dataset (4 factors and RSImod).

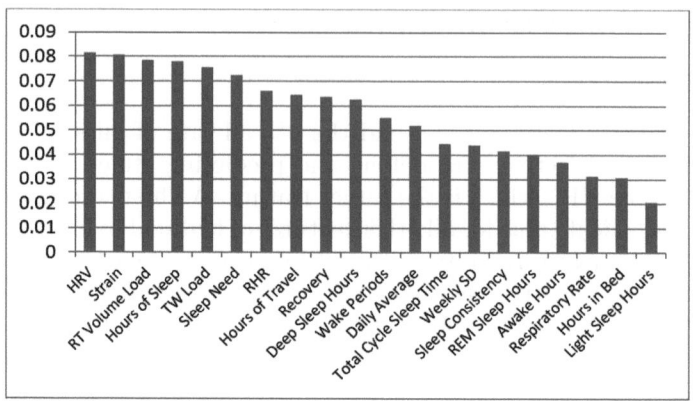

Fig. 4. Top 20 features constituting the optimal dataset.

- The features "hours of sleep", "hours in bed", "total cycle sleep time", "awake hours", "wake periods", "deep sleep hours", "REM sleep hours", "light sleep hours", "sleep consistency", and "sleep need", collected using the WHOOP strap, showed the highest correlation with *factor 0* based on their factor loadings. We computed the value for factor 0 as a linear combination of these features, using their factor loadings as weights, and named this factor "Sleep".
- The features "HRV", "RHR", "recovery", and "respiratory rate", also collected from the WHOOP strap, showed the highest correlation with *factor 1*. We calculated the value for factor 1 for each record by using the values of these features and their factor loadings as weights, naming this factor "Cardiac Rhythm".
- Similarly, features "TWLoad", "strain", "RT volume load", "weekly SD", and "daily average", collected during strength and resistance training sessions, constituted *factor 2*, which we named "Training Strain". The "hours of travel" feature formed *factor 3*, named "Travel Schedule".

We compared the MSE and adjusted R^2 scores of our approach using latent factors against those of multi-linear regression, incorporating all input features from the dataset to assess the efficacy of factor analysis in reducing redundancy without significant information loss. (Please refer to Table 2 for detailed comparisons).

Table 2. MSE and R^2 scores comparison.

Dataset	MSE	R^2
Original (41 features)	0.048	0.560
Optimal (4 factors)	0.028	0.694

Notice that the optimal dataset after factor analysis improved the model's prediction accuracy (slightly smaller MSE) and reduced variability (slightly increased adjusted R2 score). It signifies that factor analysis effectively captured the essential information in the data by eliminating redundant features, resulting in reduced data dimensionality.

Among the features categorized under "Sleep", "hours of sleep" was identified as the most significant, having the highest factor loading. For "Training Strain", the most significant feature was "strain"; for "Cardiac Rhythm", it was "HRV"; and for "Travel Schedule", it was "hours of travel". Following the silhouette score recommendation, we utilised the k-means clustering algorithm to partition the dataset into four distinct athlete clusters. Table 3 summarises the average cluster statistics.

Table 3. Average cluster statistics.

Cluster	Athletes	Sleep (hours)	Training	RSImod
Moderate Performers (Cluster 1)	4, 6, 13, 16, 17	6.78	2060	0.32
Intensive Trainers (Cluster 2)	2, 8, 10, 12, 16	6.59	2268.5	0.24
High Performers (Cluster 3)	3, 7, 9, 11	7.28	1935	0.39
Suboptimal Performers (Cluster 4)	1, 5, 15	7.28	1890	0.27

Overall, we observed an imbalance in the dataset, with 38% of the records associated with high and very high RSImod scores and 62% associated with low and moderate RSImod scores. To address this imbalance, we applied the SMOGN [6] technique for data balancing, increasing the overall sample size to 712. This adjustment ensured that 50% of the records belonged to high and very high levels of athlete readiness and the remaining 50% to low and moderate levels. We partitioned the dataset into four subsets by segregating the athlete data points corresponding to each cluster.

To validate model efficacy, we introduce missing data into the dataset by deliberately removing 213 RSImod score values, resulting in 499 records allocated for training and those 213 for testing using a traditional 70:30 training-to-test ratio. Over each of these subsets, we trained an XGBoost regressor, with RSImod as the target feature. For each record in the test dataset, we first identified the athlete's cluster affiliation using an unsupervised machine learning model. Next, we computed the distance from the record to the centroid of each cluster to obtain a similarity score. This similarity score was passed through the Softmax function to generate probabilistic weights for each cluster. The final RSImod score was imputed as a weighted average of the XGBoost regressor predictions, with the Softmax probabilities serving as weights corresponding to their respective clusters. The hybrid model achieved an MSE of 0.0102, outperforming the individual cluster-based models with MSEs of 0.0265, 0.0287, 0.0271, and 0.0197 for Clusters 1, 2, 3, and 4, respectively.

To validate the efficacy of the imputations performed by the proposed approach, it was compared against five state-of-the-art missing value imputation (MVI) techniques: KNN, XGBoost, EM, Cart, and MICE. The evaluation criteria used were Root Mean Square Error (RMSE) and Mean Absolute Error (MAE). Table 4 and 5 present the RMSE and MAE values obtained for each technique across five different features, respectively (Fig. 5).

Table 4. RMSE comparison.

Technique/Feature	Hours of Sleep	Sleep Need	TWLoad	OR	HRV
KNN	0.78	1.28	0.98	1.1	0.88
MICE	0.74	1.25	0.95	1.08	0.86
EM	0.82	1.35	1.05	1.18	0.95
Cart	0.79	1.3	1.00	1.12	0.9
XGBoost	0.81	1.32	1.02	1.15	0.92
Proposed	0.72	1.21	0.93	1.05	0.84

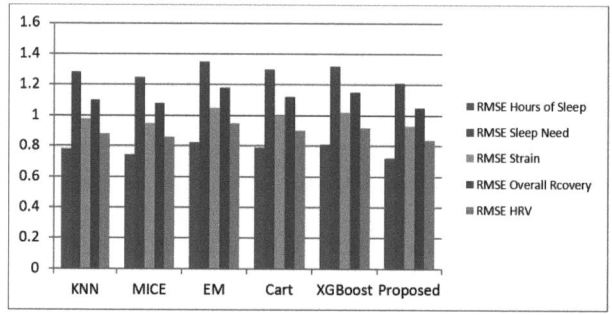

Fig. 5. RMSE comparison of proposed technique with state-of-the-art techniques.

The RMSE and MAE values for sleep features, training metrics, cognitive variables, and cardiac rhythm parameters consistently showed lower error rates with our approach compared to KNN, XGBoost, EM, Cart, and MICE. Specifically, for hours of sleep, our technique achieved an RMSE of 0.72 and MAE of 0.58, while the best-performing comparative technique, MICE, obtained an RMSE of 0.74 and MAE of 0.60. Similar trends were observed across all features, demonstrating the robustness and efficacy of our proposed method in handling missing data (Fig. 6).

To further validate the efficacy of the proposed methodology, RSImod predictions were generated using XGBoost regressor models trained on datasets

Table 5. MAE comparison.

Technique/Feature	Hours of Sleep	Sleep Need	TWLoad	OR	HRV
KNN	0.62	1.02	0.79	0.85	0.7
MICE	0.6	1	0.77	0.84	0.68
EM	0.65	1.08	0.84	0.92	0.75
Cart	0.63	1.04	0.8	0.87	0.71
XGBoost	0.64	1.06	0.82	0.9	0.73
Proposed	0.58	0.97	0.76	0.82	0.67

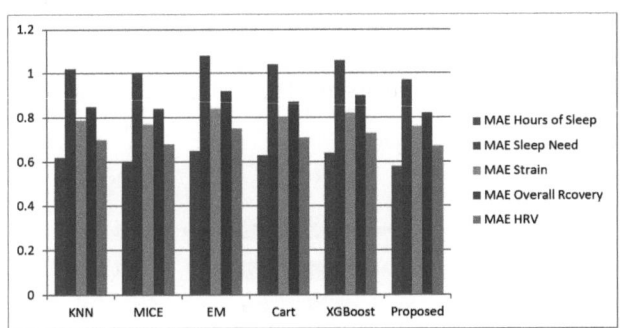

Fig. 6. MAE comparison of proposed technqiue with state-of-the-art techniques.

imputed by these MIV techniques (RSImod score imputation for 98 missing values). Table 6 compares model performances in terms of MSE and R^2 score.

The model fit over the dataset imputed using our proposed approach achieved the lowest MSE of 0.0102, substantially outperforming KNN (0.0534), MICE (0.0312), EM (0.0490), CART (0.0361), and XGBoost (0.0283). Moreover, it exhibited the highest R^2 score of 0.897, indicating superior model fit and predictive accuracy. The superior performance of our proposed technique can be attributed to several key factors. Firstly, our approach integrates a hybrid methodology that leverages feature importance analysis, factor analysis, and athlete clustering to identify and prioritize influential features for imputation. By considering both linear relationships and complex interactions through ensemble methods like XGBoost, our technique effectively captures the nuanced patterns present in the dataset, thereby improving the accuracy of imputed values. Furthermore, incorporating athlete clustering ensures that missing values are estimated based on athletes' similarities, enhancing the imputation process's relevance and precision tailored to individual performance profiles. This comprehensive approach addresses missing data challenges and optimizes predictive modelling outcomes for assessing athletic readiness and performance indicators.

Additionally, we generated SHAP (SHapley Additive exPlanations) values for the predicted RSImod values. These values provide a clear and interpretable rationale behind each imputed value [22]. As depicted in Fig. 7, SHAP values for

Table 6. MSE and R^2 comparison of RSImod score prediction over imputed datasets

MIV technique	MSE	R^2 score
KNN	0.0534	0.487
MICE	0.0312	0.680
EM	0.0490	0.512
CART	0.0361	0.734
XGBoost	0.0283	0.760
Proposed Approach	0.0102	0.897

a predicted RSImod score highlight sleep's paramount importance. Effectively managing training strain and optimizing cardiac rhythm can enhance athlete readiness and performance. Conversely, the travel schedule exhibits the least significance.

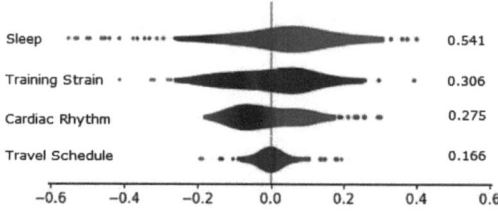

Fig. 7. SHAP value explanation of predicted RSImod score - features are listed by importance, with positive SHAP values indicating a positive impact on the prediction and negative values indicating a negative impact. The colour gradient reflects the magnitude of the impact, with darker colours representing stronger influences.

By generating SHAP values for the predicted RSImod values, our approach not only imputes missing data but also quantifies the contribution of each feature to the imputation outcome. This enables us to understand which features significantly influence the imputed RSImod score and how they interact with each other in the imputation process.

4.2 Discussion

The time complexity of the proposed methodology can be broken down as follows:

Feature Importance Analysis

– Pearson's Correlation: Calculating the correlation between two features has a time complexity of $O(N)$, where N is the number of data points. Since we calculate this for all pairs, the total complexity for Pearson's correlation for M features is $O(M^2 N)$.

- Random Forest Feature Importance: Training a Random Forest with T trees, each with depth D, has a $O(T{\cdot}D{\cdot}N{\cdot}logN)$ complexity. Extracting feature importance is usually $O(T{\cdot}M)$.
- XGBoost Feature Importance: The training complexity of XGBoost is $O(K{\cdot}N{\cdot}logN)$, where K is the number of boosting rounds. Extracting feature importance adds a complexity of $O(M)$.
- Aggregated Feature Importance Score Calculation: Combining feature importance scores from the three methods involves normalization and averaging, with a $O(M)$ complexity.

Factor Analysis. The time complexity of performing factor analysis primarily depends on the matrix operations involved. The complexity of M features and N samples is typically $O(M^3 + M^2N)$ due to the eigenvalue decomposition step.

Athlete Clustering. K-means clustering has a complexity of $O(I{\cdot}K{\cdot}N{\cdot}M)$, where I is the number of iterations, K is the number of clusters, N is the number of data points, and M is the number of features.

Training XGBoost Regressors. Training an XGBoost model has a complexity of $O(K{\cdot}N{\cdot}logN)$. Since we train one regressor per cluster, this becomes $O(C{\cdot}K{\cdot}N{\cdot}logN)$, where C is the number of clusters.

Missing Value Imputation

- Cluster Assignment: Calculating similarity scores with each cluster has a $O(C{\cdot}M)$ complexity.
- Softmax Probabilities: Computing softmax has a complexity of $O(C)$.
- Weighted Prediction: Calculating the weighted average of predictions has a complexity of $O(C)$.
- Thus, the overall time complexity is $O(M^3 + M^2N + N{\cdot}logN)$.

Table 7. Time complexity comparison.

MIV technique	Time Complexity
KNN	$O(N^2 M)$
MICE	$O(I \cdot M \cdot N \cdot \log N)$
EM	$O(I \cdot N \cdot M^2)$
CART	$O(N \cdot \log N \cdot M)$
XGBoost	$O(K \cdot N \cdot \log N)$
Proposed Approach	$O(M^3 + M^2N + N \log N)$

Table 7 compares the time complexity of the proposed approach with state-of-the-art MVI techniques. While the proposed methodology exhibits a comparable time complexity to KNN and EM techniques, it is relatively higher than CART, MICE, and XGBoost. However, these considerations are outweighed by its substantial improvements over state-of-the-art imputation techniques in terms of RMSE (up to 12.20%) and MAE (up to 10.77%). Moreover, predictions of athletic readiness (RSImod) scores over the dataset imputed using our proposed technique demonstrate enhancements, achieving up to 80.85% improvement in MSE and up to 79.99% in R^2 score compared to the current state-of-the-art. In cohort studies requiring robust imputation methods, imputation quality directly impacts the reliability of subsequent analyses and conclusions drawn. The proposed methodology's ability to achieve significantly lower error rates enhances the accuracy and trustworthiness of imputed data, thereby improving the overall validity and insights derived from cohort studies.

Additionally, the SHAP value-generated insights is crucial for coaches, trainers, and sports scientists to make informed decisions regarding athlete performance and readiness management. By providing a clear and interpretable rationale behind each imputed value, SHAP values enhance the trustworthiness and applicability of our imputation method in optimizing athletic training and monitoring strategies.

5 Conclusion

Addressing missing data in cohort studies is critical for ensuring the reliability and accuracy of outcomes. Our proposed data-driven MIV technique, integrating clustering, predictive modelling, feature sensitivity analysis, and factor analysis, significantly enhances the handling of complex datasets, leveraging benefits like inclusion of data-specificity, reduced dimensionality, quadratic speed-up, improved accuracy of imputation and enhanced interpretability.

With a dataset from collegiate basketball athletes, including diverse metrics from wearables, surveys, and physical tests, our method effectively models the impact of stressors on athletic readiness, quantified by RSImod. Notably, our technique outperforms existing methods with improvements in RMSE and MAE by up to 12.20% and 10.77%, respectively. Moreover, predictions of RSImod scores show remarkable enhancements, achieving up to 80.85% improvement in MSE and 79.99% in R^2 score compared to current standards. Leveraging SHAP values enhances interpretability, empowering coaches and practitioners with insightful decision-making tools based on comprehensive data insights.

This quality-driven approach ensures that the benefits gained from superior performance outweigh the modest increase in computational complexity, making it a preferred choice for researchers seeking precise and dependable imputation results. This methodology represents a significant advancement in addressing missing data challenges in sports science research, offering a robust framework for enhancing accuracy, reliability, and insights into athlete performance and readiness assessments. Its adaptable nature suggests potential applicability across diverse domains, promising to extend its benefits to various fields beyond sports science.

References

1. Li, J., et al.: Comparison of the effects of imputation methods for missing data in predictive modelling of cohort study datasets. BMC Med. Res. Methodol. **24**(1), 41 (2024)
2. Tenan, M.S.: Missing data in sport science: a didactic example using wearables in American football. Sports Med. **53**(6), 1109–1116 (2023)
3. Ribeiro, C., Freitas, A.A.: A data-driven missing value imputation approach for longitudinal datasets. Artif. Intell. Rev. **54**(8), 6277–6307 (2021). https://doi.org/10.1007/s10462-021-09963-5
4. D'Ambrosio, A., Aria, M., Siciliano, R.: Accurate tree-based missing data imputation and data fusion within the statistical learning paradigm. J. Classif. **29**, 227–258 (2012)
5. Pires, I.M., et al.: Improving human activity monitoring by imputation of missing sensory data: experimental study. Future Internet **12**(9), 155 (2020)
6. Senbel, S., et al.: Impact of sleep and training on game performance and injury in division-1 women's basketball amidst the pandemic. IEEE Access **10**, 15516–15527 (2022)
7. Griffin, A., et al.: Training load monitoring in team sports: a practical approach to addressing missing data. J. Sports Sci. **39**(19), 2161–2171 (2021)
8. Jiang, N., Gruenwald, L.: Estimating missing data in data streams. In: Kotagiri, R., Krishna, P.R., Mohania, M., Nantajeewarawat, E. (eds.) DASFAA 2007. LNCS, vol. 4443, pp. 981–987. Springer, Heidelberg (2007). https://doi.org/10.1007/978-3-540-71703-4_89
9. Shoaib, M., Scholten, H., Havinga, P.J.M.: Towards physical activity recognition using smartphone sensors. In: Proceedings of the 2013 IEEE 10th International Conference on Ubiquitous Intelligence and Computing. Vietri sul Mere, Italy, pp. 80–87 (2013)
10. Guo, Q., Liu, B., Chen, C.W.: A two-layer and multi-strategy framework for human activity recognition using smartphone. In: Proceedings of the 2016 IEEE International Conference on Communications (ICC), Kuala Lumpur, Malaysia, pp. 1–6 (2016)
11. Ni, D., et al.: Multiple imputation scheme for overcoming the missing values and variability issues in its data. J. Transp. Eng. **131**, 931–938 (2005)
12. Phung, L.S.: Deep learning methods for health data imputation and classification. Ph.D. thesis. Doctoral Dissertation (2021)
13. Phung, S., Kumar, A., Kim, J.: A deep learning technique for imputing missing healthcare data. In: Proceedings of the 2019 IEEE Engineering in Medicine and Biology Society, pp. 6513–6516 (2019)
14. Taber, C.B., Sharma, S., Raval, M.S., et al.: A holistic approach to performance prediction in collegiate athletics: player, team, and conference perspectives. Sci. Rep. **14**, 1162 (2024)
15. Barrientos, A.F., Deborsherr Sen, G.L., Dunson, D.B.: Bayesian inferences on uncertain ranks and orderings: application to ranking players and lineups. Bayesian Anal. **1**, 1–3 (2022)

16. Hall, M.A.: Correlation-based feature selection of discrete and numeric class machine learning. Technical report (2000)
17. Sharma, S.U., et al.: A hybrid approach for interpretable game performance prediction in basketball. In: 2022 International Joint Conference on Neural Networks (IJCNN), pp. 01–08 (2022)
18. Kroese, D.P., et al.: Data Science and Machine Learning: Mathematical and Statistical Methods. CRC Press (2019)
19. Babaee Khobdeh, S., Yamaghani, M.R., Khodaparast Sareshkeh, S.: Machine learning methods for data imputation in biomedical datasets: a comparative analysis. Comput. Biol. Med. **121**, 103800 (2020)
20. Poggio, C., et al.: Evaluating the effectiveness of imputation methods for missing athlete workload data: a systematic review. Sports Med. - Open **6**, 33 (2020)
21. Munir, T., Martinez, M., Al-Jumaily, A.: Deep learning techniques for real-time missing data imputation: applications in wearable health monitoring systems. IEEE Trans. Biomed. Circuits Syst. **14**(6), 1125–1133 (2020)
22. Serletis, D., Whittaker, J.: Improving athlete readiness predictions through hybrid machine learning models: a case study in collegiate sports. J. Appl. Sports Sci. **36**(4), 215–227 (2023)

TANS: A Chess-Inspired Notation System for Strategy Analysis of Tennis Games

Yuexi Song[1]($\boxtimes$), Chuanfei Li[2]($\boxtimes$), Hao Cao[3]($\boxtimes$), Ling Wu[4]($\boxtimes$), Huanhuan Zheng[1]($\boxtimes$), and Zhenkai Liang[1]($\boxtimes$)

[1] National University of Singapore, Singapore, Singapore
yuexi06@u.nus.edu, sppzhen@nus.edu.sg, liangzk@comp.nus.edu.sg
[2] Raffles Institution, Singapore, Singapore
Chuanfeili08@gmail.com
[3] Nanyang Technological University, Singapore, Singapore
NIE25.CH@e.ntu.edu.sg
[4] ChemT Biotechnology, Singapore, Singapore
ling.wu@chemtbio.com

Abstract. Tennis is a challenging sport that requires a comprehensive set of skills, including physical abilities, technical proficiency, and mental resilience. More importantly, tennis is also a game of strategy, demanding players to constantly evaluate the situation, calculate the shot placement and tactics, and control the flow of match. Chess, another game of strategy, has long benefited from the standard notation system, which enables deep and systematic analysis of game strategy in chess and even AI integration. In contrast, tennis currently lacks a structured and standardized framework for analyzing strategic game play. To bridge this gap, this paper presents Tennis Algebraic Notation System (TANS), a novel tennis notation system inspired by traditional chess notations. Unlike existing notation methods, TANS makes it easy for automated data extraction and analysis of game strategy and winning patterns. Using charted data with the new notation system, we can carry out analysis of match statistics and winning strategy analysis of tennis doubles matches. By summarizing tennis matches in a concise and information-rich notation system, we believe that TANS represents an important step toward enabling future systematic analysis of tennis strategy and AI-based game simulation.

Keywords: Strategy Analysis · Notation System · Tennis Analytics

1 Introduction

Tennis is a challenging sport that requires a comprehensive set of skills, including physical abilities, technical proficiency, and mental resilience. More importantly, tennis is also a game of strategy, demanding players to constantly evaluate the situation, calculate the shot placement and tactics, and control the flow of the match. Data analytics is widely used in tennis games to give an overall view of the game and players. However, there is only a limited amount of research in automated strategy analysis in tennis [14].

Chess is another game of strategy. Chess has a mature notation system for game charting and analysis [5], which also enables extensive analysis and modeling [18], including AI-based chess players. In contrast, a key missing piece in tennis is a notation system that precisely and concisely describes tennis moves. The challenge is that tennis movements are much more complex than those in chess. Tennis games are influenced by ball speed, spin, player positions, etc. It is not straightforward to include these features into a concise description.

However, we observed that *chess and tennis have many similarities in game strategy*. For example, both games have the notion of controlling space, planning moves, etc. Motivated by this similarity, we view tennis as a board game in which *one player moves other players through the placement of tennis balls*. As the tennis ball is the cause of player movements, we use the ball's position as the primary focus in our notation, which is comparable to the piece positions of the chess game.

In this paper, we define TANS, Tennis Algebraic Notation System. Inspired by the algebraic notation system [5] of chess, TANS describes detailed moves in tennis games in a concise format. TANS is designed to support both tennis analytics and strategy analysis. Based on TANS, we adopted algorithms, such as Grey Relational Analysis (GRA), to identify game strategies, showcasing the capability of TANS in enabling fine-grained strategy analysis.

In our evaluation, we focus on tennis doubles games when carrying out strategy analysis, as tennis doubles games have stronger strategy patterns. Using the data charted from matches such as Australian Open Doubles Final, we showed that TANS can support standard tennis analytics tasks. In addition, it can reveal deeper insights in winning patterns. We analyzed the effectiveness of employing the "I-formation" in tennis doubles and identified winning patterns for the serving team, which shows the strongest correlation with winning the point.

In summary, TANS records tennis matches in a concise and precise notation system. We believe that TANS represents an important step toward enabling future systematic analysis of tennis strategy and AI-based game simulation. In particular, we made the following contributions in this paper.

- We define TANS, a novel tennis notation system. Inspired by chess notation systems, we define the syntax of the notation language to record tennis moves.
- We develop applications on top of TANS to show that TANS can support deeper tennis analytics and strategy discovery.
- We evaluated the TANS language and applications using past professional games, where we discovered interesting strategy patterns from charted games.

Paper Organization. The rest of the paper is organized as follows. Section 2 introduces related work and background on chess notation systems. Section 3 introduces the design and notation of TANS. We discuss a few tennis analytics applications built on top of TANS in Sect. 4 and present the evaluation results in Sect. 5. Section 6 concludes the paper.

2 Background and Related Work

2.1 Tennis Strategy Analysis

Tennis is a highly strategic sport, and over the years, researchers have made continuous efforts to model game play and analyze winning strategies. As a start, Chiu et al. [8] developed a mathematical model and utilized math and physics simulation in identifying optimal strategy for tennis. Vis et al. [22] used descriptive variables such as stroke type, player position and used data mining for pattern discovery in tennis rallies. Martínez-Gallego et al. [16] utilized some other descriptive variables such as result of the last shot of the game, shots per game, and analyzed the game structure and point ending characteristics of tennis doubles matches. Similarly, Torres-Luque et al. [23] designed an observational instrument for obtaining objective information about singles matches for analysis. Kocib et al. [12] observed 8 sets in 18 professional men's doubles matches to examine the frequency and effectiveness of various tactical formations. Focused on tennis double games, Liu et al. [15] developed a Markov-based framework for modeling doubles game play. Given the additional complexities of doubles tennis, including player coordination and dynamic positioning, this work represents an important step toward enhancing the strategic understanding of doubles matches. These studies provided valuable insights into tennis strategy analysis. However, different research works have come up with different formulations of tennis games, requiring manual annotation of tennis games for each research work. Thus a major limitation to these studies has been the lack of comprehensive match charting data.

With the advent of the Match Charting Project, this limitation has been significantly mitigated. The project, hosted on TennisAbstract.com [7], provides thousands of data points from professional matches, charted based on their predefined charting language. Carlo, et al. [19] utilized the language and developed an AI Framework for tennis singles games. In contrast to mainly statistically analysis work done previously, their work opened a new door for utilizing AI simulation in identifying game winning strategies. However, the charting language from TennisAbstract was originally designed for ease of recording rather than for machine learning application, limiting the accuracy of the framework. Hence, a key missing component across these works is a systematic notation language specifically designed to support strategy analysis of tennis games.

2.2 Chess Notation Systems

There have been significant developments in analyzing chess game winning moves and patterns. Chess has a well structured notation system which allows powerful computational tools to analyze chess games and dive deep into strategic exploration. A groundbreaking development in this area is AlphaZero [21], which is a chess AI that mastered the game through reinforcement learning in self-play. Later, Ruoss et al. [20] optimized performance through reinforcement learning to generate moves without relying on explicit search. A basis for these advancements is the algebraic notation [5], the official notation system of chess. It includes the following components.

1. **Chessboard Coordinates.** The board has files (columns) labeled a–h (left to right). The ranks (rows) are numbered 1–8 (bottom to top), shown in Fig. 1.

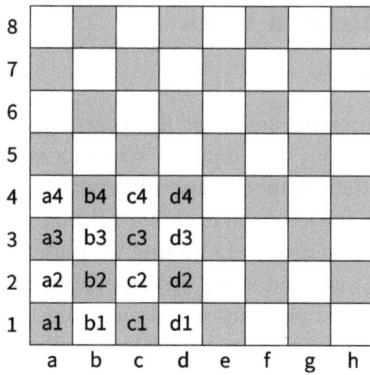

Fig. 1. Chessboard and coordinates.

2. **Piece Abbreviations.**
   - K: King
   - Q: Queen
   - R: Rook
   - B: Bishop
   - N: Knight
   - No letter for pawns (just the destination square)
3. **Move Notation.**
   Write the piece abbreviation followed by the destination square. For example, e4 (pawn to e4); Nf3 (knight to f3); Bb5 (bishop to b5).
   Captures. Use x between the piece and the destination. For example, Nxe5 (knight captures on e5); Rxd7 (rook captures on d7); exd5 (pawn from file e captures on d5).
   Special Moves. *Castling*: Kingside: O-O, Queenside: O-O-O. *Pawn promotion*: e8=Q (pawn reaches e8 and promotes to a queen). *En passant*: exd6 e.p.
   Extra notations If there are two of the same piece that can move to the same square, specify which one moved by adding: The file it came from (Rea8 means the rook from the "e" file moved to a8). If both pieces are on the same file, use the rank instead (R1d3 if rooks are on d1 and d5, and the one from d1 moves to the d3 square).
   Check and Checkmate. *Check*: Add + behind the piece abbreviation and destination square, e.g., Qg5+.
   Checkmate: Add # behind the piece abbreviation and destination square, e.g., Qh8#.
4. **Game End Notations.**
   1-0 (White wins).
   0-1 (Black wins).
   $1/2 - 1/2$ (Draw).

The algebraic notation is the standard representation to describe moves. Additional information, such as player names, dates, are included as tags at the beginning of the description to form the Portable Game Notation (PGN) [10] for chess. PGN is widely used by chess software and chess game databases.

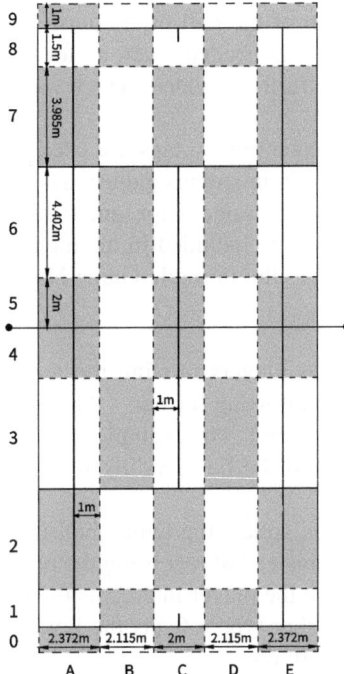

Fig. 2. Tennis court coordinate in TANS.

3 Tennis Algebraic Notation System

Inspired by chess notation systems, we developed a tennis algebraic notation system (TANS) to enable systematic analysis of tennis strategies.

The detailed description of the language is as follows:

1. **Court Coordinates.**
   A naive method to assign court coordinate is to equally divide the court, similar to the chess board. However, unlike in a chess board, the regions in the tennis court are of different importance in tennis strategy. Existing studies such as [17] have shown that ball landings occur more frequently in certain areas of the court, and winning shots are often associated with specific regions, such as near the net or close to the sidelines. Therefore, we made a coordinate system based on the *strategic meaning* of tennis courts.
   TANS Coordinates. As illustrated in Fig. 2, the solid lines are the court partitions where the tennis court is divided into a grid of ten rows (ranks) 0–9 by five columns (files) A-E, forming 50 distinct sections. Each section is uniquely labeled with letters and numbers, such as C4. In particular, Row 0 and Row 9 are reserved for player positioning behind the baseline where players prepare to serve, receive, or respond to long shots.
   This coarse-grained court partition is designed for easier charting by humans. It

captures the overall position of the ball on a court, such as left/right or center. In case a finer grained partition is needed, such as for computerized processing, the court columns (files) can be further divided, which we illustrate in the Appendix B.

2. **Player Abbreviations.**
   To identify shots made by each player, tags are used to label the players on court. Players are labeled as X and Y in singles games and as X, Y, Z, and W in doubles games, where X and Y belong to one team, and Z and W belong to the opposing team. The players' names can be included in the metadata section at the beginning of the charting data, similar to the Portable Game Notation (PGN) file format used in chess.

3. **Shot and Move Notation.**
   We view the tennis game as players moving the tennis ball strategically. Hence, each shot is encoded in the following format: [Player tag] + [Stroke] + [Ball landing position]. For example, WfB6 stands for Player W hitting a forehand ground stroke and the ball landing in the court position B6 (notation explained below).

   In certain situations, such as volley, the tennis ball is intercepted by the opponent before it lands. In this case, instead of recording the position where the ball lands in, we record the position where the player intercepts the ball.

   Shot Type Encoding. The shot type reflects the kind of stroke used by the player. The encoding is shown as follows:
   - s: serve
   - f: forehand ground stroke
   - b: backhand ground stroke
   - v: volley
   - c: slice
   - o: overhead (smash)
   - l: lob
   - d: drop shot

     Extra Notations. In the case where the ball is intercepted by the volleyer, the intended direction of the ball may also be encoded as the character(s) after the shot type encoding, before the ball landing position, e.g., WbxB6. This is particularly useful for analysis in the scenario where the ball is intercepted by the opponent. This field is optional as it can be inferred from the landing position of the ball. The encoding is shown as follows, which is after the shot type encoding:
   - i: down the line
   - x: cross court
   - m: down the middle

     Note at our main focus of this paper is the spatial distribution of tennis shots and related player movements. Other properties of the game, such as *ball speed*, can be added as part of extra notations in the future, e.g., observational instruments highlighted in Torres-Luque et al. [23].

4. **Player Positions.** The position of the players is an important part of the game, especially in tennis doubles, where the placement of each player is crucial to the strategic game play. And before every shot, the position of the player may change, hence, the

player position is recorded along with each shot. To capture this, we use a tuple of the form (X, Y) for singles games and (X, Y, Z, W) for doubles games, where each element represents the court section occupied by the corresponding player right before the time of the shot.

5. **Point Ending.**
   The ending of a point is encoded in the format of [Winning Player/Team] + [Outcome].
   Player/Team: X (Player X in singles games); Y (Player Y in singles games); X (Team X/Y in doubles games); Z (Team Z/W in doubles games).
   Outcome: V (Winner) E (Forced error of opponent); U (Unforced error of opponent). For example, ZV means Team Z/W wins with a winner point. In the case of forced or unforced error, the type of the error may also be encoded as follows: n (net); w (wide); h (long).

Summary of the Overall Language

Every point comprises of several shots and players' positions may be changing throughout the game point. For every point, it is encoded as [Shot Encoding] followed by optional [Player Position Encoding] in chronological order, where the player position refers to the players' positions immediately before the shot is played. At the end of each sequence, the result of the point is noted.

The following is an example charting of a doubles point where player X begins the with a serve, and the team Z/W ultimately wins the point:

XsC6, (B0, D4, B6, E9)
WfB6, (A1, C4, B6, D9)
XbD6, (A1, D4, B6, E8)
WfE1, (A0, C4, B6, E9)
ZV

4 Strategy Analysis

In this section, we discuss a few analyses enabled by our description language.

4.1 Winning Shot Sequence Identification

Our language effectively describes a tennis game as a sequence of moves in string format. Therefore, similar to chess winning move sequences, such as the Scholar's Mate [6], tennis games also have patterns for winning. When the game is described by strings, such move sequences will be reflected as patterns of regular expressions.

Algorithm 1: Grey relational analysis for winning strategy identification

Input: X: Encoded and normalized array of shot sequences where
$X[i] = (x_i(1), \ldots, x_i(c))$
n: Total number of shot sequences
c: Number of shots in each sequence
Output: Ranked shot sequences based on average grey relational degree
for $i \leftarrow 1$ **to** n **do**
    $\xi \leftarrow 0.5$;
    let $ref \leftarrow X[i]$;
    // Set reference sequence as $X[i]$
    for $j \leftarrow 1$ **to** n **do**
        for $k \leftarrow 1$ **to** c **do**
            $\Delta_j(k) \leftarrow |ref(k) - x_j(k)|$;
        end
    end
    $M \leftarrow \max_j \max_k \Delta_j(k)$;
    $m \leftarrow \min_j \min_k \Delta_j(k)$;
    for $j \leftarrow 1$ **to** n **do**
        for $k \leftarrow 1$ **to** c **do**
            $\gamma_{0j}(k) \leftarrow \frac{m + \xi \cdot M}{\Delta_j(k) + \xi \cdot M}$;
        end
        $\gamma_{0j} \leftarrow \frac{1}{c} \sum_{k=1}^{c} \gamma_{0j}(k)$;
        // Compute grey relational coefficient between
            reference sequence and sequence j
    end
    $\gamma_i \leftarrow \frac{1}{n-1} \sum_{\substack{j=1 \\ j \neq i}}^{n} \gamma_{0j}$; // Compute average grey relational degree
        for reference sequence i excluding self-comparison
end
Sort $(\gamma_i, X[i])$ pairs in descending order based on γ_i;
return *The sorted shot sequences based on γ_i;*

In computer science, there are existing approaches to solving the *common substring of multiple strings* problem. One such algorithm [11] identifies the longest substring that appears in at least k strings. Building on this idea, we adopted the algorithm to identify the substring that is common to the most number of sequences as a winning pattern. We will report a few cases we found in the evaluation section.

4.2 Grey Relational Analysis

Beyond simple pattern matching, the notation language also enables a deeper analysis of the winning strategy. As our notation language captures matches at a finer granularity and encodes more detailed information, we are able to extract meaningful winning patterns from a much smaller set of data points.

As common winning strategies will have relatively high occurrences in games, the corresponding shot sequence described in TANS will also have a high correlation among the points charted in a game. We employed Grey Relational Analysis (GRA), which measures the similarity between data sequences by comparing the geometric shapes of their curves, allowing us to assess the closeness of their relationships. It is a method well-suited for small sample sizes and capable of revealing significant patterns with a small amount of data.

Data Preparation and Normalization

As grey relational analysis mainly deals with numerical data, we first encode each shot in sequences into numerical data using a shot encoding function. This function parses the shot into three components: the shot type, the area of the court, and a numeric spot identifier. These components are mapped to integers using predefined dictionaries. The encoded value of each shot is then calculated using a weighted formula that ensures uniqueness across combinations. Let $X_i = (x_i(1), x_i(2), \ldots, x_i(n))$ be our encoded data sequence, then each data point is mapped to its interval image(D_3 operator) using the following formula [13]:

$$x_i(k)d_3 = \frac{x_i(k) - \min_k x_i(k)}{\max_k x_i(k) - \min_k x_i(k)}; \quad k = 1, 2, \ldots, n \quad (1)$$

Grey Relational Degree Calculation

Given a sequence $X_0 = (x_0(1), \ldots, x_0(n))$ as the reference sequence and sequences $X_i = (x_i(1), x_i(2), \ldots, x_i(n))$, $i = 1, 2, \ldots, m$, their grey correlation degree is calculated through the following formula [13]:

$$\gamma(x_0(k), x_i(k)) = \frac{\min_i \min_k |x_0(k) - x_i(k)| + \xi \max_i \max_k |x_0(k) - x_i(k)|}{|x_0(k) - x_i(k)| + \xi \max_i \max_k |x_0(k) - x_i(k)|} \quad (2)$$

$$\gamma(X_0, X_i) = \frac{1}{n}\sum_{k=1}^{n} \gamma(x_0(k), x_i(k)) \quad (3)$$

where $\gamma(\mathbf{X_0}, \mathbf{X_i})$ is known as Deng's grey relational degree between X_0 and X_i, where ξ is the distinguishing coefficient [9].

Each winning sequence is used as a reference sequence to compute the grey correlation with all the other winning sequences. Then the average of the correlation with all the other winning sequences is taken as a standard to identify the best winning pattern. The summarized algorithm is given in Algorithm 1.

Table 1. Top 5 winning shots.

Shot Sequence	Frequency
vC6	7.4468%
vD5	7.4468%
vB5	6.3830%
oB5	6.3830%
oD5	4.2553%

5 Evaluation

Utilizing the language we developed, we charted the full 2022 Australian Open Men's Doubles Final match [2], 2024 Australian Open Men's Doubles Final match [3] and also US Open Women's Doubles Final match [1], excluding points lost due to unforced errors. This resulted in charted data of 129 points. More information about the data and availability is discussed in Appendix A.

With this data, we performed different types of evaluations to show the capability of TANS. First, we show that TANS can be used to carry out traditional tennis analytics; Second, we identify common winning patterns in tennis games; Finally, we apply grey-relational analysis to identify the best winning strategy for a specific match.

5.1 Match Statistics

Traditional match statistics analysis is carried out from manual notations. As TANS does not have data loss, it supports tennis match statistics. For example, when we examined the impact of serving in doubles, we found that 68% of winning points were generated by the serving team in the 2024 Australian Open's match. This indicates the importance of controlling the serving game well. And among these, 37% points started with *sC6, which refers to serving down the "T". In addition, we analyzed the opening positions of the serving team that led to winning the point. We identified that 74% of these winning points began with the server positioned at C4 and the volleyer at C4, which actually corresponds to the well-known "I-formation" in tennis doubles strategy.

Moreover, we can use a program to automatically translate TANS into other description languages, such as the notation language used by The Match Charting Project [7]. For example, our charted data of 1. XsD6, (C0, D9) 2. YbcB2, (C4, D9), YF can be translated into 5b# in the language defined by TennisAbstract while also providing additional information. It can also be translated to English description: 1st serve down the "T"; backhand return cross court; forced error of opponent.

5.2 Common Winning Pattern

We focused on ball landing or striking positions during the rallies of each point and extracted them from the charted data. We first analyzed the most common winner shot of both matches. From Table 1, we can see that 7% of the winners are generated by volleying to position C6. While C6 seems to be easy to cover, it often becomes an exposed area in doubles, particularly for teams lacking coordination and on-court synergy. This highlights the tactical advantage of targeting weak spots that emerge from insufficient teamwork. Moreover, around 24% of the winner shots are generated by overhead smashes or volleys to position B5 or position D5, and this again shows the importance of strong net control in doubles.

Table 2. Top 5 Final two-shot placement patterns.

Shot Sequence	Frequency
C4, D5	5.3191%
C4, B5	5.3191%
C4, C6	4.2553%
C4, D6	4.2553%
B3, A8	3.1915%

Table 3. Top 5 Final three-shot placement patterns.

Shot Sequence	Frequency
E6, C4, C6	3.1915%
D6, D4, D5	2.1277%
C6, C4, A6	2.1277%
B6, B4, E5	2.1277%
B6, C4, D5	2.1277%

We further checked for the most common patterns in the last two or three shots of the winner points charted. From Table 2, the most common two-shot patterns are C4, B5 and C4, D5, each with percentage of around 5%. By symmetry of B5 and D5, we can treat them as variations of one shot where the volleyer volleys cross court. Then this pattern appears in over 10% of the winning patterns. This again highlights the effectiveness of having the net player control the center and execute high-quality cross-court volleys. Furthermore, from Table 3, the most frequent three-shot winning sequence is E6, C4, C6, which reflects a strategic play: hitting a wide cross-court shot, forcing the opponent to return to the middle, and then the volleyer finishes the point with a short, well-placed volley, targeting the same player.

5.3 Winning Strategy from GRA

Using the GRA method in Sect. 4.2, we aim to identify the frequently used strategies for the serving side to win a point. In our evaluation, we calculated the correlation among the winning sequences in the charted men's doubles matches. Most of the winner points

Table 4. Top 10 winning sequences.

Sequence			Grey correlation degree
*sB6,	*bB4,	*vE5	0.7626
*sB6,	*bB4,	*vD6	0.7626
*sC6,	*fE1,	*fE6	0.7621
*sD6,	*fE3,	*fD7	0.7603
*sB6,	*fB4,	*vD6	0.7598
*sC6,	*fC4,	*vE5	0.7584
*sB6,	*fB4,	*vD5	0.7584
*sD6,	*bD4,	*oD5	0.7522
*sD6,	*fD4,	*vE5	0.7504
*sC6,	*1B2,	*oD6	0.7429

(a) sB6: Serve to B6

(b) bB4: Backhand to B4

(c) vE5: Volley to E5

Fig. 3. Screenshot of the winning pattern sB6, bB4, vE5.

are short, requiring only three shots, in cases where the point is longer, we take the first three shots for analysis. We also masked the player with *, so that the algorithm can focus on the serving team, regardless of the specific players involved.

The sequences with the top 10 correlation scores are listed in Table 4. Among these, the sequence *sB6, *bB4, *vE5 has the highest average correlation, 0.7626.

This refers to the following strategy: The serving player serves down the "T" toward opponent's backhand. Then returning player returns the ball towards the middle. Positioned near the middle of the net, the volleyer on the serving team intercepts the ball and strikes cross court, resulting in a winning point. The detailed point is shown in Fig. 3, using screenshot from the match [3].

Similar strategies were observed in women's doubles tennis. In the final match of the 2019 US Open, the patterns *sC6, *bmC4, *oB6 and *sE6, *bmC4, *oE5 exhibited high correlation scores of 0.8581 and 0.8539, respectively. Both of these patterns involve a powerful serve to the returner's backhand, prompting a return to the volleyer, who then hits it back to the returner.

6 Conclusion

In this paper, we present the Tennis Algebraic Notation System (TANS), a novel tennis notation system inspired by the algebraic notation system of chess. TANS is developed as a formal description to lay the foundation for comprehensive analysis of game strategies. In our evaluation, TANS has been demonstrated to be effective for supporting match statistics analysis, winning pattern discovery and winning strategy identification. As the future work, we aim to leverage the advancements in tennis video analysis, chess analytics, and AI solutions for chess to enable deeper analysis of tennis strategies.

Acknowledgments. We thank the anonymous reviewers for their valuable comments to improve this paper. We dedicate this paper to the Tennis Team of the Peking University Alumni Association (Singapore) for the collaborative spirit and unwavering motivation. This paper is in part supported by the Grant A-8002737-00-00 from the School of Computing, National University of Singapore.

A Tennis Game Data Used in Evaluation in TANS

We used TANS format to chart several tennis games for our evaluation, including the game between Bolelli/Vavassori and Bopanna/Ebden in Australian Open 2024 Men's Doubles Final [3], the game between Azarenka/Barty and Mertens/Sabalenka in US Open 2019 Women's Doubles Final [1] and the game between Sabalenka and Gauff in Mutua Madrid Open 2025 Women's Singles Final [4].

The charted data for our evaluation is available at our project's supporting website: https://tennis-ans.github.io/. We will use the space to host datasets and other related documents and tools for this project.

B Extension of Tennis Court Partition

An example of fine-grained partition is shown in Fig. 4. Column A is subdivided into Column 1 and Column a; Column C is subdivided into Column p and Column q; c is be used for marking the position along the center line; Column E is subdivided into Column e and Column r; Column B and Column D remain to be Column b and Column d. So the fine-grained columns labels are: l, a, b, p, c, q, d, e, r. This notation is reserved for future computerized processing, which is not used in this paper.

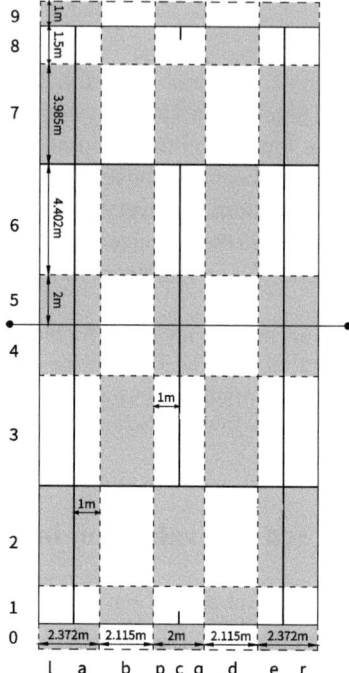

Fig. 4. Extension of Tennis court coordinate in TANS.

References

1. US open 2019 women's doubles final (2019). https://www.youtube.com/watch?v=xbxPVXOpoA8
2. Australian open 2022 men's doubles final (2022). https://www.youtube.com/watch?v=iZYCl7xq4TI
3. Australian open 2024 men's doubles final (2024). https://www.youtube.com/watch?v=mC_pSNURbzU&t=3933s
4. Mutua Madrid open 2025 women's singles final (2025). https://www.youtube.com/watch?v=_O-4_IJFIjQ
5. Chess algebraic notation. https://www.chess.com/article/view/chess-notation. Accessed April 2025
6. Scholar's mate. https://www.chess.com/terms/scholars-mate-chess. Accessed April 2025
7. Tennisabstract.com. https://www.tennisabstract.com/. Accessed April 2025
8. Chiu, C.H., Tsao, S.Y.: Mathematical model for the optimal tennis placement and defense space. Int. J. Sport Exer. Sci. **4**(2), 25–36 (2012)
9. Deng, J.: Grey Control Systems. Press of Huazhong University of Science and Technology (1985)
10. Edwards, S.J.: Portable game notation specification and implementation guide (1994). https://www.saremba.de/chessgml/standards/pgn/pgn-complete.htm
11. Chi, L., Hui, K.: Color set size problem with applications to string matching. In: Apostolico, A., Crochemore, M., Galil, Z., Manber, U. (eds.) CPM 1992. LNCS, vol. 644, pp. 230–243. Springer, Heidelberg (1992). https://doi.org/10.1007/3-540-56024-6_19

12. Kocib, T., Carboch, J., Cabela, M., Kresta, J.: Tactics in tennis doubles: analysis of the formations used by the serving and receiving teams. Int. J. Phys. Educ. Fitness Sports **9**(2), 45–50 (2020)
13. Liu, S.: Grey Systems: Analysis, Methods, Models and Applications. Series on Grey System, 2nd edn. Springer, London (2016)
14. Liu, Z., Jiang, K., Hou, Z., Lin, Y., Dong, J.S.: Insight analysis for tennis strategy and tactics. In: Proceedings of the 2023 IEEE International Conference on Data Mining (ICDM), pp. 1175–1180. IEEE (2023)
15. Liu, Z., Dong, C., Wang, C., Dong, T.Y., Jiang, K.: Exploring team strategy dynamics in tennis doubles matches. In: Proceedings of the International Symposium on Advances in Computing and Engineering (ISACE) (2024)
16. Martínez-Gallego, R., Crespo, M., Ramón Llin, J., Micó, S., Guzmán Luján, J.F.: Men's doubles professional tennis on hard courts: game structure and point ending characteristics (2020)
17. Martínez-Gallego, R., Ramón-Llin, J., Crespo, M.: A cluster analysis approach to profile men and women's volley positions in professional tennis matches (doubles). Sustainability **13**, 6370 (2021)
18. McIlroy-Young, R., Wang, R., Sen, S., Kleinberg, J., Anderson, A.: Learning models of individual behavior in chess. In: Proceedings of the 28th ACM SIGKDD Conference on Knowledge Discovery and Data Mining. ACM (2022)
19. Nübel, C., Dockhorn, A., Mostaghim, S.: Match point AI: a novel AI framework for evaluating data-driven tennis strategies (2024)
20. Ruoss, A., Delétang, G., Medapati, S., Grau-Moya, J., Wenliang, L.K., Catt, E.: Grandmaster-level chess without search. arXiv preprint arXiv:2304.11415 (2023)
21. Silver, D., et al.: Mastering chess and shogi by self-play with a general reinforcement learning algorithm. Nature **550**(7676), 354–359 (2017)
22. Terroba, A., Kosters, W.A., Vis, J.K.: Tactical analysis modeling through data mining: Pattern discovery in racket sports. In: Proceedings of the International Conference on Data Mining (ICDM) (2008)
23. Torres-Luque, G., Fernández-García, A.I., Cabello-Manrique, D., Giménez-Egido, J.M., Ortega-Toro, E.: Design and validation of an observational instrument for the technical-tactical actions in singles tennis. Front. Psychol. **9**, 2418 (2018)

Agentic Generative AI for Media Content Discovery at the National Football League

Henry Wang[1]($^\boxtimes$), Md Sirajus Salekin[1]($^\boxtimes$), Jake Lee[1]($^\boxtimes$), Ross Claytor[1]($^\boxtimes$), Shinan Zhang[1]($^\boxtimes$), and Michael Chi[2]

[1] Amazon Web Services, Seattle, USA
yuanhenw@amazon.com
[2] National Football League, New York, USA

Abstract. Generative AI has unlocked new possibilities in content discovery and management. Through collaboration with the National Football League (NFL), we demonstrate how a generative-AI based workflow allows media researchers and analysts to query relevant historical plays using natural language, rather than using traditional filter and click-based interfaces. The agentic workflow takes a user query in natural language as an input, dissects the query into different elements, and then translates these elements into the underlying database query language. The accuracy and latency of retrieval are further improved through carefully designed semantic caching. The solution performs with over 95-percent accuracy and reduces the average time of finding relevant videos from 10 min to 30 s, significantly increasing the NFL's operational efficiency and allowing users to focus more on producing creative content and engaging storylines.

Keywords: National Football League · Next Gen Stats · Agentic System · Natural Language Query · Media Content Search

1 Introduction

The demand for sports content grows every year, as shown by the increasing sizes of sports leagues' multi-year media rights deals valued at billions of US dollars [2,15]. The National Football League (NFL)[1], one of the most popular professional sports leagues in the world, produces and distributes content all year-round to meet this demand and to engage its 184 million domestic and 100 million international fans. The NFL manages petabytes of content, primarily videos of game footage, and distributes it for use across broadcasts, TV segments, digital platforms (sites, apps), and social media. In a sports entertainment landscape that is becoming increasingly more digital, producing captivating content

---

[1] https://www.nfl.com/.

Supplementary Information The online version contains supplementary material available at https://doi.org/10.1007/978-3-032-06167-6_20.

remains at the core of what the NFL delivers to fans around the world. However, it has become increasingly difficult for media teams to find and manage the right content in the NFL's ever-growing library of digital media.

The NFL partnered with our team to develop an agentic solution - powered by generative AI - that enables media and production teams to efficiently search for game videos using natural language queries[2]. The solution enables users to spend more time on the creative aspects of their roles to create more engaging content for fans.

2 Related Works

The advent of generative AI has revolutionized traditional content retrieval paradigms, transforming how we interact with and extract information from vast data repositories. With the power of Large Language Models (LLMs) [23], we can now perform text-to-SQL and Retrieval-Augmented Generation (RAG) operations that convert natural language queries into corresponding intermediate representations of the SQL queries, vector embeddings, or other structured formats, thus, enabling more intuitive and effective content retrieval. Apart from the general in-context learning capability of LLMs, recent text-to-SQL approaches can solve different complex text-to-SQL tasks using chain-of-thoughts [14], few-shot strategy [17], decomposing steps [18], self correction [20], and multi-turn steps [21]. On the other hand, RAG [3] frameworks can retrieve relevant context for any particular query. Both text-to-SQL and RAG systems can retrieve domain-specific content based on SQL or vector embeddings.

In sports, media content retrieval is a common task when preparing highlights or writing digital posts. Recently, LLMs have shown their potential in transforming how sports organizations manage, analyze, retrieve, and discover content within their vast media archives. LLMs are being used for applications including: expert commentary generation [1,4,8,9], action spotting [12], sports understanding [6,7,10,16,19,22], and video assistant refereeing systems [5]. Natural language queries are making sports content more accessible to analysts, fans, and broadcasters. Related work in soccer (football) has shown rather than relying on structured search parameters or specific keywords, users can now query and retrieve soccer information using RAG [8,13] or GraphRAG [11].

While LLMs introduce the capability to perform natural language queries and to retrieve relevant content, the extent to which LLMs can be utilized to understand and streamline complex data and media platforms (such as the NFL's Next Gen Stats platform) is yet to be explored. Specifically, it would be worthwhile to explore whether LLMs and natural language prompts can be utilized to automate and streamline the complex backend API calls used to retrieve desired media content.

---

[2] https://www.wired.com/sponsored/story/will-the-nfls-push-into-genai-transform-how-we-see-sports/.

3 Data

The NFL's Next Gen Stats (NGS) platform[3] is a comprehensive player and ball tracking system that captures and stores real-time data on every play from every NFL game since 2016. It stores large volumes of information, including information on players, teams, historical performances, and relevant advanced statistics. This data is captured through sensors in players' pads and the ball.

To build and test our agentic media search solution, our team used different statistical data such as team/player/play details, glossaries, question-answer pairs for media search, and backend OpenSearch API call syntax from the NGS platform. Our team also worked closely with NFL Subject Matter Experts (SMEs) to prepare and validate the ground truths and to retrieve search results during solution development.

3.1 Schema

Due to highly granular nature of the NFL's NGS data, the data is organized into different schemas (such as defense vs offense, passing vs rushing) to help with extracting stats and facilitating data organization. The NGS data is stored in OpenSearch and each schema consists of its own unique OpenSearch keys and fields. Table 1 & 2 (Appendix) contains descriptions of key fields to demonstrate what the schemas look like. These schemas are passed into LLMs as context to help them understand the data types and definitions of each field.

3.2 Question and Answer Pairs

240 Question and Answer (QA) pairs were collected to represent user queries and their ground truth number of plays. The questions varied in complexity and were representative of questions that would be typically asked by NFL's target end users such as research and production analysts - 30% are easy questions that can be answered by straightforward API calls, 50% are medium questions that require multiple filtering conditions and the remainder 20% are complex questions that not only require deep football knowledge but also may involve multiple schema. 100 of the questions were set aside for the test set and the rest were used as development data to construct few-shots examples and enhance our prompting strategy. Examples of the QA pairs are shown in Table 3 (Appendix).

4 Methodology

We built an end-to-end, natural-language "agentic" search experience - the system interprets a user's query, asks clarifying questions when necessary, then composes and executes OpenSearch calls to fetch the matching plays and their media links. The workflow runs in AWS, is orchestrated with LangGraph and is surfaced through a React front-end (screenshots in the Appendix). Figure 1 gives the high-level flow. Details of the design are described below.

---

[3] https://nextgenstats.nfl.com/.

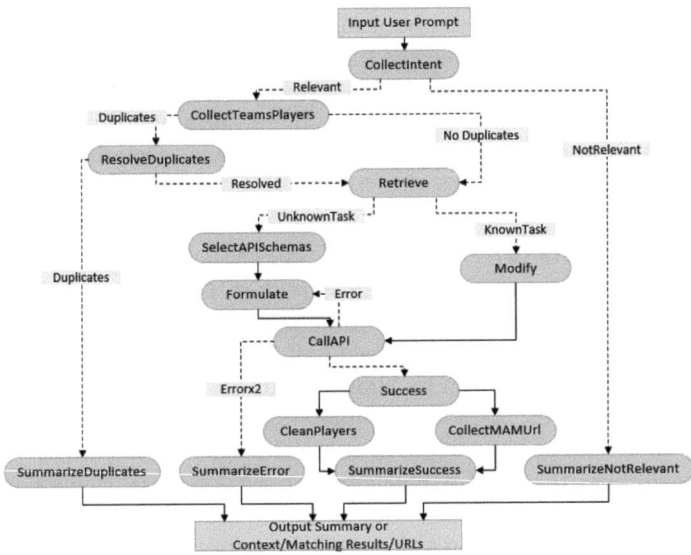

Fig. 1. Overall diagram of the Agentic workflow

4.1 Agentic Workflow

The distinctly different schemas warrant the need for an agentic workflow that interprets which tables and fields are needed when constructing the API calls.

Assess User Intent. A low-latency model (such as Claude 3 Haiku) classifies the prompt. Non-football questions receive a polite rejection; valid football queries trigger the graph. This guardrail prevents out-of-scope inputs from derailing later steps.

Extract Entities, Actions, and Conditions. The LLM breaks down the prompt into Entities (players/teams), Actions (stats of interest), and Conditions. For a query such as "Find all plays where Patrick Mahomes throws a touchdown farther than 10 yards", "Patrick Mahomes" is the Entity, "touchdown throw" is Action, ">10 yards" is the Condition. When situations with duplicate names appear (it's not uncommon for players in NFL to have same last names), the LLM prompts the user to confirm their player of interests and store confirmed player/team id for the session.

Select Relevant API Schema: Based on extracted entities, actions and constraints, the system first conduct a semantic similarity search to identify if similar queries have been encountered in the past:

- If not, the system will pin down relevant context. A router LLM picks the most relevant NGS schema(s)—e.g. *passing, rushing, team defense, team offense,* etc. Each schema contains relevant fields that LLM can use to construct the API call. For example, the "*passYards*" field corresponds to the yardage of

passing, and "touchdown" is a boolean indicator of whether the play resulted in a touchdown. Only the relevant schemas are passed in as context to reduce overall context length and reduce the chance LLM picks inaccurate fields.
- If similar queries are addressed before, the system will directly leverage the information instead of reformulating the problem. Details are described in **Semantic caching** below.

Formulate and Execute API Calls: The API-formulation LLM receives team/ player IDs, reference schema, and few-shot demonstrations on how APIs are constructed. It reasons step-by-step, maps extracted actions and conditions to corresponding fields and emits the final API call with values assigned to each field. Figure 2 from Appendix shows a sample of formulated API call and reasoning LLM provides. A python runner executes the API call and an LLM summarizes the returned results into natural language responses to the user; Upon syntax failure, the LLM will attempt auto-correct using error messages. If the final API call still fails after three tries, the model will prompt the user to rephrase original query and run the workflow again.

Conversational Experience: Follow-up conversations inherit prior context. After "*Find me all the plays where Patrick Mahomes throws a touchdown farther than 10 yards,*" a user may ask "*What about all the throws that were intercepted?*" The LLM implicitly keeps "Patrick Mahomes" as the entity, and will rephrase the user's prompt to "*Find all the plays where Patrick Mahomes throws are intercepted.*"

Semantic Caching: We optimize latency and accuracy by storing prior queries and responses with player/team names redacted and replaced by placeholders (e.g. "[PLAYER]", "[TEAM]"). When a new query gets submitted, the system first applies a vector search to find similar queries in the past. If a high score is achieved (meaning highly-similar queries have been addressed in the past), the cached API calls with placeholders will be replaced with the fresh IDs and executed directly, allowing the system to bypass expensive reformulation.

Link Plays to Assets in Media Asset Management(MAM) Solution: Finally, the play IDs are serialized and sent to NFL's MAM system, which will retrieve unique URLs for each asset. The React UI embeds the returned URLs. See Fig. 2 for an example where a user can click on any of the returned NGS Media Links to access the media content directly in a MAM. At this point, the end-to-end content searching experience with natural language is complete.

4.2 System Infrastructure

The entire agentic workflow is implemented in the AWS ecosystem. There are 3 major components of this system. These are: 1) LLM model using Amazon Bedrock converse API, 2) Memory using Redis and 3) Workflow orchestration using LangGraph. The overall architecture diagram of the infrastructure is included in Appendix Fig. 5.

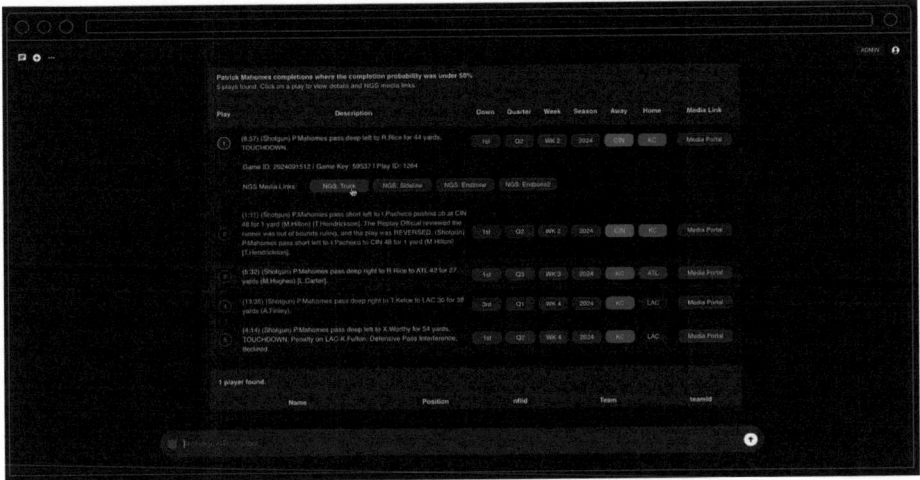

Fig. 2. Links to retrieved assets embedded in React frontend

5 Results

To develop the solution, our team used 100 QA pairs to refine our prompts and iteratively collected feedback from NFL SMEs to create a smooth user experience. The answer is considered accurate if the constructed APIs contain the correct filter&values and the count of records matches with ground truth value. During testing, the solution achieved over 95% accuracy on the 100 QA pairs. The solution has yielded notable process improvements across the NFL's media teams and 250 users were onboarded within the first month after its deployment. Users reported significant time savings: With free-form natural language queries, users can now search for and retrieve relevant video in an average of 30 s, down from over 10 min per query, giving them more time to focus on finding creative insights. They can also ask follow-up questions if they deem certain results returned interesting, so they can maintain the research flows without navigating between different tools.

6 Conclusion

In this paper, we discussed the implementation and efficacy of LLM-driven agentic workflows in transforming natural language queries into complex API calls within the NFL's Next Gen Stats system. Today, this solution is used daily by all of the NFL's media editors. For simple searches that are expected to return videos for only a few specific plays, users can find relevant videos nearly 6 times faster than before. This time savings for finding relevant videos scales up to 20 times faster as search complexity increases and more game plays match the user's prompt. In the future, there are opportunities to further enhance the capabilities of the tool and create fan-facing version of the solution, providing richer contents and offerings to hundreds of millions of avid football fans worldwide.

References

1. Cook, A., Karakuş, O.: Llm-commentator: novel fine-tuning strategies of large language models for automatic commentary generation using football event data. Knowl.-Based Syst. **300**, 112219 (2024)
2. Elberse, A., Warner, E.: The nfl's $110-billion media rights deals (2022). https://www.hbs.edu/faculty/Pages/item.aspx?num=62434. Accessed: 2025-04-17
3. Fan, W., et al.: A survey on rag meeting LLMs: towards retrieval-augmented large language models. In: Proceedings of the 30th ACM SIGKDD Conference on Knowledge Discovery and Data Mining, pp. 6491–6501 (2024)
4. Ge, K., et al.: Scbench: A sports commentary benchmark for video LLMs. arXiv preprint arXiv:2412.17637 (2024)
5. Held, J., Itani, H., Cioppa, A., Giancola, S., Ghanem, B., Van Droogenbroeck, M.: X-vars: introducing explainability in football refereeing with multi-modal large language models. In: Proceedings of the IEEE/CVF Conference on Computer Vision and Pattern Recognition, pp. 3267–3279 (2024)
6. Jiang, T., Wang, H., Salekin, M.S., Atighehchian, P., Zhang, S.: Domain adaptation of vlm for soccer video understanding. In: Proceedings of the Computer Vision and Pattern Recognition Conference, pp. 6111–6121 (2025)
7. Lee, C., Lin, T., Pfister, H., Zhu-Tian, C.: Sportify: question answering with embedded visualizations and personified narratives for sports video. IEEE Trans. Vis. Comput. Graph. (2024)
8. Li, X., et al.: Multi-modal large language model with rag strategies in soccer commentary generation. In: 2025 IEEE/CVF Winter Conference on Applications of Computer Vision (WACV), pp. 6197–6206. IEEE (2025)
9. Sarfati, N., Yerushalmy, I., Chertok, M., Keller, Y.: Generating factually consistent sport highlights narrations. In: Proceedings of the 6th International Workshop on Multimedia Content Analysis in Sports, pp. 15–22 (2023)
10. Schilling, A., et al.: Querying football matches for event data: towards using large language models. In: International Sports Analytics Conference and Exhibition, pp. 216–227. Springer (2024)
11. Sepasdar, Z., Gautam, S., Midoglu, C., Riegler, M.A., Halvorsen, P.: Soccergraphrag: Applications of graphrag in soccer. In: International Workshop on Graph-Based Approaches in Information Retrieval, pp. 1–10. Springer (2024)
12. Shin, Y., Park, S., Han, Y., Jeon, B.K., Lee, S., Kang, B.J.: Soccer-clip: vision language model for soccer action spotting. IEEE Access **13**, 44354–44365 (2025)
13. Strand, A.T., Gautam, S., Midoglu, C., Halvorsen, P.: Soccerrag: multimodal soccer information retrieval via natural queries. In: 2024 International Conference on Content-Based Multimedia Indexing (CBMI), pp. 1–7. IEEE (2024)
14. Tai, C.Y., Chen, Z., Zhang, T., Deng, X., Sun, H.: Exploring chain-of-thought style prompting for text-to-sql (2023). arXiv preprint arXiv:2305.14215
15. The Guardian: NFL finalizes blockbuster $113bn media rights deal through 2033 season. The Guardian (2021). Accessed: 2025-04-17
16. Xia, H., et al.: Sportu: a comprehensive sports understanding benchmark for multimodal large language models. arXiv preprint arXiv:2410.08474 (2024)

17. Xie, X., Xu, G., Zhao, L., Guo, R.: Opensearch-sql: enhancing text-to-sql with dynamic few-shot and consistency alignment. Proceedings of the ACM on Management of Data **3**(3), 1–24 (2025)
18. Xie, Y., et al.: Decomposition for enhancing attention: Improving LLM-based text-to-SQL through workflow paradigm. arXiv preprint arXiv:2402.10671 (2024)
19. Yang, Z., Xia, H., Li, J., Chen, Z., Zhu, Z., Shen, W.: Sports intelligence: assessing the sports understanding capabilities of language models through question answering from text to video. arXiv preprint arXiv:2406.14877 (2024)
20. Yuan, H., Tang, X., Chen, K., Shou, L., Chen, G., Li, H.: Cogsql: a cognitive framework for enhancing large language models in text-to-SQL translation. In: Proceedings of the AAAI Conference on Artificial Intelligence, vol. 39, pp. 25778–25786 (2025)
21. Zhang, H., Cao, R., Xu, H., Chen, L., Yu, K.: Coe-SQL: In-context learning for multi-turn text-to-SQL with chain-of-editions. arXiv preprint arXiv:2405.02712 (2024)
22. Zhang, J., Han, D., Han, S., Li, H., Lam, W.K., Zhang, M.: Chatmatch: exploring the potential of hybrid vision-language deep learning approach for the intelligent analysis and inference of racket sports. Comput. Speech Lang. **89**, 101694 (2025)
23. Zhu, Y., et al.: Large language models for information retrieval: a survey. arXiv preprint arXiv:2308.07107 (2023)

Action Sequence Modeling for Tactical Training in Handball

Luke Wildman[1]([✉])[iD], Roland Nemes[2]([✉])[iD], and Zhe Hou[3]([✉])[iD]

[1] Brisbane Handball Club, Brisbane, Australia
[2] Hosei University, Tokyo, Japan
nem_roland@hosei.ac.jp
[3] Griffith University, Brisbane, Australia
z.hou@griffith.edu.au

Abstract. Handball is a highly dynamic and complex team sport, characterized by continuous player interactions, rapid transitions between attack and defense, and frequent decision-making under pressure. These factors create significant challenges for formal tactical modeling and performance analysis, as highlighted in previous systematic reviews of match analysis and action sequence complexity in handball. Unlike more discretized sports like baseball or even football, handball's fluidity demands advanced methods to capture and simulate strategic behaviors effectively. This study investigates a novel approach for analyzing handball tactical sequences by applying Probabilistic Model Checking (PMC) to model player actions, decisions, and outcomes. Using Markov Decision Processes (MDPs) and the Process Analysis Toolkit (PAT), we construct probabilistic simulations of handball attacks to evaluate how incremental improvements in player performance—such as passing accuracy, shooting effectiveness, or decision timing—impact overall team success rates.

Keywords: Handball · Sports analytics · Probabilistic model checking

1 Introduction

Handball (also known as European Handball and Team Handball) [16] has been an Olympic sport since 1972 and is estimated to be played by 30 million players by the International Handball Federation [8] and as such is considered to be one of the most popular team sports in the world. Handball action sequences have been studied by Tilp and others [20–22]. Action sequences are combinations of actions performed by players or teams during offensive or defensive phases of the game. Action sequences are influenced by tactical and situational variables such as the type of defence, the score difference, the game period, and the quality of the opponent. A systematic review of Handball Match Analysis was performed by Ferrari *et al.* in 2019 [4].

This paper considers the use of model-checking technology (widely used in the analysis of mission critical systems) to take a probabilistic model (a Markov

Decision Process) of handball play to calculate the probability of scoring. Model-checking technology can also be used for witness and counter-example generation to generate example action sequences for inspection. A Markov decision process (MDP) is a mathematical model for sequential decision-making under uncertainty. It consists of a set of states, a set of actions, and transition probabilities that depend on the current state and action. MDPs have been used to model and solve control problems for stochastic systems, such as robotics, planning, reinforcement learning, and so on. They have also been applied to model sports action sequences, such as possessions in football, baseball, tennis, and other sports where the goal is to measure the contribution of each action to the final outcome.

Our intended use of the model is to investigate the possible benefits of making incremental marginal improvements to the decision making and performance characteristics of the team players (through targetted training activities). We follow the approach of using MDP to model decisions and performance described in [13] and [9]. Building on initial work with a junior women's team preparing for a minor tournament, we have now expanded the model's validation by integrating datasets from matches played by two EHF Champions League teams (one male, one female) and a female Youth World Championship team. This expanded dataset allows comparison of tactical modeling across different age groups, genders, and performance levels, offering new insights into how decision-making patterns evolve with skill and experience.

The broader goal of the project is to develop methods that leverage machine learning and formal probabilistic modeling to help coaches identify strengths and weaknesses in both their own teams and their opponents. By simulating various tactical scenarios and player combinations, coaches can make informed decisions to maximize efficiency, adapt training activities, and increase the team's winning probability. Furthermore, the framework moves toward analyzing and understanding player decision-making processes—a major frontier in sports analytics—by making action sequences and tactical options explicitly visible and quantifiable. Our findings demonstrate that Probabilistic Model Checking offers a highly transparent, explainable alternative to conventional machine learning models. It enables simulation-driven strategy development without requiring massive datasets, making it practical even for teams with limited analytical resources. Future work will aim to incorporate dynamic defensive adjustments, expand the model's action space, and integrate automated video analysis pipelines to enable large-scale deployment.

2 Related Work in Decision Making in Handball

Decision-making in handball operates at the intersection of rapid perception, tactical cognition, and high-stakes interaction, presenting unique challenges for formal modeling. Unlike other team sports with broader spatial-temporal margins, handball compresses cognitive demands into milliseconds, requiring players to anticipate, decide, and act in dynamically shifting contexts [1,6]. Defensive decisions, in particular, rely heavily on the real-time coupling between outfield defenders and goalkeepers, who must interpret subtle kinematic cues (e.g., shoulder tilt, arm movement) to predict shot direction and intent before execution [5,11]. These decision sequences are rarely isolated—they emerge from fluid, adaptive interplay between contextual variables such as game score, fatigue, and opponent tendencies [15,19]. While machine learning models can effectively capture structured game behavior and probabilistic state transitions (e.g., pass networks, spatial densities), the modeling of internal decision-making processes is hindered by a lack of labeled perceptual-cognitive data and ground truth for "intent" [3,12]. Therefore, we defer the modeling of decision-making components until more granular multimodal datasets—such as eye-tracking, biomechanical markers, and real-time emotional state indicators—become available. In addition to the work on decision-making in handball, there is a large body of work on the application of machine learning to handball. Ichimura et al. [7] have applied machine learning to the prediction of handball player performance. Mizuno et al. [18] have applied random forest to the prediction of handball player performance. Marczinka et al. [17] consider technical elements in defence focusing on the differences between positions and genders.

3 PCSP# Language Overview

To model the handball play, we use the modelling language Probabilistic Communicating Sequential Programs (PCSP#) [14] and the Process Analysis Toolkit (PAT) [24] as the model checker. Given a model of the desired system expressed in *PCSP#* and its desired properties, PAT will automatically and exhaustively search all possible cases to verify if the system satisfies the desired property. When given a model which includes probabilistic choice and a probabilistic reachability property, PAT will calculate the probability of reaching the desired states [23]. A system is modelled using a set of variables and processes. A variable is usually an integer or an enumerable type within a certain range to ensure the number of states is countable and finite. When any of the variables is assigned a different value, the system is considered to have transited into a different state. More information on the use of model checkers and the algorithms used may be found in [14] and the references above. Another widely used alternative model-checking tool for probabilistic system is Prism [10]. We now introduce the relevant aspects of the modelling language via a handball example.

4 Basic Model of a Handball Attack

Handball is played on a $20 * 40$ m court with $2 * 3$ m goals at both ends. Around each goal is a 6 m line, in a D-shape marking the goal area (which is inclusive of the line). There is also a 9 m dashed line which simply signifies the distance of 3 m required for free throws, and a 7 m mark used for taking penalties. See Fig. 2. There are normally 7 players on court for each side, made up by 6 field players and 1 goal keeper. There are also several substitutes for each side with substitution over the side-line occurring any time and allowed multiple times per player. Only the goalkeeper is allowed to play in the goal area. Field players cannot play the ball while standing in the goal area, or gain advantage by moving through the goal area. Shots on goal must be taken from outside the goal area or while in the air over the goal area before touching down. Similarly defenders may not defend from inside the goal area. As for other invasion games, the usual modes of attack are fast breaks (in the case that the defence is disorganised) or attacks on an organised zone defence around the goal area. The attacking players are usually organised in a D-shape around the defence and the usual positions are labelled accordingly: *left wing, left back, centre back, right back, right wing* and lastly the *line player* or *pivot* who plays in the middle, among the defence on the goal area line (Fig. 1).

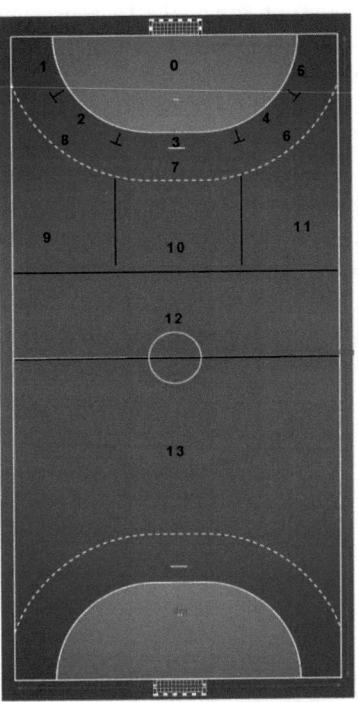

Fig. 1. Handball court with positions marked

A very simple model of a handball attack around the goal area (or 'D') is now provided for further introduction to the modelling language. We model the progress of the attack by the location of the players and ball relative to 14 zones. Zones 1 to 5 represent sequential positions around the goal area line from *left-wing* to *right-wing*, zones 6, 7, and 8 represent 3 back positions between the goal line and the 9-metre line, for *left back, centre back* and *right back* respectively, and zones 9, 10, and 11 represent three back positions behind the 9-metre line. Finally, zone 0 is inside the goal and represents scoring a goal. Zone 12 is used to represent an attacking position behind the backs to the half-way line from where it is very difficult to score by a direct shot on goal. Zone 13 represents the other half of the court, which from the point of view of the attacking team represents the defensive zone. Zone 14 is used to represent states where the attacking team loses the ball (causing a turn-over) by the ball being played outside of the playing court or missing a shot, or the ball being intercepted by the defence (including the goal keeper by a save), or otherwise being turned over because of an attacking foul or other reason. We use a single variable called *ball* to represent the zone occupied by the ball. The location of the players (in the order LW LB CB RB RW PV) is captured in the array variable *aloc*. For example, $aloc = [1, 6, 7, 8, 5, 3]$ models the normal positions occupied by the attacking players when facing a 6-0 defence where the defensive players are lined up around the goal area and $ball = 7$ represents it being held by the centre back, given the attacking team is in possession of the ball.

We model the defence positions $d1$ to $d5$ representing the five zones around the goal area and the variable *gap* to capture whether there is gap in one of those zones. For example, $gap[d1] == 1$ represents a gap in zone 1. Note this model is adequate for a simple 6-0 defence but will need development for more complex defence arrangements (See Sect. 7.2).

Finally, we keep track of the number of passes in the attack with the variable *pass*. This is due to the rules regarding *passive play* where the referees may force a turnover if it appears that the attacking team is not consistently attacking the goal. This call is made by the referees based on their read of the play, but as a general rule the number of passes in an attack is less than 15 [25]. We define a constant *maxpass* to represent this. The predicate *inplay* describes the states where the ball remains in play. Predicates are logical expressions involving the model variables and may be used in conditional expressions in the model. In this case *inplay* is true if the ball is not in the goal ($ball\ !=0$) or not lost/turned over ($ball\ !=14$), and that the referee has not called passive play ($pass < maxpass$).

Given the above very simple model of the state, we model the attacking play by describing each player as a process. The CB process below models the probabilistic choice of gameplay decisions made by the centre back faced with specific situations. The first line models the condition that the centre back may only play the ball if it is *inplay*. This is expressed as a conditional process **if** (*cond*) { P } **else** { Q } that behaves as P if the condition *cond* holds or otherwise behaves as Q. In this case Q is the special process $Skip$ which just terminates because the ball is not in play.

$CB = $ **if** $(inplay)$ { **case** { $aloc[cb] == 7$ && $gap[d3] == 0$: *pcase* {
30 : $cb\_\ lb \rightarrow short\_\ pass(cb,\ lb);\ LB$
30 : $cb\_\ rb \rightarrow short\_\ pass(cb,\ rb);\ RB$
10 : $cb\_\ pv \rightarrow pivot\_\ pass(cb,\ pv);\ PV$
10 : $cb\_\ lw \rightarrow long\_\ pass(cb,\ lw);\ LW$
10 : $cb\_\ rw \rightarrow long\_\ pass(cb,\ rw);\ RW$
5 : $cb\_\ m10 \rightarrow move\_\ wball(cb,\ 10);\ CB$
5 : $cb\_\ shot \rightarrow shot(cb)$
} $default : undef \rightarrow Skip$
}} **else** {$Skip$};

The situation described by this process is that the centre back occupies location 7 on the field and there is no gap in defensive position $d3$ directly in front of them ($gap[d3] == 0$). We only describe one situation, but in general many situations may be described where each situation is captured as a branch of a **case** process. In general the case process is written **case** { $b_1 : P_1\ b_2 : P_2\ ...\ default : P_d$ } where the combined case process behaves as a subprocess P_i if the associated condition b_i holds. The conditions are evaluated one by one until a true one is found and if no conditions hold, the case process behaves as the default P_d. In this basic situation we have estimated a probability of 30% that the CB passes the ball to the left back (LB), a 30% probability that they pass the ball to the right back (RB), and so on, with a 5% chance of shooting. This is modelled by a probabilistic choice **pcase** $\{\ p_1 : P_1\ p_2 : P_2\ ...\ p_n : P_n\ \}$ where the probabilities (p_i) are normalised to sum up to 1. Each branch of the probabilistic process (P_i) in the above example identifies events (e.g., cb_lb, cb_rb); an action to be executed; and then the next player taking control of the ball (and the overall attack process). For example, the process description states that there is a 10% chance of passing to the pivot, which is modelled by the event cb_pv followed by an action $pivot\_pass$ from the centre back to the pivot, and then if the pass is successful, the PV process takes over. The PV process models what happens when the ball is received by the pivot. The probabilities related to the pivot actions are different to the centre back. That is, in the situation where the pivot receives that ball on the line and there is a gap at that defensive position, then the pivot will almost certainly shoot and a very small percentage of the time they may do something else such as pass the ball to the centre back.

$$PV\ =\ if\ (inplay)\ \{\ case\ \{\ loc[pv] ==\ \&\&\ gap[d3] == 1 :$$
$$pcase\ \{$$
$$\quad 99\ :\ pv\_shot\ \rightarrow\ shot(pv)$$
$$\quad 1\ :\ pv\_cb\ \rightarrow\ short\_pass(pv, cb)\ ;\ CB$$
$$\}$$
$$\}\}\ else\ \{Skip\};$$

Other player positions are described similarly. Decisions are dependent on the court situation: player position, ball position, defence position, and whether the attacker has an unhindered path to the goal line to shoot. They may also depend on the performance of defence (e.g., the goal keeper may be particularly good at saving goals from the wing making the winger less likely to take a shot from zones 1 or 5).

5 Player Performance

We capture the performance of the players with arrays recording the probability that a short pass, long pass, or pass to the pivot succeeds (there are N players). This success rate is calculated based on statistics gathered in matches and scaled against elite player performance [25]. Values are elided for formatting purposes.

$$var\ short\_pass\_succ[N]\ =\ [98, ..., 98];$$
$$var\ long\_pass\_succ[N]\ =\ [85, ..., 84];$$
$$var\ pivot\_pass\_succ[N]\ =\ [60, ..., 50];$$

Process $short\_pass(pl, rp)$ below models the success of passing a ball from player pl to receiver rp. If the pass is successful (event $spass.pl.rp$), then the ball goes to the location of the receiving player and the number of passes increments by 1. If the pass fails (event $to.pl$), then the ball goes to location 14 representing a turn over. The probability of failure is 100 minus the probability of success.

Processes *long_pass* and *pivot_pass* are defined similarly and are not shown in this paper.

$$short\_pass(pl, rp) = pcase \{$$
$$short\_pass\_succ[pos[pl]] : spass.pl.rp \{ball = loc[rp]; pass\text{++}\} \rightarrow Skip$$
$$100 - short\_pass\_succ[pos[pl]] : to.pl \{ball = zout\} \rightarrow Skip\};$$

The effectiveness of a shot is captured in a similar way for each player and each shooting position in the array variable *shot_effect*. Values are elided for formatting purposes. The effect of the shot, if successful (event *goal*), is represented by setting the ball location to 0 Otherwise the event *miss* is represented by setting the ball location to zone 14.

$$shot(n) = pcase \{$$
$$shot\_effect[ball - 1][pos[n]] : goal\{ball = 0\} \rightarrow Skip$$
$$100 - shot\_effect[ball - 1][pos[n]] : miss\{ball = zout\} \rightarrow Skip \};$$

6 Model Simulation

The handball model is simulated by providing an assertion to be checked and application of the PAT model checker. Assuming that an attack starts with the centre back, we define the process *Play* as *CB*. We define the proposition *scoregoal* as the state where the *goal* variable is equal to 0. The assertion to be checked is then defined as follows.

$$Play = CB;$$
$$\#define\ scoregoal\ ball == 0;$$
$$\#assert\ Play\ reaches\ scoregoal\ with\ prob;$$

When executed, PAT returns the probability that the assertion is valid with minimum and maximum values. E.g. $[0, 0.4832]$. A minimum probability of 0 represents the fact that it is possible that the play fails to score.

Another useful aspect of model checking is the generation of witness sequences and counter examples. Consider the following reachability assertion:

$$\#assert\ Play\ reaches\ scoregoal;$$

PAT returns that the assertion is valid and provides the following trace (with some simplifications) as a witness.

$$< init \rightarrow [\ if\ (inplay)] \rightarrow [(loc[cb] == 7)] \rightarrow [(gap[d3] == 1)] \rightarrow$$
$$0.5 \rightarrow cb\_shot \rightarrow 0.4 \rightarrow goal >$$

The above example illustrates some of the basic concepts used in the model of the handball attack. However, there are many limitations to the model and we progressively deal with these in the following sections.

7 Extensions

In our previous work [26], we focused on modeling simple attacks such as fast breaks. The current paper extends this by considering more complex attack patterns and basic defensive structures. The attack patterns analyzed are derived from matches in the EHF Champions League and Youth World Championship, while the defensive modeling is based on observations from the EHF Champions League. Importantly, the model framework we have developed is readily extensible to incorporate additional attack and defence patterns of increasing complexity.

7.1 Attack

We extend the model by incorporating several additional attack patterns: the simple cross, the empty cross/"Jugo" style attack, the second pivot attack pattern, and pivot crossing movements. These represent common tactical variations seen in high-level handball matches (see [7] for more details on common tactical variations in handball). The attack patterns are introduced into the model by encoding the behaviour as in $PCSP\#$. We have modified the process description for Center Back behaviour for the CB when they are in zone 10 as follows. First it is checked that the CB, pivot and left and right backs are in the correct positions. That is, the CB is in zone 10, the pivot is in zone 4, the left back is in zone 9 and the right back is in zone 11. The process then encodes the decision to initiate the different styles of attacks with probabilities derived from the frequencies observed in the EHF Champions League.

The event $pvxr$ encodes the decision to initiate the pivot cross. Assuming it starts 17% of the time (embedded in a pcase) it starts by the pivot moving to zone 10 followed by the CB passing to the pivot. Note that this represents movement of the pivot without the ball. In practice, the pivot usually actually starts moving prior to CB receiving the ball, anticipating the pass from the LB, and the pivot is in zone 10 when the CB receives the ball. However we have simplified the process for the purpose of this explanation. The process then evolves the $PVXR$ process which is described further below. Similarly, if the empty cross on the left (or Jugo left) ($cbexl$) is chosen then the CB moves to zone 11 and passes to the LB. The process then evolves the $CBEXL$ process which is also described further below. The $CB10$ process also includes the $CBEXR$ process which is the same as $CBEXL$ but with the CB moving to zone 9 and passing to the RB, as well as other possible passes, movements and shots as described earlier for basic model.

$CB10 = [aloc[cb] == z10 \;\&\&\; ball == z10]$
   $gap[d3] == 0 \;\&\&\; aloc[pv] == z4 \;\&\&\; aloc[lb] == z9$
   $\&\&\; aloc[rb] == z11 : pcase \{$
   ...
   $17 : pvxr \rightarrow move(pv, z10); cb\_pv \rightarrow short\_pass(cb, pv); PVXR$

The $PVXR$ process is then defined as follows. First the state is checked to see if the ball is in play. This checks that the previous pass was successful and did not result in a turn over due to a bad pass or other attacker error. (Note this is very unlikely, however it is a possibility must be accounted for). Next, we implement the role switch between the cb and the rb. This represents that the CB role is now being played by player sw who was previously the RB. By describing the process in terms of roles we are able to reuse the same process descriptions for the CB and the RB without having to write player-specific processes. This is done by calling the *switch* process with the cb and rb roles as arguments. This is followed by the new cb moving to zone 10 (completing the empty cross) and evolves into the $PVL10$ process.

$$PVXR = if(inplay) \{sw.cb.rb\{call(switch, cb, rb)\} \\ \rightarrow cbm10 \rightarrow move(cb, z10) \ ; \ PVL10\};$$

The switch function is defined as follows. It simply swaps the players in the two roles.

$$\#define \ switch(pl, \ rv) \ \{var \ t = pos[pl]; pos[pl] = pos[rv]; pos[rv] = t\};$$

The $PVL10$ process is defined similarly to PV in the basic model, but with the pivot being in zone 10 and Pivot moving to the left back lb with the ball. In this case it is mostly likely that the pivot will pass to the LB but the description allows for other options such as when there is an open shooting opportunity or the PV decides to pass it elsewhere. (Note, there is obviously a number of possible other variation, but we have only modelled those observed in practice.)

The $CBEXL$ (Jugo to the left) process is defined similarly to the Pivot cross process. As part of the initiation described above in $CB10$, the centerback moves with the ball to zone 11. This is continued in $CBEXL$ which after checking that this has not resulted in a turn over due to an attacker error, the ball is passed to the LB (this is a long pass given that the CB is in zone 11), switches roles with the rightback rb, the new CB moves into zone 10, and evolves into the LB process, noting that the LB has the ball.

$$CBEXL = if \ (inplay) \ \{long\_ \ pass(cb, \ lb); \\ sw.cb.rb\{call(switch, cb, rb)\} \ \rightarrow \ cbm10 \\ \rightarrow \ move(cb, z10) \ ; \ LB\};$$

A slightly different move is the 2nd pivot attack pattern. This starts similarly to the pivot cross but instead of pivot, the wing (left wing in this case) moves out to zone 10 and the CB passes the ball to the LW the LW passes to the RB while the CB executes an empty cross with the LB (switching roles.

$$LW2PV = if(inplay) \ \{sw.cb.lb\{call(switch, cb, lb)\} \\ \rightarrow \ cbm10 \ \rightarrow \ move(cb, z10) \ ; \ LWR10\};$$

What is different here is that the LW changes into a second pivot and moves to zone 4. That is, there is a change of role for that player from a wing role to a second pivot (PV2) role. This is encoded in the following event sequence. Following the *LW* passing to the *RB*, the *LW* transforms to the 2nd pivot role and moves to zone 4. The process then evolves into the *RB* process as the *RB* now has the ball.

$$lw\_rb \rightarrow short\_pass(lw, rb)\,;\ tr.lw.pv2\ \{call(trans, lw, pv2)\} \rightarrow$$
$$pv2m4 \rightarrow move(pv2, z4);\ RB$$

The *trans* function transforms the player at role *pl* to the role *rv* as well as their location. The first role (left wing in this case) is no longer being used so we set the role to *pnull* and the player's location to *zout* (out of play).

$$\#define\ trans(pl,\ rv)\ \{pos[rv] = pos[pl]; pos[pl] = pnull;$$
$$aloc[rv] = aloc[pl]; aloc[pl] = zout\};$$

7.2 Defence

In this section, we extend the model to include defensive formations and behaviors. Specifically, we incorporate two common defensive systems: the 6-0 defense (where all defenders are positioned along the 6-meter line) and the 5-1 defense (with five players on the 6-meter line and one advanced defender).

6-0 Defence. The 6-0 defence is modeled using a simplified representation focused on defensive gaps rather than explicit defender positions. We consider six potential gaps in the defensive formation: five gaps around the goal area (D) and one gap in front for the 5-1 formation. Each gap is represented as a binary state (0 or 1) where 1 indicates the presence of a defensive gap.

The five gaps around the goal area are positioned as follows:

- d1: A gap on the left wing, occurring when either the wing defender is out of position or the adjacent defender fails to provide coverage
- d2: A gap on the left side, arising when the second defender is displaced and neither the first nor third defender compensates
- d3: A gap in the center, representing a significant area where the middle defenders have failed to maintain their formation
- d4: A gap on the right side, following similar principles to d2
- d5: A gap on the right wing, analogous to d1

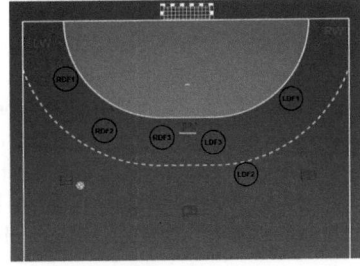

Fig. 2. Handball court with positions marked

Additionally, $d6$ represents a gap at zones 7 and 10, which would allow the center back to attempt shots from these positions. This abstraction captures a key defensive principle: the defence typically only exposes gaps on the side opposite to the ball's location. However, when defenders step out to challenge shots from the backs, gaps may emerge if other defenders fail to adjust their coverage appropriately. The probabilities for gap formation can be calibrated using empirical data collected from different variations of the 6-0 defense (designated as 6-0A, 6-0B, and 6-0C), combined with tactical analysis of defensive behavior.

$$var\ gap[6] = [0,0,0,0,0,0];$$

The function *CloseGap* models a simplified version of a sliding 6-0 defence with four possible outcomes with the following probabilities:

- 0.1: The defence closes all gaps.
- 0.5: The defence closes a gap in front of the player and leaves a gap on the furthest wing.
- 0.2: The defence opens a gap in front of the player and closes a gap on the furthest wing.
- 0.2: The defence opens a gap in front of the player and leaves a gap on the furthest wing.

$$
\begin{aligned}
CloseGap(x) = \ &pcase\ \{ \\
&[0.1]\ :\ gca.x\ \{resgap5\} \rightarrow DEF60 \\
&[0.5]\ :\ gcwo.x\ \{resgap5;\ call(wgap,x)\} \rightarrow DEF60 \\
&[0.2]\ :\ go.x\ \{resgap5;\ gap[admap[aloc[x]]] = 1\} \rightarrow DEF60 \\
&[0.2]\ :\ gowo.x\ \{resgap5;\ call(wgap,x) \rightarrow DEF60\ \};
\end{aligned}
$$

Defence Process. The defense is implemented as a separate process that interacts with the attacking processes through defined communication channels. These channels facilitate the exchange of information about player movements, and passing actions. The defensive process takes inputs from the attacking processes, produces appropriate responses through its outputs, and is composed with the attacking processes to create the complete game model.

Channels are a *PCSP#* feature that allows processes to communicate with each other. We use channels of length 0 to indicate that the communication is synchronous, as illustrated by the following.

$$channel\ spass\ 0;$$

Channels are defined for passing and movement events.

We synchronize the defence actions with the events modelling the performance of an action rather than the event describing the decision to perform the action. That is, the defence process is triggered by the observation of the action rather than having read the mind of the attacking player.

Input to the channels is defined using the ! operator. For example, the values *pl* and *rp* are input to the short pass channel *spass* in the following update of the short pass process. The values of *pl* and *rp* are composed to form the value *pl.rp* which is passed to the defence process via the channel.

$$short\_pass(pl, rp) = pcase \{$$
$$short\_pass\_succ[pos[pl]] : spass!pl.rp \rightarrow \{ball = aloc[rp]; pass\texttt{++}\} \rightarrow Skip$$
$$1000 - short\_pass\_succ[pos[pl]] : to.pl \{ball = zout\} \rightarrow Skip \};$$

The defence process uses the output of the channel. The value is read form the channel using the ? operator. In the following example, the value *y.x* is read from the short pass channel *spass*.

This example is for the 6-0 defence. The process DEF60 calls the process CloseGap with the value *x* which is the player that received the pass. The CloseGap process is defined further below.

$$DEF60 = spass?y.x \rightarrow CloseGap(x) \;\square$$
$$lpass?y.x \rightarrow CloseGap(x) \;\square$$
$$ppass?y.x \rightarrow CloseGap(x) \;\square$$
$$movb?y.x \rightarrow CloseGap(y) \;\square$$
$$mov?y.x \rightarrow CloseGap(y);$$

5-1 Defence. The 5-1 defence is a simple extension of the same idea. The difference is that additional defensive zone is added in front of the 6-0 line.

Composition of Attack and Defence. The attack and defence processes are composed using the ||| operator. This is a *PCSP#* operator that allows the composition of processes that communicate via channels. The syntax is as follows.

$$AD = CB \;|||\; DEF60;$$

The attack begins with the process *CB* and evolves as described in the process description. The defence process interleaves with the attack process by synchronising on the channel events.

8 Application

We have applied the model to the analysis of handball matches from the European Handball Federation (EHF) Champions League and the IHF U-18 World Championships. The four matches cover sex, age and tier. The European Handball Federation (EHF) Champions League, established in 1956, is Europe's premier club competition in handball, encompassing both men's and women's tournaments. Hungarian clubs have been prominent in both competitions. Telekom

Veszprém HC has reached the men's final four times, each time finishing as runner-up, while MOL-Pick Szeged is the reigning Hungarian Cup holder. In women's handball, Győri Audi ETO KC dominates with six Champions League trophies, and FTC-Rail Cargo Hungaria, current Hungarian champions and cup winners, have also achieved runner-up status in the competition. These high-stakes encounters not only fuel intense rivalries but also provide valuable data for analyzing tactical sequences and player decision-making.

Event Logging. Two analysts watched match in a media-player displaying and entered every attacking episode into spreadsheets. Each atomic or combination event was logged with time, zone and outcome.

Atomic actions (28)	– Ball control: receive, dribble, pick-up, fake-shot, fake-pass; – Passes: parallel, cross/X-over, long diagonal, return, pivot feed, bounce, wing; – Shots: wing, back-court, pivot, 6 m breakthrough, fast-break, lob, jump, standing; – Defence/keeper: block, steal/intercept, goalkeeper save, foul earned, 7 m earned; – Outcomes: goal, miss, turnover, ball-out.
Multi-action combinations (18)	Wing-Yugo, Yugo, Pivot cross, PvSlide, Rb/Lb -Pv 2-2, Lw/Rw -Rover, Cb-Rb/LbX, RunP, LongP, RetP, PvBlock, Lw/Rw 2ndPv, Ten Pv- Lw/RwX

We also recorded the occurrences of passing and movement events and shooting locations and success rates. The results are provided in the appendix.

Probabilistic Verification. The data was hand-coded into the $PCSP\#$ model and simulated in the Process Analysis Toolkit (PAT). Four reachability queries were run for every team similar to the following examples. The first example is the probability of eventually scoring. The second example is the probability of eventually scoring given that the attack is a pivot cross (left or right). Due to the size of the model, the reachability queries were run to maximum depth of 5 passes. This is not considered a significant limitation due the fact that the number passes that actually impact the outcome of an attack is often less than 5.

#assert AD reaches scoregoal with prob;
#assert AD $\models$ F(pvxl $\parallel$ pvxr) && X F(scoregoal) with prob;

Results. Table 1 shows the results of the matches with the fast breaks excluded and a summary of the predicted number of goals and success rate. We have excluded the fast breaks from the analysis as they are not currently part of the model. (Note we covered gast breaks in the earlier paper.)

The maximum probability of the complex combinations are presented in Table 2.

Table 1. Analysis results (A = Actual, P = Predicted)

Tier	Competition	Sex	A (attacks/goals/success%)	P (goals/success%)
T1	EHF Champions League	♂	Szeged.A (46/23/50)	(19/42)
			Veszprém.A (57/22/39)	(26/45)
T1	EHF Champions League	♂	Szeged.B (63/32/52)	(21/33)
			Veszprém.B (52/39/59)	(17/33)
T2	EHF Champions League	♀	Győri ETO (40/21/53)	(16/40)
			FTC (39/15/38.5)	(15/38)
T4	IHF U-18 Worlds	♀	Netherlands (46/17/37.0)	(12/26)
			China (42/8/19.0)	(11/26)

Table 2. Success Rates by Play Pattern and Team (A = Actual, P = Predicted)

Play	SzA		VeB		Ch		Ne		FTC		ETO		SzB		VeB	
	A	P	A	P	A	P	A	P	A	P	A	P	A	P	A	P
P2	73	40	33	24	50	19	-	-	-	-	-	-	80	27	100	33
PX	43	75	0	26	20	21	60	20	50	29	-	-	100	38	100	35
EX	-	-	29	20	-	-	33	20	20	29	38	34	50	22	37	27
X	60	22	57	19	0	23	40	20	50	30	75	30	80	26	60	29

8.1 Discussion

Key Findings. Across four elite and youth matches the probabilistic model correctly identified the winning team in two cases and produced a mean absolute error of 6.8% points in goal-probability prediction.

Interpreting Discrepancies. The youth mismatch (Netherlands vs China) exposes two boundaries of the current framework. First, player-level parameters were calibrated on senior data; youth error rates in passing and shooting are higher and more variable, amplifying divergence. Second, the model assumes a standard 6-0 or 5-1 defensive structure with fixed gap-closing probabilities. Video review shows that China defended with a high-pressure 6-0 variant (6-0C), stepping aggressively into first-line gaps and contesting most passes. Those pressure cues induce state transitions—especially forced turnovers—outside the calibrated range, inflating forecast error. These observations confirm that model fidelity may depend less on sample size than on defensive intensity and style representativeness.

The 2nd (EHF Champions League) match between Szeged and Veszprem predicts the opposite outcome to the actual result (Szeged 32, Veszprem 39). This is due to the fact that the model does not cover fast breaks. In this particular match, Szeged suffered an extraordinary high number of turnovers (18) resulting in 16 fast breaks for Veszprem whereas there were only 2 fast break for

Szeged from 7 Veszprem turnovers. However the model does correctly predict the outcome of the non-fastbreak play.

The maximum probabilities of the complex combinations are not indicative of the actual success rate. Further investigation is required to understand the reasons for this.

Practical Value for Coaches. Despite sparse data, the framework already yields actionable insights. PAT counter-examples reveal that Szeged's most profitable sequence was the Wing-to-Second-Pivot (success = 73 %), yet it accounted for only 20 % of attacks. Conversely, Veszprem's Pivot-Cross produced zero goals in seven attempts; reallocating those possessions to X-cross would have yielded an estimated +0.9 goals per game. Such diagnostics let coaches target training time without exhaustive video coding.

Limitations. The manual analysis of each game takes significant time and is error prone, and as such we were only able to analyse a small number of matches. As such the Data is confined to 6 teams and only a small number of atomic labels; rare tactics (e.g., 7 v 6 empty-goal) are absent. Because we did not retain a systematic double-coded sample, we cannot report a formal inter-rater statistic (e.g., Cohen's k); quantifying annotation uncertainty is therefore left to future work. Defensive reactions are currently modelled as static zone gaps; dynamic elements such as stepped-out first-line pressure and adaptive goalkeeper positioning remain abstracted away. The present implementation is also episodic and ignores fatigue, score effects, and stochastic time-outs. Davis *et al.* [2] discuss these and other limitations of sports analytics models.

9 Further Work and Conclusion

The details presented above define a framework for the consideration of playing conditions; description of decision options (as probabilistic choices); and the description of some performance characteristics such as passing and shooting (also probabilistic).

While our initial results are promising, there are several key areas for future development of this work. A primary limitation is the relatively small dataset currently available for analysis. The model would also benefit from including goalkeeper success rates to better reflect defensive capabilities. Player fatigue effects on performance metrics should be considered, as these can significantly impact game outcomes over time. Additionally, analyzing defensive adjustment patterns across multiple games would provide insight into tactical adaptations, especially in the case of superiority or inferiority due to exclusions. The impact of timeouts and substitutions on attack effectiveness represents another important avenue of investigation. These extensions would create a more comprehensive framework for analyzing handball tactics and performance.

References

1. Bonnet, G., et al.: Toward a better understanding of decision-making in handball: a systematic review. Mov. Sport Sci. **108**, 5–20 (2020)
2. Davis, J., et al.: Evaluating sports analytics models: challenges, approaches, and lessons learned (2022)
3. Espoz-Lazo, S., Hinojosa-Torres, C.: Modern handball as a dynamic system: orderly chaotic. Appl. Sci. **15**(7), 3541 (2025)
4. Ferrari, W., Sarmento, H., Vaz, V.: Match analysis in handball: a systematic review. Monten. J. Sports Sci. Med. **8**, 63–76 (2019)
5. Fontaine, J.: The role of emotions in intuitive decision-making: a study with expert handball coaches. In: Naturalistic Decision-Making Conference Papers (2024)
6. Hinz, M., et al.: Differences in decision-making behavior between elite and amateur team-handball players in a near-game test situation. Front. Psychol. **13**, 854208 (2022)
7. Ichimura, S., Moriguchi, T., Nemes, R.: Development of data model for attacking behavior in handball. In: 7th European Handball Federation Scientific Conference "Sustainability in Handball - Circle of a Handball Life", Porto. Conference Coordinators: João Monteiro. Beata Kozlowska, Noémi Szécsényi (2024)
8. IHF. International handball federation (2023). Accessed 6 Aug 2023
9. Kan, J.: Sports strategy analytics using probabilistic model checking and machine learning (2023)
10. Kwiatkowska, M., Norman, G., Parker, D.: PRISM: probabilistic symbolic model checker. In: Proceedings of Tools Session of Aachen 2001 International Multiconference on Measurement, Modelling and Evaluation of Computer-Communication Systems (2001)
11. Le Menn, M., et al.: Handball goalkeeper intuitive decision-making: a naturalistic case study. J. Hum. Kinet. **70**, 249–260 (2019)
12. Lenzen, B., et al.: Situated analysis of team handball players' decisions: an exploratory study. J. Teach. Phys. Educ. **28**(3), 327–343 (2009)
13. Liu, Z., Jiang, K., Hou, Z., Lin, Y., Dong, J.S.: Insight analysis for tennis strategy and tactics. In: 2023 IEEE International Conference on Data Mining (ICDM), Los Alamitos, CA, USA, pp. 1169–1174. IEEE Computer Society (2023)
14. Liu, Z., Ma, M., Jiang, K., Dong, J., Hou, Z., Shi, L.: PCSP# denotational semantics with an application in sports analytics, vol. 14900, pp. 71–102. Springer, Cham (2024)
15. Magnaguagno, L., et al.: Cognitive representations of handball tactic actions in athletes: the function of expertise and age. PLoS ONE **18**(4), e0284941 (2023)
16. Marczinka, Z.: Playing handball: a comprehensive study of the game. Trio Budapest (1993)
17. Marczinka, Z.: Analysis of technical elements in defence: focusing on the differences between positions and genders. In: 7th European Handball Federation Scientific Conference "Sustainability in Handball - Circle of a Handball Life", Porto: Conference Coordinators: João Monteiro. Beata Kozlowska, Noémi Szécsényi (2024)
18. Mizuno, H., Nemes, R.: Examination of scoring efficiency in men's handball using random forest. In: 7th European Handball Federation Scientific Conference "Sustainability in Handball - Circle of a Handball Life", Porto: Conference Coordinators: João Monteiro. Beata Kozlowska, Noémi Szécsényi (2024)
19. Nicolosi, S., et al.: Situational analysis and tactical decision-making in elite handball players. Appl. Sci. **13**(15), 8920 (2023)

20. Perl, J., Tilp, M., Baca, A., Memmert, D.: Neural networks for analysing sports games. In: Routledge Handbook of Sports Performance Analysis, pp. 237–247 (2013)
21. Rudelsdorfer, P., Schrapf, N., Possegger, H., Mauthner, T., Bischof, H., Tilp, M.: A novel method for the analysis of sequential actions in team handball. Int. J. Comput. Sci. Sport (IJCSS) **13**, 69–84 (2014)
22. Schrapf, N., Tilp, M.: Action sequence analysis in team handball. J. Hum. Sport Exerc. **8**(3proc), S615–S621 (2013)
23. Song, S., Gui, L., Sun, J., Liu, Y., Dong, J.S.: Improved reachability analysis in DTMC via divide and conquer. In: IFM, pp. 162–176 (2013)
24. Sun, J., Liu, Y., Dong, J.S., Pang, J.: PAT: towards flexible verification under fairness. In: Bouajjani, A., Maler, O. (eds.) CAV 2009. LNCS, vol. 5643, pp. 709–714. Springer, Heidelberg (2009). https://doi.org/10.1007/978-3-642-02658-4_59
25. Vurgun, N., Bilge, M., Eler, S., Eler, N., Şentürk, A.: Current developments in handball game analysis. J. Popul. Ther. Clin. Pharmacol. **30**(12), 421–435 (2023)
26. Wildman, L.: Probabilistic model checking of handball action sequences. In: 2023 IEEE 28th Pacific Rim International Symposium on Dependable Computing (PRDC), pp. 332–336 (2023)

Author Index

A
Angelova, Maia 1

B
Bandara, Ishara 1
Barot, Vishal 235
Bianco, Mattia Donna 209
Brecht, Tim 162
Breytenbach, Gerhardt 18

C
Cao, Hao 253
Chi, Michael 268
Claytor, Ross 268
Clinckemaillie, Winter 36

D
Demir, Ege 53
Divakaran, Srikrishnan 235
Dwyer, Daniel B. 1

G
Grobler, Jacomine 18

H
Hashimoto, Sakiko 99
Hong, Jin-i 116
Hou, Zhe 276
Huang, Tao 69
Huang, Yangyi 69
Hur, Nathan 76

I
Iaboni, Evan 162
Ishibe, Kai 99

J
Javadpour, Leili 92
Jiang, Kan 126

K
Kase, Yuto 99
Kaya, Tolga 218, 235
Khazaeli, Mehdi 92
Kim, Eun-jin 1
Kim, Jongbae 116
Kim, Jongsung 116
Kondapalli, Ashwinth Reddy 92
Kwon, Yeong-hun 116

L
Lee, Jake 268
Lee, Yun-hwan 116
Li, Chuanfei 253
Li, Zizhen 126
Liang, Zhenkai 253
Lin, Jun 69
Liu, Zhaoyu 126, 142
Lodhi, Fauzan 162

M
Maddox, Eva 218
Malhotra, Ruchika 155
Manuri, Federico 209
Meena, Shweta 155
Mor, Subodh 155
Moskovitch, Robert 193

N
Nemes, Roland 276

P
Pitassi, Miles 162

R
Rajasegarar, Sutharshan 1
Ranasinghe, Pamudu 179
Ranasinghe, Pasindu 179
Raval, Mehul S. 218, 235
Reeves, Adam 92

© The Editor(s) (if applicable) and The Author(s), under exclusive license to Springer Nature Switzerland AG 2026
J. Dong et al. (Eds.): ISACE 2025, LNCS 15925, pp. 293–294, 2026.
https://doi.org/10.1007/978-3-032-06167-6

Rize, Denis   193
Rossi, Luca Francesco   209

S
Şahin, Yusuf H.   53
Saldanha, Paulo   193
Salekin, Md Sirajus   268
Sanna, Andrea   209
Shah, Dhairya   218
Sharma, Srishti   235
Shelyag, Sergiy   1
Slembrouck, Maarten   36
Song, Yuexi   253
Soulsby, Jonathan   76
Su, Shenyi   142
Sun, Jing   76

T
Taber, Christopher B.   235
Taber, Christopher   218

U
Üre, Nazım Kemal   53

V
Vanhaeverbeke, Jelle   36
Verstockt, Steven   36

W
Wadhwa, Bimlesh   155
Wang, Henry   268
Wang, Kun   69
Washida, Yudai   99
Wildman, Luke   276
Wu, Ling   253

X
Xia, Zehan   69

Y
Yasuda, Ryoma   99

Z
Zhang, Shinan   268
Zhao, Zixiao   76
Zheng, Huanhuan   253
Zheng, Jiaxin   69

MIX
Papier aus verantwortungsvollen Quellen
Paper from responsible sources
FSC® C105338

If you have any concerns about our products,
you can contact us on
ProductSafety@springernature.com

In case Publisher is established outside the EU,
the EU authorized representative is:
**Springer Nature Customer Service Center GmbH
Europaplatz 3, 69115 Heidelberg, Germany**

Printed by Libri Plureos GmbH
in Hamburg, Germany